Plant Physics

Plant Physics

KARL J. NIKLAS AND
HANNS-CHRISTOF SPATZ

THE UNIVERSITY OF CHICAGO PRESS   CHICAGO AND LONDON

The University of Chicago Press, Chicago 60637
The University of Chicago Press, Ltd., London
© 2012 by The University of Chicago
All rights reserved. Published 2012.
Paperback edition 2014
Printed in the United States of America

23 22 21 20 19 18 17 16 15 14   2 3 4 5 6

ISBN-13: 978-0-226-58632-8 (cloth)
ISBN-13: 978-0-226-15081-9 (paper)
ISBN-13: 978-0-226-58634-2 (e-book)
DOI: 10.7208/chicago/9780226586342.001.0001

Library of Congress Cataloging-in-Publication Data

Niklas, Karl J.
    Plant physics / Karl J. Niklas and Hanns-Christof Spatz.
        p. cm.
    Includes bibliographical references and index.
    ISBN-13: 978-0-226-58632-8 (cloth : alk. paper)
    ISBN-10: 0-226-58632-4 (cloth : alk. paper)   1. Plant physiology. 2. Botanical
chemistry.   I. Spatz, Hanns-Christof.   II. Title.
    QK711.2.N54 2012
    571.2—dc23

                                                                        2011024765

Contents

Preface

It is not essential that a concept should be perfectly clear. It is not essential that you should be able to distinguish a chair from a stool that closely resembles a chair. As a matter of fact, there is always a fuzziness about a concept, and one of the main differences between ordinary life and science is that scientific concepts are less fuzzy. But they are fuzzy nevertheless. It is only in mathematics that we find clear-cut concepts, and it is probably this inhuman characteristic that makes the subject repellent to many people.—J. L. Synge, *Talking About Relativity*

This book has two interweaving themes—one that emphasizes plant biology and another that emphasizes physics. For this reason, we have called it *Plant Physics*. The basic thesis of our book is simple: plants cannot be fully understood without examining how physical forces and processes influence their growth, development, reproduction, and evolution.

The history of science has shown that botanists can learn much from collaborations with physicists, mathematicians, and engineers. While writing the 1682 edition of his *The Anatomie of Plantes*, Nehemiah Grew (1641–1712) sought the advice of the engineer-inventor and author of *Micrographia* Robert Hooke (1655–1703) to explain how cells expand and why some tissues are stiff and strong while others are not. The philosopher and engineer Herbert Spencer (1820–1903) first proposed that "internal mechanical stress itself" caused the formation of "strong bonds" in plant cell walls, whereas the Barba-Kick law, which helped Gustave Eiffel (1832–1923) to construct his tower, was essential to the work of Alfred G. Greenhill (1847–1927), who sought to define mathematically the physical limits to the height of a tree. And without the elegant dimensionless equation developed in 1885 by the engineer and mathematician Osborne Reynolds (1842–1912), which permits modeling of the complex behavior of fluid flow around different-sized objects, the field of bio-fluid mechanics would be impossibly "fuzzy" (Vogel 1981).

Physical scientists can also learn much from studying plants. In 1638, Galileo Galilei (1564–1642) used the hollow stalks of grass to illustrate the idea that peripheral rather than centrally located construction materials provide most of the resistance to bending forces. He also developed the concept of geometric self-similarity, which presaged J. S. Huxley's (1887–1975) study of allometry and biological scaling principles (Huxley 1932) by comparing the mechanics of small and large oak trees. Leonardo da Vinci's (1452–1519) interest in fluid mechanics was inspired by observing the cross-sectional areas of tree trunks and noting that they roughly equal the sum of the cross-sectional areas of branches above any point. Likewise, his drawings illustrating the concept of a parachute and an autogyroscopic propeller are alleged to be based on his study of the dandelion's pappus and the maple tree's samara (Richter 1970). Over two hundred years later, in a book published after his death, Duhamel du Monceau (1700–1782) compared the skeletons of animals with the wood in trees (du Monceau 1785). In 1811, T. A. Knight (1759–1838) studied the effects of mechanical perturbation on the growth and morphology of trees. And in 1868, Julius von Sachs (1832–1897) wrote about the role of turgor (the hydrostatic pressure exerted on cell walls by their living protoplasts) in stiffening plant tissues and organs—a concept based on his observations of balloons.

This book explores these and many other insights that emerge when plants are studied with the aid of physics, mathematics, engineering, and chemistry. Much of this exploration dwells on the discipline known as solid mechanics because this has been the focus of much botanical research. However, *Plant Physics* is not a book about plant solid mechanics. It treats a wider range of phenomena that traditionally fall under the purview of physics, including fluid mechanics, electrophysiology, and optics. It also outlines the physics of physiological processes such as photosynthesis, phloem loading, and stomatal opening and closing. These and other aspects of plant biology demand attention and are treated to the best of our abilities in this book.

By its nature, any attempt to understand plant physics requires a considerable familiarity with mathematics, quantitative analyses, and computational methods. Our experience as teachers has shown that these occupations are best presented first using words and analogies and only later with formulas and mathematical derivations. Therefore, the majority of mathematical derivations and quantitative examples are placed in "boxes" to improve the flow of the text and the communication of concepts. A list of the most frequently used mathematical symbols (and their definitions and units) is provided at the beginning of the book. Less frequently used symbols are defined when they are used in the text to avoid confusion when reading individual chapters out of sequence.

This book is organized so as to present basic concepts first and, when appropriate, to deal with them in greater detail in subsequent chapters as new material is presented and explained. For this reason, we advise readers unfamiliar with the topics treated in this book to read the text linearly, from the beginning to the end. For those readers who are already familiar with some or all of the material presented in this book, we have subdivided each chapter into numbered sections that are cross-referenced throughout the book so that a topic treated in different chapters can be traced as it is developed further. Equations in the text and in the boxes are treated similarly. In this way, topics such as photosynthesis, the cell wall, and the mechanics of wood, which are dealt with at different levels of complexity, can be followed throughout the book as new details or concepts are introduced. As an additional aid, a glossary is provided at the end of the book to assist those who may be unfamiliar with some of the more technical terms used in botany, engineering, chemistry, or physics.

The first chapter covers such fundamental topics as the importance of plant life, the relationship between organic form and function, plant reproduction and development, the importance of multicellularity, and the developmental basis of the basic plant body plans—topics that establish a conceptual framework for much of the material presented in the following chapters. The second chapter offers a general outline of how fundamental physical principles and processes affect plant growth and ecology. Its purpose is to introduce the reader to environmental phytophysics. Many of the concepts introduced in these first two chapters are elaborated in chapters 3 7, wherein we present the physical and chemical principles required to understand plant water relations, solid and fluid mechanics, electrophysiology, and optics in relation to plant form, function, and ecology. Chapter 8 attempts to synthesize this information by emphasizing how different plant materials and processes are juxtaposed to function as a single organic entity. The last two chapters describe and discuss many of the experimental tools and modeling approaches that have been used to discover new things about plant physics. It is our hope that this organization allows *Plant Physics* to be used in the classroom as well as by professional scientists who find our perspectives on particular topics useful.

In terms of citations and the literature, even a casual inspection of the many currently available textbooks shows that there are many philosophies and practices concerning references. However, based on our teaching experiences, we believe that references are optimally placed at the end of individual chapters, rather than compiled into a single bibliography, which can be cumbersome and difficult to use. Additionally, we believe that references should be limited to those that are accessible by means of ISBN numbers. In this way, a reader can find the more extensive literature, both old and new, that is typically widely scattered in a variety of journals. Clearly, it would be pretentious to claim an extensive knowledge of this literature. Fortunately, bibliographies are far less necessary than in the past. Electronic tools to search for even the older literature using keywords have become extremely efficient and generally accessible.

We hope this book inspires future generations of students to study plant physics. Scientists have learned a great deal about how physical principles and processes influence plant growth, behavior, and evolution. However, there is still very much more to learn. It is to those future generations that we dedicate this book.

Literature Cited

Du Monceau, D. 1785. *La physique des arbes* ... I. Paris, France.

Galilei, G. 1638. *Discorsi e dimonstrazioni matematiche, intorno a due nuove scienze.* Leida, Italy: Appresso gli Elsevirii.

Grew, N. 1682. *The antomie of plantes* ... . 2nd ed. London, England.

Huxley, J. S. 1932. *Problems of relative growth.* New York: MacVeagh.

Knight, T. A. 1811. On the causes which influence the direction of the growth of roots. *Phil. Trans. Roy. Soc. London* 1811:209–19.

Reynolds, O. 1885. An experimental investigation of the circumstances which determine whether the motion of water shall be direct or sinuous, and the law of resistance in parallel channels. *Phil. Trans. Roy. Soc. London* 174:935–82.

Richter, J. P. 1970. *The notebooks of Leonardo da Vinci (1452–1566), compiled and edited from the original manuscripts.* New York: Dover.

Sachs, J. 1868. *Lehrbuch der Botanik.* Leipzig: Engelmann.

Vogel, S. 1981. *Life in moving fluids: The physical biology of flow.* Boston: Willard Grant.

Acknowledgments

This book would not have been possible without the support and professionalism of Christie Henry, the University of Chicago Press acquiring editor who agreed to publish *Plant Physics*, and her assistant Amy Krynak, who oversaw the entire project. We the authors also want to acknowledge the courtesy, efficiency, and professionalism of Michael Koplow (manuscript editor), Andrea Guinn (designer), David O'Connor (production manager), and Micah Fehrenbacher (promotions manager). We are especially grateful to Norma Sims Roche (copyeditor) for her meticulous attention to detail and to Adrianna Fusco (who designed the cover of the paperback edition). Many colleagues contributed to this book with suggestions and advice, notably Professors Leonid Fukshansky, Rainer Hertel, Bruno Moulia, Wolfgang Merzkirch, Thomas Owens, and Randy Wayne, as well as two anonymous reviewers. We also thank Edward Cobb (Cornell University) for many of the photographs used in our book. Finally, we gratefully acknowledge the Alexander von Humboldt Stiftung whose support ultimately made this project possible.

Recommended Reading

Denny, M. W. 1988. *Biology and the mechanics of the wave-swept environment.* Princeton, NJ: Princeton University Press.

Ennos, R. 2011. *Solid mechanics.* Princeton, NJ: Princeton University Press.

Gates, D. M. 1980. *Biophysical ecology.* New York: Springer Verlag.

Gibson, L. J., M. F. Ashby, and B. A. Harley. 2010. *Cellular materials in nature and medicine.* New York: Cambridge University Press.

Gordon, J. E. 1976. *The new science of strong material or why you don't fall through the floor.* 2nd ed. London: Penguin.

———. 1978. *Structures or why things don't fall down.* London: Penguin.

McGhee, G. R. Jr. 1999. *Theoretical morphology: The concept and its application.* New York: Columbia University Press.

Monteith, J. L. 1973. *Principles of environmental physics.* New York: Elsevier.

Nobel, P. S. 2005. *Physicochemical and environmental plant physiology.* 3rd ed. Amsterdam: Elsevier.

Raven, P. H., R. F. Evert, and S. E. Eichhorn. 2010. *The biology of plants.* 7th ed. New York: Freeman.

Ruse, M. 2003. *Darwin and design: Does evolution have a purpose?* Cambridge, MA: Harvard University Press.

Stephens, R. C. 1970. *Strength of materials.* London: Edward Arnold.

Vincent, J. F. V. 1990. *Structural biomaterials.* Princeton, NJ: Princeton University Press.

Vogel, S. 1981. *Life in moving fluids: The physical biology of flow.* Boston: Willard Grant.

———. 1988. *Life's devices.* Princeton, NJ: Princeton University Press.

———. 1998. *Cats' paws and catapults.* New York: Norton.

Wainwright, S. A., W. D. Biggs, J. D. Currey, and J. M. Gosline. 1976. *Mechanical design in organisms.* New York: Wiley and Sons.

Wayne, R. 2009. *Plant cell biology.* Amsterdam: Elsevier.

Zimmermann, M. H. 1983. *Xylem structure and the ascent of sap.* Berlin: Springer Verlag.

Frequently Used Symbols

Basic units: L, length (m); M, mass (kg); t, time (s)
Note: SI units are used throughout.

A	Cross-sectional area or surface area [m^2]
E	Elastic modulus in tension or compression [N/m^2]
EI	Flexural stiffness [N m^2]
F	Force [N or kg m s^{-2}]
g	Acceleration due to gravity [m/s^2] (g = 9.80665 m/s^2)
G	Torsional modulus [N/m^2]
GJ	Torsional stiffness [N m^2]
I_x, I_y	Second moments of area of a cross section in the x and y axes [m^4]
J	Polar moment of inertia of a cross section [m^4]
M	Bending moment [N m]
N	Newton [kg m s^{-2}] (1 Newton = 1 kg m s^{-2})
P	Pressure [N/m^2]
Re	Reynolds number [dimensionless]
T	Torque [N m]
U	Velocity [m/s]
V	Volume [m^3]
w	Weight per unit length [N/m]; width [m]
x, y, z	Rectangular coordinates [dimensionless]
δ	Deflection from a reference coordinate axis [m]
ε	Strain (unit distortion) [dimensionless]
$\varepsilon_x, \varepsilon_y, \varepsilon_z$	Strain in x-, y-, and z-directions [dimensionless]
μ	Dynamic viscosity [kg m^{-1} s^{-1}]

ν	Poisson's ratio [dimensionless]
θ, ϕ	Angle [degree or radian]
ρ	Density [kg/m^3]
$\sigma_x, \sigma_y, \sigma_z$	Normal stress components parallel to x, y, and z axes [N/m^2]
τ	Shearing stress [N/m^2]
$\tau_{xy}, \tau_{xz}, \tau_{yz}$	Shearing stress components in rectangular coordinates [N/m^2]
υ	Kinematic viscosity [m^2/s]
$\Delta\psi$	Potential difference [volt]
ψ_w	Water potential [N/m^2]

An Introduction to Some Basic Concepts

When you have eliminated the impossible, whatever remains, however improbable, must be the truth.—Sir Arthur Conan Doyle, *The Sign of Four*

Our goal in this chapter is to introduce some basic concepts in the study of plant life and biophysics, concepts that might be unfamiliar to physicists, engineers, or mathematicians interested in learning about plants or to biologists who want to learn more about physics.

The topics discussed in this chapter can be thought of as a philosophical prolegomena to the rest of the book. They include the limits of natural selection, the role of endosymbiosis in the evolution of plants, and some practical and philosophical issues, among which the value and pitfalls of reductionism and modeling are important launching pads for interpreting the concepts introduced in other chapters.

To begin with, it is important to recognize that the biological and physical sciences have much in common. Both can be used to explore the relationships that exist between form and function. Both help us to recognize that these relationships are contingent on local environmental conditions (i.e., the "working place" of the engineered artifact and the "habitat" of the organism). Both are experimental sciences that can be used to achieve great quantitative rigor. Both have rich theoretical frameworks on which to draw. And both are very practical sciences in the sense that physicists, chemists, engineers, and biologists recognize that the phenomena they study often resist the tidy, elegant analytical solutions that appear in the pages of many textbooks.

Despite these parallels in perspective, the physical and biological sciences are not entirely compatible. Typically, the physical scientist does not encounter systems that can alter form and substance in response to environmental changes. Nor does he or she deal with systems that can reproduce, mutate, or evolve. In the physical sciences, form-function relationships are known in advance when a machine or structure or synthetic enzyme or material is constructed. In addition, the building materials are specified in advance. In contrast, the biologist must deduce form-function relationships, an activity that, with very few exceptions, tends to be a risky adventure in speculation. The biologist must also deal with organic shapes, geometries, structures, and materials that have few, if any, counterparts in the physical sciences. Indeed, the distinction between a "structure" and a "material" as traditionally defined by engineers often becomes blurred when we examine the ultrastructure of a plant, animal, or some other life-form.

1.1 What is plant physics?

The title of this book reflects the juxtaposition of plant biology and classical physics to better understand the physical factors that have helped to shape plant form-function relationships, ecology, and the broad evolutionary patterns we see in the fossil record. This approach uses physical laws and processes, engineering principles, and mathematical tools to discover how an organism functions, grows, and reproduces. It also uses these tools to explore adaptive evolution by means of natural selection. The fundamental premise of any biophysical enquiry is that organisms cannot obviate physical laws and processes, which must therefore influence the course of organic evolution. That is not to say that all of biology can be reduced to mathematics and physical phenomena and processes, nor does it mean that evolutionary history is prefigured in the same way that Newton believed the universe was deterministic. Rather, the approach presented here merely assumes that much of biology can be understood by taking a reductionist approach and that whatever remains must be approached from a strictly biological perspective, one that fully acknowledges the important roles played by random events and historical contingency during the course of life's long and complicated history on earth.

For this reason, a biophysical approach can explain why certain hypothetical organisms are physical impossibilities, but it cannot explain why certain kinds of organisms exist. The course of organic evolution is influenced by random processes such as mutation, genetic recombination, and extinction events just as it involves the participation of nonrandom processes, among which natural selection is extremely important (Futuyma 1998). The concept of natural selection is complex and often misunderstood, so much so that it is sometimes accused of having circular logic; that is, "that which is fit survives, that which survives is fit." Indeed, many essays and symposia have been dedicated to the topic of natural selection, which can be defined in a variety of ways (e.g., Sober 1984). For example, natural selection can be defined as the process by which genetic variants in populations are winnowed to eliminate those variants that are less suited to the environment. This simply means that the traits exhibited by successfully reproducing individuals in any population are not identical to those of the entire population of potentially reproducing individuals. The disparity between reproductively successful individuals as a group and the rest of the population *is* natural selection. Accordingly, when natural selection is said to result in the adaptation of individuals to their environment, what is really being said is that a continual contrast exists between parent and offspring and that this contrast is generally advantageous to survival and reproduction for at least some individuals under particular environmental conditions.

Evolution by means of natural selection has no "purpose," and so it can have neither foresight nor intent. The traits that allow individuals in one generation to reproduce successfully may not be those that allow future generations to reproduce successfully, especially if the environment changes abruptly and unpredictably. Nevertheless, evolutionary trends that appear "directional" are not uncommon. Trends in the fossil record are discernible for both plants and animals. For example, the geological record shows that during periods of stable environmental conditions, the size range in many fossil lineages increased. Tree-sized lycopods and horsetails flourished and adapted to the comparatively stable and luxuriant coal swamps of the Carboniferous period (Taylor et al. 2009). Today, the descendants of these plants are herbaceous and small. In contrast, the size range of many lineages was reduced during periods of global attrition or environmental instability. The paleoecologist might suggest that stable environments permit plants to evolve longer maturation times and thus achieve greater size whereas, in contrast, environmental instability or attrition selects against organisms that require long periods to achieve reproductive maturity. Certainly, these are reasonable hypotheses that echo the classical distinction between the environmental regimes that favor K- or r-selection. The concept of K- and r-selection regimes deals with the trade-offs between the degree to which environmental conditions are stable or unpredictable and the fecundity and precocity of the species that cope with them. Habitats with environmental conditions that are predictable permit species to exist that require longer times to reach reproductive maturity (i.e., K-selection). Such species typically cannot maintain viable populations in habitats characterized by conditions that are unpredictable because individuals die before they reach reproductive maturity. Thus, natural selection in unpredictable environmental conditions favors the existence of species that grow rapidly, reach reproductive maturity quickly, and produce numerous progeny (i.e., r-selection). To a certain extent, K-selection is really not "selection" at all because it does not exclude a priori species that are very fecund and reproductively precocious.

Regardless of whether species experience K- or r-selection, throughout the vicissitudes of Earth's long history, they experience the results of physical laws and processes that are invariant and ubiquitous. For this reason, evolutionary history has other equally strong directional components or, at least, "historical signals" that reflect the inextricable interconnectedness of organic form, function, and environment. The first land plants were small by present-day standards and lacked specialized tissues for the conduction of water and sap (Taylor et al. 2009). They also possessed none of the organographic distinctions among leaves, stems, and roots that characterize the vast majority of land plants today. Yet within a comparatively short time (by geological standards), the surface of the earth was colonized and made green by organisms that manufactured wood with which to elevate specialized leaves and reproductive structures to heights that rival those of modern trees. Importantly, the fossil record shows that many of the form-function relationships observed in one ancient plant lineage are mirrored in other plant lineages that evolved independently but along the same morphological or anatomical pathways. Thus, the capacity to form wood evolved in the lycopods, the horsetails, and the progymnosperms independently. Within each of these three lineages, roots and large leaves also evolved, albeit in very different ways. And within each of these lineages, plants evolved the capacity to produce spores that produce unisexual gametophytes (heterospory). These and other examples of convergent evolution attest to the strong bond between form, function, and environmental context.

Plant physics provides the tools to explore this triumvirate quantitatively and to learn how organic form functions in the environmental setting of organisms that exist today and—by inferences drawn from the fossil record and what we know about biology in general—organisms that are now long extinct. By doing so, it sheds light on present-day ecology and the evolutionary history of every form of life, past and present and—yes, in theory—life-forms that have yet to evolve.

1.2 The importance of plants

Over 90% of all visible living matter is plant life—the substance that cleans the air and provides food, wood, fibers for clothing, important pharmaceuticals, the coal that fueled the Industrial Revolution, and many model organisms with which to explore genetics and development. In addition, plants have evolved into the largest life-forms on earth. Consider the largest extant animal, the magnificent blue whale (*Balaenoptera musculus*), which can weigh as much as 136 metric tons and measure 34 m in length, thus surpassing what may have been the largest dinosaur, *Argentinosaurus*, which is estimated to have weighed 60 metric tons with a body length of 35 m. Yet, however impressive these body sizes may appear, they pale in comparison to modern plants. For example, the brown alga *Macrocystis pyrifera* can grow 60 m in length annually, whereas the General Sherman tree (a specimen of *Sequoia sempervirens*) is estimated to weigh 1,814 metric tons and measures over 84 m in height.

It is surprising, therefore, that plants receive comparatively little attention in many textbooks devoted to biology or evolution. Apparently, attempts to drive home their importance can fail miserably. But consider food production, for example—or, more precisely, the annual production of organic carbon by plants—and its ecological consequences in terms of oxygen and carbon dioxide processing.

Aside from a small amount of organic carbon produced by chemoautotrophic organisms, plants provide virtually all of the organic carbon used by heterotrophs as food. The magnitude of organic carbon produced annually by land plants is on the order of 25 billion tons, as can be shown by a simple calculation (box 1.1). Regardless of how large this number seems, it pales in comparison to the estimated amount of organic matter produced annually by algae! Naturally, not all of this organic matter is available as food for humans. Roughly 70% of the organic carbon produced annually is in the form of cellulose and lignin, both of which contain carbon and neither of which is digestible by humans, although both are consumed in the form of paper and wood products.

BOX 1.1 **The amount of organic carbon produced annually**

The calculation of the amount of organic carbon produced annually by terrestrial plants rests on three facts:

1. The transfer of 12 g of carbon (one gram-atom) in the form of carbon dioxide into organic matter requires 469 kJ of energy (i.e., the production of 1 ton of organic carbon requires roughly 4×10^7 kJ).
2. Carbohydrates constitute roughly 80% of all organic matter.
3. The annual solar energy flux at the outer boundary of the earth's atmosphere is about 5×10^{21} kJ/yr (= 8.4 J of solar energy cm^{-2} min^{-1}).

We begin this calculation with the observations that only 40% of annual solar energy flux at the outer boundary of the atmosphere reaches the earth's surface (i.e., 2×10^{21} kJ/yr) and that only roughly one-half of this solar radiation is in the form of photosynthetically active radiation (PAR). In addition, only about 60% of this light energy is absorbed by land plants, of which only 1% is used to convert carbon dioxide into organic carbon (~6×10^{18} kJ/yr). The rest of the light energy is dissipated as heat or reflected back into the atmosphere. Assuming that 20% of the earth's surface is covered by plants, we estimate that 1.2×10^{18} kJ/yr is available for photosynthesis. Accordingly, about 0.02% of the annual solar energy flux at the atmosphere's outer boundary is converted by terrestrial plants into organic carbon.

With this amount of energy, roughly 3×10^{10} tons of carbon are fixed per year, which translates into an annual processing of about 8.3×10^{10} tons of oxygen and about 1.3×10^{11} tons of carbon dioxide. Even if we assume that 15% of the annual available energy for this processing is lost through respiration, we still come up with an estimated annual productivity on the order of 25 billion tons of organic carbon.

1.3 A brief history of plant life

We have used the word *plant* rather glibly so far. Yet the meaning of this word is sometimes ambiguous because it can be used in at least two ways: the traditional way, which groups organisms based on their shared characteristics (the grade level of organismal construction), and the phylogenetic way, which groups organisms based on their evolutionary histories (the clade level of evolutionary ancestor-descendant relationships).

The traditional way defines any eukaryotic organism as a plant if it carries out photosynthesis and possesses cell walls. This definition groups all of the unicellular and multicellular algae together with the more familiar nonvascular and vascular land plants, collectively called the embryophytes. This definition has some advantages. For example, it draws attention to a shared metabolism that requires the acquisition of photosynthetically active radiation, carbon dioxide, water, and minerals to support growth and reproduction. It also highlights the physical presence of an external layer of materials that provides the protoplasm of cells with mechanical rigidity and protection. By doing so, the traditional definition focuses our attention on convergent evolution among otherwise dissimilar evolutionary lineages. In turn, this convergence invites us to explore whether these and other shared features confer adaptive advantages.

Nevertheless, the traditional definition has two drawbacks. Unless qualified in some way, it excludes nonphotosynthetic organisms that have evolved from photosynthetic ancestors, such as the "fungus" *Saprolegnia* and the parasitic flowering plant *Monotropa*. More important, it neglects a complex evolutionary history that shows us that photosynthetic eukaryotes possessing cell walls have evolved independently many times during earth's history—which is the basis for affirming that convergent evolution has occurred in the first place!

Evidence for convergent evolution comes from the many detailed comparative studies using DNA sequence, biochemical, ultrastructural, molecular, and morphological data that reveal the polyphyletic nature of the organisms we call algae (Palmer et al. 2004). These studies indicate that there are at least five separate algal lineages (table 1.1). In contrast, similar phylogenetic analyses reveal that the embryophytes are a monophyletic group of organisms that includes the mosses, liverworts, hornworts, lycopods, ferns, horsetails, gymnosperms, and angiosperms. Collectively, all of these land plant groups trace their evolutionary history back to a common ancestor that was shared with the closest living relatives of the land plants, the modern charophycean algae.

Phylogenetic analyses also show that the multiple evolutionary origins of the algal lineages (including the green algal lineage that ultimately gave rise to the embryophytes) were the consequence of primary, secondary, and even tertiary endosymbiotic events (see table 1.1). In a very real sense, therefore, the history of plants, as traditionally defined, is reticulate by means of extensive lateral gene transfer (Kutschera and Niklas 2004, 2005).

TABLE 1.1 **Chlorophyll composition and postulated origin of plastids in some plant lineages**

Group	Chlorophylls	Plastid origin
Embryophytes	*a* and *b*	Primary
Charophytes	*a* and *b*	Primary
Chlorophytes	*a* and *b*	Primary
Glaucophytes	*a* and *b*	Primary
Rhodophytes	*a* and *c*	Primary
Euglenoids	*a* and *b*	Secondary (green)
Cryptomonads	*a* and *c*	Secondary (red)
Stramenopiles	*a* and *c*	Secondary (red)
Haptophytes	*a* and *c*	Secondary (red)
Dinoflagellates	*a*, various	Tertiary (various)

Source: Graham and Wilcox (2000).

As the word implies, *endosymbiosis* refers to the evolution of symbiotic relationships among different kinds of organisms in which one or more attained the physiological status of being an organelle in a host cell. For example, it is now widely accepted that plant plastids (of which the chloroplast is one) evolved when an ancient heterotrophic or chemoautotrophic prokaryote engulfed a photosynthetic prokaryote and evolved a mutually beneficial endosymbiotic relationship with it. This primary endosymbiotic event led to the evolution of the red algae (Rhodophyta), the green algae (Chlorophyta), the charophycean algae (Charophyta), and ultimately, the embryophytes. Phyletic molecular analyses indicate that the ancestral proto-plastid was very much like modern cyanobacteria, which liberate oxygen as a by-product of photosynthesis. Similar studies based on DNA sequences indicate that the first mitochondria probably evolved from prokaryotes very much like extant free-living α-proteobacteria (for a review, see Kutschera and Niklas 2004, 2005).

Secondary endosymbiotic events also occurred. These events are believed to have resulted in the evolution of other algal lineages, such as the euglenoids and the stramenopiles (which include the brown algae and the diatoms) (Graham and Wilcox 2000). Secondary endosymbiotic events occur when a eukaryotic heterotroph engulfs a photosynthetic unicellular alga that subsequently assumes the physiological role of a chloroplast within the host cell. The chlorophyll compositions of these secondarily acquired chloroplasts suggest that they are the remnants of unicellular green or red algae (see table 1.1). The corresponding ancestral host cell was some sort of heterotrophic eukaryote (i.e., an animal). Tertiary endosymbiotic events have also occurred, but these events are comparatively

rare and appear to have been limited to a group of algae called the dino-
flagellates.

Various lines of evidence exist for the evolution of organisms called
plants by means of endosymbiotic events. We will mention only eight:

1. The presence of organelle-specific DNA that is nonhistonal (as in the cytoplasm
   of prokaryotes)
2. The high degrees of sequence homology between the DNA of chloroplasts and
   cyanobacteria and between the DNA of mitochondria and proteobacteria
3. Organelle ribosomes that are similar to those of prokaryotes (70S ribosomes) but differ from those found in the cytoplasm of eukaryotic cells (80S
   ribosomes)
4. The observation that the 70S ribosomes of prokaryotes and organelles are both
   sensitive to the antibiotic chloramphenicole, whereas 80S ribosomes are not
5. The initiation of messenger RNA translation in prokaryotes and in organelles
   by means of a similar mechanism
6. The lack of a typical (cytoplasmic) actin/tubulin system in both organelles and
   prokaryotes
7. Fatty acid biosynthesis in plastids via acyl-carrier proteins (as in certain
   bacteria)
8. The presence of a double membrane surrounding plastids and mitochondria

How, then, should we use the word *plant*? Our view is that both the
traditional definition and the phylogenetic definition have their place in
this book, provided that the distinction between the two is made whenever the meaning is potentially ambiguous. Our reason for this is simple.
All photosynthetic eukaryotes share many metabolic, ultrastructural,
and structural features (e.g., photosynthesis, plastids, cell walls) that profoundly influence how these organisms make their living and thus much
of their ecology, regardless of their unique evolutionary history. In turn,
these shared features necessitate similar solutions to environmental conditions. For example, the foliar leaves of vascular plants have functional
analogues among many multicellular algae, most notably the kelps (brown
algae; Phaeophyta). This convergence in form reflects similar functional
obligations; for example, both organisms require large surface areas for
capturing sunlight and exchanging mass or energy with a fluid medium.
Because the blades of kelps perform many of the same functions of the
foliar leaves of ferns and other vascular plants, these organic structures
look very much alike.

By the same token, convergent evolution cannot be affirmed without a well-grounded and reasonably accurate phylogeny. That is, we cannot say that two or more lineages have converged independently on similar form-function relationships unless we can be sure these groups evolved independently. The phylogenetic perspective is therefore required whenever we attempt to address the nature of adaptive evolution. A word of caution is required in this context because phylogenetic (cladistic) hypotheses depend on the taxa included in an analysis and on the characters (traits) and character states employed to construct phylogenetic relationships. Numerous studies also demonstrate that the inclusion of fossil taxa can alter analyses based exclusively on extant species. In this sense, every phylogenetic hypothesis is just that, a hypothesis that may or may not be valid. Hypotheses about algal phylogenies are particularly "volatile" because of extensive lateral gene transfer resulting from endosymbiotic events in which the nuclear genomes of host cells acquired some of the genetic information originally contained in the genomes of their endosymbionts.

1.4 A brief review of vascular plant ontogeny

In preparation for our discussion of the physical properties of specific plant tissues (see chap. 8), it will be useful to review briefly some of the developmental relationships among the various plant tissue systems because these relationships have a profound effect on the kinds of physical forces different plant tissues typically experience. In this context, we focus on the developmental biology of a stereotypical vascular plant (fig. 1.1). This digression into what may be viewed as a strictly "botanical" topic is necessary if we are to fully appreciate the differences between the primary tissues, such as the epidermis and primary vascular tissues (which mechanically operate very much like hydrostatic devices; see section 8.5), and the secondary tissues, such as wood (which mechanically behave very much like a category of materials called cellular solids; see section 8.7).

Primary and secondary tissues can be distinguished using a variety of different criteria. However, from a strictly developmental perspective, the primary tissues trace their origins to the activities of apical meristems, whereas the secondary tissues are produced by cells derived from the two lateral meristems (Esau 1977; Beck 2005): the vascular cambium and the cork cambium (also called the phellogen). The initial cells in apical and

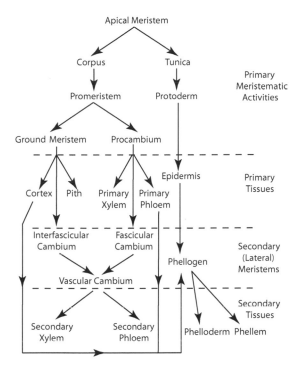

FIGURE 1.1. Flow diagram showing the developmental relationships among the primary and secondary tissues that develop in a stereotypical dicot stem. Note that the phellogen (cork cambium) can be traced back to one of three sources depending on the species or individual circumstances. (Adapted from Esau 1977 and Niklas 1992.)

lateral meristems give rise to undifferentiated cells (called derivatives) that subsequently differentiate and mature into the various types of tissues, tissue systems, and organs. The apical meristems provide stems and roots with the potential to increase in length indefinitely by means of primary growth. Lateral meristems permit stems and roots to increase in girth by means of the accumulation of secondary tissues.

The apical meristems give rise to the three primary tissue systems in stems, leaves, and roots: the dermal tissue system (epidermis), the primary vascular system (primary xylem and primary phloem), and the ground tissue system. In the case of stems, the epidermis traces its developmental origin back to a region of the apical meristem called the tunica, which produces meristematic cells (called protoderm) that develop into mature epidermal cells. The cells within the primary vascular and ground tissues systems trace their origins ultimately back to cells produced by the corpus, a region in the apical meristem that gives rise to the promeristem, which gives rise to the procambium and the ground meristem. The ground tissues (cortex and pith) in a typical dicot stem are composed predominantly of comparatively undifferentiated cell types, whereas arguably the most highly specialized and complex cell types differentiate in the vascular tissue system. The developmental origins of the three primary tissue systems in roots are far more complex than those in stems, in part because a variety of root apical meristematic arrangements exist across different species (Esau 1977; Beck 2005).

With the advent of secondary growth, the anatomical differences that may exist between the stems and roots of vascular plants gradually disappear, although subtle differences persist in wood anatomy that can be used to distinguish even between old portions of roots and stems. A noteworthy feature of the ontogenetic transition from primary to secondary growth is the replacement of the bulk of the three primary tissue systems with secondary tissues that take on their functional roles. As secondary xylem accumulates within the center of stems and roots, the cortex and primary phloem (as well as the oldest secondary phloem) are radially displaced and become compressed against the epidermis, which can expand to accommodate the increase in tissue volume to only a limited extent before it dies. With the continued accumulation of wood, the epidermis is eventually shed, while the other more external primary tissues are crushed and either completely or incompletely absorbed. As a consequence of secondary growth, the functional roles of the epidermis are taken over by the outer bark (phellem), which is produced and continuously replaced by the meristematic activities of the cork cambium (phellogen). Thus, the aeration of living tissues within older portions of stems and roots (such as the vascular cambium), which was originally provided by stomata, is supplied by the spongy aerenchyma-like cellular architecture of the lenticels in the outer bark. Likewise, the protective role of the cuticular membrane is taken over by the suberized walls of the phellem. And, for some species, the outer bark may also function as an "elastic skin" that can afford mechanical support, much as the cuticular membrane and outer and inner periclinal walls of the epidermis help to rigidify primary stems.

We will return to these and other aspects of vascular plant ontogeny later when dealing with growth stresses and reaction wood (sections 8.8 and 8.9). However, for the time being, we draw attention to the fact that the bulk of the cells produced by primary growth behave as hydrostatic mechanical devices (section 8.5).

1.5 Plant reproduction

The topic of plant reproduction may seem out of place in a book devoted to plant physics. To some, it may seem that reproduction of any sort cannot be studied with the tools of engineering, physics, or mathematics. Yet many aspects of plant reproduction are amenable to rigorous quantitative analyses using these tools, such as the dispersal of pollen by wind and the aerodynamics of wind-dispersed seeds and fruits (see chap. 6).

If it is true that plants are often given little attention in introductory biology textbooks, it is also true that they are often misunderstood in terms of their reproductive biology. This misunderstanding comes largely from the fact that the typical land plant life cycle differs from that of vertebrates and many other kinds of animals. This difference emerges clearly when we compare the sexual life cycle of a typical mammal with the reproductive life cycle of a typical land plant, such as a cabbage.

In the mammalian life cycle, the organism referred to as the *individual* is multicellular, diploid, and unisexual. It produces cells specialized for sexual reproduction that undergo meiosis to produce haploid gametes, sperm or egg cells. Copulation results in the fertilization of eggs by sperm, the formation of diploid embryos from zygotes, and the birth of new multicellular individuals. Notice that the only haploid cells in this life cycle are sperm or egg cells. It is called the haplobiontic-diploid (H-d) life cycle because it contains only one kind of multicellular individual, and that individual is diploid. The H-d life cycle also occurs in some algae. For example, the brown alga *Fucus* has a life cycle that, in this sense, is fundamentally like that of any mammal. The multicellular individual is a diploid organism, which produces gametes. One difference between the life cycles of *Fucus* and mammal, however, is that some species of *Fucus* are bisexual and thus produce both sperm and egg cells.

The H-d life cycle differs profoundly from that of the majority of algae and the embryophytes, which are characterized by a diplobiontic (D) life cycle. In the D life cycle, a multicellular diploid plant produces haploid spores by meiotic cell divisions, which subsequently divide mitotically to develop into multicellular haploid plants, which produce sperm or egg cells or both. The plant generation producing spores is called the sporophyte (which is technically asexual since it is neither male nor female); the plant generation producing sperm or eggs cells, or both, is called the gametophyte. Although the distinction between unisexual and bisexual gametophytes is important when dealing with population genetics, inbreeding depression, and strategies for minimizing self-fertilization, a more important detail of the D life cycle for the sake of the present discussion is that it contains two multicellular *individuals* that alternate in their growth and development to complete the cycle of sexual reproduction. For this reason, the diplobiontic life cycle is also called *the alternation of generations*.

The extent to which the gametophyte or the sporophyte dominates the diplobiontic life cycle differs between the nonvascular and vascular land plant lineages. Among the nonvascular land plants (mosses, liverworts, and hornworts), the gametophyte generation is the larger and seasonally persistent individual in the life cycle. In contrast, the gametophyte generation is small and short-lived in the life cycle of the vascular land plants, in which the sporophyte generation is massive and long-lived in comparison. Thus, when we think of a moss, we conjure up its gametophytic form; when we see a cabbage, we see its sporophyte. Yet in either case, the moss or any other species with a diplobiontic life cycle has two life-forms that can have very different morphologies and anatomies as well as very different ecological and physiological requirements for growth and survival.

There are other plant life cycles, such as the haplobiontic-haploid (H-h) life cycle seen among the charophycean algae, in which the only multicellular individual is haploid and the only diploid cell in the life cycle is the fertilized egg (the zygote). However, from a biophysical perspective, every plant life cycle is similar in three respects: plants do not mate consciously, gametes must come together to produce viable embryos, and spores or zygotes are typically transported by a biotic or abiotic vector. Collectively, these attributes can be summed up by one word, *dispersal*, which falls under the purview of biophysics.

Indeed, the methods by which plants disperse reproductive structures are as manifold as they are sophisticated and complex. Consider just a few examples: the air-filled bladders of pine pollen, which facilitate wind dispersal and add buoyancy when pollen grains float in a pollination droplet; the parachute-like pappus (a modified calyx) of the dandelion fruit; the autogyroscopic wings of the maple fruit and pine seed; the hygroscopic awns that help wheat fruits to burrow underground; the hygroscopic spore wall elaters that expand and contract to push spores out of horsetail sporangia; the triggerlike stamen filaments of mountain laurel (*Kalmia*) that "dust" visiting insects with pollen; and the explosive fruits and seed integuments of the exploding cucumber and wood sorrel.

1.6 Compromise and adaptive evolution

One of the many lessons learned from studying biology is that even very simple unicellular organisms perform many tasks simultaneously to survive, grow, and reproduce successfully. One of the many lessons learned from studying engineering is that the construction of a multitasking device typically requires compromises in design because the best ways to perform some tasks structurally or mechanically conflict with the best ways to perform other tasks. These two lessons meet head on in the study of plant physics because, in some respects, organisms are like multitasking machines and most, if not all, forms of life manifest compromises in their design.

We can illustrate what is meant by a biological design compromise by means of a simple schematic for the process of photosynthesis:

$$CO_2 + H_2O \quad \xrightarrow[\text{chlorophyll}]{\text{radiant energy}} \quad (CH_2O)_n + O_2.$$

In words, this schematic says that carbon dioxide and water are converted in the presence of chlorophyll into simple carbohydrates, denoted by $(CH_2O)_n$, and that oxygen is released as a by-product. The acquisition of CO_2 from water can pose a problem for aquatic plants even if they have an active uptake mechanism (see section 2.3). But the acquisition of water poses little problem, which permits aquatic plants to have very large surface area-to-volume ratios. Terrestrial plants, however, always run the risk of dehydration. One of the major adaptive innovations that permitted plants to populate the land was the evolution of a cuticle. This external covering provides a barrier to the passive diffusion of water by virtue of an outer waxy layer and a subtending cuticular membrane that may penetrate the primary cell walls of the epidermis and, depending on the species, one or more hypodermal cell layers (see section 8.1). Although the chemical composition of the cuticle reduces the loss of water from surfaces exposed to the air, it also provides a barrier to the passive diffusion of CO_2, which is essential for photosynthesis. The function of the cuticle, therefore, involves a trade-off, or "design compromise," because neither nature nor the best engineer has produced a substance that completely impedes the passage of water but permits the passage of CO_2 or O_2.

Naturally, this trade-off has been circumvented in a number of ways. One was the evolution of stomata, which provide pores through which atmospheric gases can diffuse and enter the plant body; the diameters of these pores are regulated by changes in the turgor pressure of pairs of flanking guard cells. Land plants that lack stomata, such as many moss, liverwort, and hornwort species, are confined ecologically to moist microhabitats. Among these species, adaptive evolution has resulted in metabolic ways to suspend metabolic activity when individuals become dehydrated (as well as ways to repair the effects of cell damage once water becomes available again). Another example of a "design constraint" is illustrated in box 1.2.

BOX 1.2 **Photosynthetic efficiency versus mechanical stability**

An example of a "design constraint" is illustrated by means of the formula for the bending moment exerted on a vertical or cantilevered cylindrical beam with radius R and length L by virtue of its own weight:

(1.2.1) $M = 0.5g\rho\pi R^2 L^2 \sin\emptyset,$

where M is the bending moment, g is the acceleration due to gravity ($g = 9.807$ m/s^2), ρ is the bulk density of the beam, and $\emptyset$ is the angle measured between the vertical and the beam's longitudinal axis. Note that the mass of the beam is $\rho\pi R^2 L$, which becomes weight (and thus a force) when multiplied by g. Also note that the bending moment is a product of a weight-force and a lever arm. Regardless of these details, this formula shows that the maximum bending moment is achieved when $\emptyset = 90°$ (when the beam is oriented parallel to the ground) and that the bending moment equals zero when $\emptyset = 0°$ or $180°$ (when the beam is perfectly vertical). Thus, if gravity-induced bending moments were our only concern, we would predict that cylindrical plant stems and leaves should be oriented vertically to minimize their bending moments. However, computer simulations show that the best orientation for a cylindrical photosynthetic stem or leaf to receive sunlight occurs when either organ is oriented parallel to the ground (i.e., when $\emptyset = 90°$). Thus, a trade-off exists between the ideal mechanical and photosynthetic orientations for cylindrical green stems and leaves. Perhaps for this reason, the lateral branches that bear more or less horizontally oriented leaves on many tree trunks and those of shrub or herbaceous species tend to be oriented at angles that are more than $0°$ but less than $90°$.

It is important to bear in mind that organic structures are not *designed* in any purposeful, goal-oriented way (Ruse 2003). As noted earlier in this chapter, adaptive evolution by means of natural selection has neither a plan nor a grand design. And mass extinction events in the history of life demonstrate unequivocally that previously successful forms of life can be exterminated when their environments change suddenly. Throughout this book, words and phrases such as "design," "architecture," "trade-off," and "compromise" are used as shorthand to describe form-function relationships that can be approached quantitatively using a biophysical perspective. Thus, we can say that the *architecture* of a leaf is *designed* to deal with the physiological *trade-offs* that exist when photosynthesis occurs, but only if we continue to remind ourselves that the word "designed" implies nothing more than that the morphology and anatomy of a leaf are capable of reconciling the many physiological and structural requirements imposed on any organ that carries out photosynthesis.

1.7 Elucidating function from form

Our discussion about design compromises and trade-offs thus far illuminates two important lessons to keep in mind when studying plant physics: that it is possible to optimize, but rarely ever maximize, the performance of many tasks performed simultaneously; and that as the number of tasks that a thing performs increases, both the number of trade-offs and the efficiency with which each task is performed typically decrease. These lessons, which emerge from studying engineering as well as biology, show us that adaptive evolution by means of natural selection can never result in a "perfect" organism. The most that the processes of random mutation and natural selection can ever achieve are organisms that work reasonably well.

These features raise another important aspect of the study of plant physics, which can be framed as a deceptively simple question: How do we know the function(s) that an organic structure performs? Indeed, can we be sure that a structure performs *any* function(s)? The answers to these questions are far from easy.

Every biophysical approach assumes a priori that organic form-function relationships exist and that we are intellectually equipped to identify them correctly. In many cases this presumption presents no problem. When we observe a tree trunk or branch, it is reasonable to assume that one of the functions of these structures is to provide mechanical support, and thus to explore their geometry and material properties in the context of solid mechanics. Even though we know that these stems also operate hydraulically by conducting water and sap in their xylem and phloem, we can adduce a priori that any self-supporting structure must sustain its own weight. But the elucidation of function from form is not always that simple.

Consider the prickles on a rose stem, which are typically and incorrectly called thorns. A prickle is a lenticular outgrowth of the cork cambium, and a thorn is a modified lateral stem. Even a casual inspection of a prickle suggests that this structure might function to protect stems from herbivores (or florists). Rose prickles can be sharp, brittle, and nasty structures. Thus, they *can* function defensively. In turn, these observations might lead to the hypothesis that prickles are an evolutionary consequence of a continued conflict between the ancestors of modern roses and the creatures that would eat them. This hypothesis is certainly tenable. Yet rabbits and deer are frequently seen to devour rose stems profusely covered with prickles, suggesting that "the war of the roses" can be lost or won, but not necessarily as a result of the effectiveness of prickles. Perhaps rose prickles serve some other function. The prickles of many rose cultivars, particularly those with a climbing growth habit, curve downward. It is possible, therefore, that prickles function like grappling hooks, allowing stems to cling to and climb over neighboring objects, much like the trichomes (plant hairs) on the leaves of *Galium* assist this climbing weed to attach to and shade the leaves of "host" plants (Bauer et al. 2010). This hypothesis is also tenable. Yet some rose cultivars lack prickles, but are as successful at climbing as their prickle-bearing counterparts.

The point to be made here is not to affirm or reject alternative hypotheses about the function(s) performed by rose prickles or trichomes. Candidly, we are not sure that rose prickles currently perform *any* function. It is possible that these structures are vestigial, in the sense that they were once functional during the evolution of some ancient rose species and that they have been retained by some descendants because there was no negative selection pressure against their formation on stems. It is equally possible that prickles perform two or more tasks, but that their absence on some cultivars results in no detectable detriment. Rather, our point is to caution against the notion that function can be invariably deduced from form and to caution against what have been called "somewhat-so-stories"—a predilection to affirm that every form-function relationship is a direct consequence of adaptive natural selection without benefit of rigorous quantitative analyses and solid evidence from the fossil record.

1.8 The basic plant body plans

If we keep in mind the extensive morphological and anatomical convergent evolution that has taken place among the various plant lineages, distinctions among the various plant body plans can be far more easily and profitably drawn on the basis of how plants achieve their organized growth and, if present, their basic tissue constructions. This approach identifies only four basic body plans: unicellular, colonial, siphonous, and multicellular. These body plans can be distinguished on the basis of a few basic developmental processes or events (fig. 1.2):

1. The presence or absence of vegetative cytokinesis, which determines whether the plant body is based on a uninucleate or multinucleate cellular plan (e.g., the algae *Chlamydomonas* and *Bryopsis*, respectively)
2. The separation of cell division products or their aggregation by means of a common extracellular matrix, which determines whether the body plan is unicellular or colonial (e.g., the algae *Calcidiscus* and *Phaeocystis*, respectively)
3. The indeterminate growth of the multinucleate cell, which results in the siphonous body plan (e.g., the algae *Bryopsis* and *Caulerpa*)
4. The establishment of symplastic continuity among cells during and after cell division by means of "cytoplasmic bridges," plasmodesmata, and so forth, which establishes the multicellular body plan (e.g., the alga *Volvox* and the moss *Polytrichum*)

The multicellular body plan has three basic variants that can be described morphologically, albeit not mechanistically, in terms of the number of planes of cell division (see fig. 1.2). When cell division is restricted to one plane or orientation, unbranched filaments can be formed (e.g., the alga *Spirogyra*); when confined to two orientations, cell divisions can give rise to branched filamentous, monostromatic, or pseudoparenchymatous tissue constructions (e.g., the algae *Stigeoclonium*, *Volvox*, and *Ralfsia*, respectively); and, when cell division occurs in all three planes, a multicellular body layout becomes possible, which can simultaneously manifest a filamentous and parenchymatous construction (e.g., the alga *Fritschiella*). Since each of these three multicellular variants can involve cell divisions taking place in different parts of the plant body, a large number of multicellular body plan variants exist. Although a number of growing-point (meristematic) characteristics collectively influence whether a variety of morphological and anatomical features are achieved in a particular body part or plan (e.g., the duration of activity and the number of cells involved

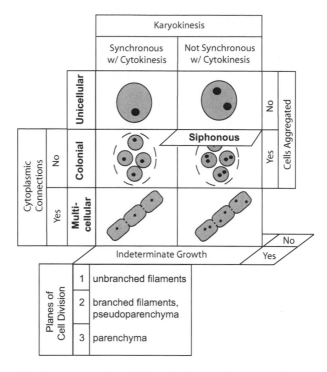

FIGURE 1.2. Graphic representation of the four developmental processes (i.e., the presence or absence of correlated karyokinesis and cytokinesis, the presence or absence of cell aggregation, the presence or absence of cytoplasmic continuity among adjoining cells, and presence or absence of indeterminate growth) that result in one of four plant body plans (unicellular, colonial, siphonous, or multicellular).

at each location, as well as the extent to which cells, tissues, or organs differentiate or differ in symmetry, number, etc.), none of these features is especially relevant to the fundamental distinctions that can be drawn among the four basic plant body plans.

1.9 The importance of multicellularity

Despite the vast diversity in the appearance and internal structure of plant body plans, the vast majority of plants must perform five basic functions:

1. intercept sunlight
2. exchange mass and energy with the external environment
3. transport materials within the plant body

4. cope with externally applied mechanical forces, such as gravity and the movement of air or water
5. reproduce sexually if adaptive evolution is to occur

With the exception of parasitic plant species, which have lost the capacity for photosynthesis and thus do not require sunlight, plant species must perform all five functions simultaneously. All of these functions, including the internal transport of materials, are performed by unicellular plants, regardless of whether they are comparatively small, like the green alga *Chlamydomonas*, or exceptionally large, like the siphonous green alga *Caulerpa*, which grows indefinitely in size over its lifetime.

In some respects, the unicellular organism is the penultimate multitasking life-form. This body plan is the dominant life-form among the majority of algal lineages. However, as we will also see, the multicellular plant body has evolved independently many times (Niklas 2000; Knoll 2011). This convergence suggests that multicellularity confers some adaptive advantages, presumably because it allows for the differentiation and specialization of tissues, tissue systems, and specialized reproductive cells.

Like a building that can provide ventilation, plumbing, living space, and mechanical support, the multicellular plant body is structurally reinforced by internal cell walls, which can function mechanically as an interconnected system of struts, beams, columns, and plates. In many respects this infrastructure functions like an endoskeleton (Moulia et al. 2006), albeit one that lacks a neuromuscular component. Although infrequently recognized, this cell wall infrastructure does not result a priori in the complete compartmental isolation of cell protoplasts. Much to the contrary, the multicellular plant's protoplasm may be continuous among adjoining cells, tissues, or tissue systems by virtue of cytoplasmic threads, called plasmodesmata, that pass through the walls of adjoining cells. Perhaps the best example of this supracellular cytoplasmic continuity is seen in angiosperm phloem. Within this tissue, cells called sieve tube members are stacked end to end to form sieve tubes. The protoplasts of adjoining sieve tube members (and their companion cells) are all interconnected by numerous clusters of plasmodesmata (see section 3.9). This configuration results in a supracellular transport system of living matter that can communicate physiologically from the growing tips of roots to the growing tips of stems and leaves. A functional analogue to the phloem is found in the trumpet cells of some large marine brown algae, which can transport metabolites over many meters through a continuous supracellular protoplast.

These examples of supracellular protoplasmic systems, which are by no means exceptional, are relevant to whether one subscribes to the cell theory, which implicitly considers every multicellular organism to be a "republic" or loose "confederation" of cells, or the organismal theory, which views the organism as an integrated whole that can become cellularized under some circumstances. Although many aspects of growth and development can be best studied at the level of individual cells, the cell theory has fostered the idea that the concept of the cell and the concept of the organism are logically interchangeable. It is true that a multicellular and a unicellular plant are not very different at one level: a single living protoplast resides in each, because the multicellular plant's protoplast is highly dissected, but not completely dissected into isolated compartments. It is equally true that morphological complexity can be achieved without benefit of multicellularity, a fact easily validated when one examines unicellular, siphonous green algae such as *Bryoposis*, *Codium*, and *Caulerpa*.

However, the principal propositions of the cell theory differ fundamentally from those of the organismal theory (table 1.2). According to

TABLE 1.2 **Developmental and phylogenetic corollaries of the cell and organismal theories**

Cell theory
   Developmental corollaries
      All living things are made up of cells
      Each cell is an individual of equal morphological rank
      Each multicellular organism is an aggregate of cells
      The properties of the organism are the sum of its cells
      Ontogeny is the cooperative effort of cells
   Phylogenetic corollaries
      Unicellular organisms are ancient and "elementary"
      Elemental units formed colonial organisms
      Multicellular organisms evolved from colonial organisms

Organismal theory
   Developmental corollaries
      Ontogenesis is the property of the whole organism
      Ontogenesis is the resolution of the whole into parts
      Growth and differentiation are properties of the protoplasm
      Cell division may or may not involve separation of the protoplasm
      If separation occurs, cells are subordinate to the organism
   Phylogenetic corollaries
      Unicellular organisms are nonseptate individuals
      Multicellular organisms are septate individuals
      Unicellular and multicellular organisms are homologous
      Colonial organisms are derived, not ancient
      Cellularization obtains division of labor/mechanical benefits

the developmental and phylogenetic corollaries of the cell theory, the cell is the basic structural and physiological unit, multicellular organisms are aggregates of cells, the properties of individual cells collectively dictate the properties of the multicellular organism, and multicellularity evolved from a colonial body plan configuration. In contrast, the organismal theory views growth and development as properties of the whole organism and unicellular and multicellular organisms as homologous individuals. Thus, contra Theodor Schwann's proclamation (i.e., *Omnis cellula e cellula*—each cell from a cell), a multicellular plant should be viewed according to the dictum of Heinrich de Bary (i.e., *Die Pflanze bildet Zellen, nicht die Zellen bilden Pflanzen*—plants make cells, cells do not make plants).

Regardless of one's perspective on the cell theory versus the organismal theory, from a biophysical perspective, the cell wall infrastructure is particularly important because the mechanical and hydraulic properties of most plant tissues and organs are correlated with the chemistry, size, shape, and geometry of cell walls. Numerous biophysical studies have demonstrated correlations between the mechanical and hydraulic properties of tissues and organs and the chemical and physical properties of the cell walls within them. For example, some plant tissues are composed of interlacing cells with thin walls and very little wall-to-wall contact area, which thereby create large intercellular spaces filled with air (e.g., aerenchyma). Tissues with this composition tend to be mechanically weak, but they contribute little weight per unit volume and typically serve to ventilate and cool leaves, stems, and roots. In contrast, tissues such as sclerenchyma and wood (secondary xylem) are dominated by cells with very thick walls and have very large wall-to-wall contact areas. These tissues are exceptionally rigid and strong and thus provide mechanical support. In addition, the thick, lignified walls of xylary cell types are devoid of protoplasts when mature and, by virtue of wall perforations, serve as conduits for the mass transport of water (i.e., tracheids and vessel members).

Literature Cited

Bauer, G., M.-C. Klein, S. N. Gorb, T. Speck, D. Voigt, and F. Gallenmüller. 2010. Always on the bright side: The climbing mechanism of *Galium aparine*. *Proc. Roy. Soc. London*, ser. B, doi:10.1098/rspb.2010.2038.

Beck, C. B. 2005. *An introduction to plant structure and development.* New York: Cambridge University Press.

Esau, K. 1977. *Anatomy of seed plants.* 2nd ed. New York: John Wiley.

Futuyma, D. J. 1998. *Evolutionary biology.* 3rd ed. Sunderland, MA: Sinauer Associates.

Graham, L. G., and L. W. Wilcox. 2000. *Algae.* Upper Saddle River, NJ: Prentice Hall.

Knoll, A. H. 2011. The multiple origins of complex multicellularity. *Ann. Rev. Earth Planet. Sci.* 39: 217–39.

Kutschera, U., and K. J. Niklas. 2004. The modern theory of biological evolution: An expanded synthesis. *Naturwissenschaften* 91:255–76.

———. 2005. Endosymbiosis, cell evolution, and speciation. *Theory in Biosciences* 124:1–24.

Moulia, B., C. Coutand, and C. Lenne. 2006. Posture control and skeletal mechanical acclimation in terrestrial plants: Implications for mechanical modeling of plant architecture. *Amer. J. Bot.* 93:1477–89.

Niklas, K. J. 1992. *Plant biomechanics: An engineering approach to plant form and function.* Chicago: University of Chicago Press.

———. 2000. The evolution of plant body plants: A biochemechanical perspective. *Ann. Bot.* 85: 411–38.

Palmer, J. D., D. E. Soltis, and M. W. Chase. 2004. The plant tree of life: An overview and some points of view. *Amer. J. Bot.* 91:1437–45.

Ruse, M. 2003. *Darwin and design.* Cambridge, MA: Harvard University Press.

Sober, E. 1984. *The nature of selection: Evolutionary theory in philosophical focus.* Cambridge, MA: MIT Press.

Taylor, T. N., E. L. Taylor, and M. Krings. 2009. *Paleobotany: The biology and evolution of fossil plants.* New York: Academic Press.

CHAPTER TWO

Environmental Biophysics

In the course of centuries a huge and cohesive network of interconnecting and partially overlapping facts has been established which we can call the scientific knowledge of nature. It is,
however, a net not without holes, and some of the meshes are weaker than others.
—Erwin Chargaff, *Serious Questions*

The goal of this chapter is to illustrate the concepts underlying environmental biophysics, an area of research that treats aspects of form-
function relationships quantitatively from the perspective of the physical
laws and processes that define abiotic phenomena. This area of enquiry
is so important to every aspect of biology that it deserves an extensive
discussion early on in any general treatment of plant physics. This chapter also gives us an opportunity to introduce an assortment of tools that
will be drawn on in future chapters. Among these tools are the transport
laws—Newton's law of viscosity, Fick's diffusion law, and Fourier's heat-
transfer law—which, as we will see, all have the same basic mathematical
form.

The examples used in this chapter emphasize the five functional tasks
that every photosynthetic eukaryote must perform to survive, grow, and
reproduce: light interception, exchange of mass and energy with the external environment, transport within the plant body, coping with mechanical forces, and sexual reproduction (see section 1.5). Naturally, the
performance of these functional tasks cannot be treated synoptically. Our
objective here is merely to illustrate how each of these tasks can be explored with some of the basic tools of environmental biophysics. While
doing so, it is useful to remember that all of these tasks are performed
simultaneously, and that the manifold functional obligations of even the
morphologically simplest plant require optimization because unavoidable

trade-offs exist between the performance of two or more tasks with conflicting design requirements. In addition, because the aquatic and terrestrial environments differ profoundly in their physical properties, and because the performance of each functional task is influenced by these physical properties, we have elected to compare and contrast these two environmental extremes and largely neglect what might be considered intermediate environmental conditions.

Finally, we want to draw attention to an important message emerging from the topics treated in this chapter: although organisms cannot avoid the consequences of physical laws and processes, they cope with them adaptively by virtue of their ability to change size, shape, or geometry as they grow and develop. This ability can be seen by examining each of the formulas presented in this chapter. Within each formula, one or more variables are physically invariant; that is, their magnitudes depend on the abiotic environment and cannot be changed. However, we also find one or more variables whose magnitudes can change as a function of changes in the size, shape, or geometry of an individual plant. In this sense, organisms push against their environment as much as the environment pushes back.

2.1 Three transport laws

We begin our exploration of environmental biophysics by examining some transport laws that express important physical phenomena. Fortunately, these transport laws are relatively simple to understand intuitively, as can be seen with three "thought experiments."

In the first experiment, imagine a drop of dye placed in a glass of still water. Observe how it diffuses rapidly outward from its initially high concentration, only to diffuse progressively more slowly thereafter. From this observation, we might quickly intuit that the rate of diffusion is dependent on the dye's instantaneous concentration gradient. Indeed, as we shall see, the equation for the passive diffusion of a substance such as a dye in water contains an expression for a concentration gradient. In the second experiment, imagine that you are sliding your hand over a deck of cards placed on a table. Notice how the cards move with respect to one another and with respect to the surface on which they rest. Notice that the cards move in the same direction but to different degrees with respect to the table's surface. By repeating this experiment at different speeds, you might easily conclude that the extent of shearing depends on a velocity gradient—and indeed, that is correct. Finally, consider heat transport. Imagine that you place your hand on a cold glass windowpane. The cooling you immediately sense is the transfer of heat from your warm hand to the colder glass. Given enough time, the temperature difference between your palm and the glass will be reduced and the rate of heat loss will decrease.

These three thought experiments illustrate that the rate at which mass or energy is transported depends on a driving force (e.g., a concentration gradient, a velocity gradient, or a thermal gradient). Dimensional analysis also reveals that each driving force must be multiplied by a proportionality factor with the appropriate units. This principle is illustrated by each of the three transport laws that figure prominently in environmental biophysics: Fick's diffusion law, Newton's law of viscosity, and Fourier's heat-transfer law.

Fick's first law of diffusion in its one-dimensional form is given by the formula

(2.1) $$J_i = -D_i \frac{dC_i}{dx},$$

where J_i is the flux of the diffusing substance i, D_i is the molecular diffusivity of i, and dC_i/dx is the concentration gradient of i. The negative sign on the right-hand side of equation (2.1) is a convention that indicates that the flux is in the positive direction when the gradient is negative. Together with the continuity equation,

(2.2) $$\frac{dJ_i}{dx} = -\frac{dc_i}{dt},$$

equation (2.1) yields the well-known formula for Fick's second law:

(2.3) $$\frac{dc_i}{dt} = D_i \frac{d^2c_i}{dx^2}.$$

It is noteworthy that the numerical values of D_i for many substances diffusing in water or in air typically differ by as much as four orders of magnitude (see the table in box 2.1). This difference indicates that passive diffusion in water is a much slower (and thus biologically difficult) process.

Newton's law of viscosity is given by the formula

(2.4) $$\tau = \mu \frac{dU}{dx},$$

TABLE 2.1 **Some physical properties of air and water at 20°C**

Physical property	Air	Water
Density (ρ)	1.205 kg/m^3	998 kg/m^3
Dynamic viscosity (μ)	1.808 × 10^{-5} Pa s	1.003 × 10^{-3} Pa s
Kinematic viscosity (v)	15 × 10^{-6} m^2/s	1.0 × 10^{-6} m^2/s
Thermal conductivity (k)	0.0257 W m^{-1} °C^{-1}	0.599 W m^{-1} °C^{-1}
Surface tension (σ)	—	0.0729 N/m

where τ is the shear stress generated between two layers of a moving fluid with dynamic viscosity μ and velocity gradient dU/dx. As in the case of molecular diffusivity, the numerical values of μ differs between fluids (table 2.1).

Finally, Fourier's law of heat-transfer is given by the formula

$$\textbf{(2.5)} \qquad\qquad H = -k\frac{dT}{dx},$$

where H is the heat flux in a body with thermal conductivity k and temperature gradient dT/dx. Notice the use of the negative sign, which indicates, once again, that the flux is in a positive direction when the gradient is negative.

There are a number of reasons to have all three of these transport laws expressed in the same way, and it is not difficult to do so. Consider that all of the "diffusivities" have the same units (m^2/s) and, for gases, the same magnitudes, as well as similar temperature and pressure dependencies, which can be derived from kinetic theory. Indeed, mathematical analogies for "diffusivity" and "concentration gradient" exist for equations (2.4) and (2.5). For example, Newton's law of viscosity becomes

$$\textbf{(2.6)} \qquad\qquad \tau = v\frac{d(\rho U)}{dx}$$

because kinematic viscosity v (which is dynamic viscosity divided by fluid density ρ) is an analogue of "diffusivity" and because the product of fluid density and velocity U can be thought of as "the concentration of momentum."

Likewise, Fourier's heat-transfer law can be converted into a diffusion formula by multiplying temperature T by the volumetric heat capacity of the surroundings, denoted by ρC_p, to obtain a "heat concentration" and to derive a "thermal diffusivity" D_T, which equals $k/\rho C_p$. In this conversion, equation (2.5) takes the form

(2.7)
$$H = -D_T \frac{d(\rho C_p T)}{dx}.$$

The numerical value of the volumetric heat capacity used in equation (2.7) depends on the fluid through which heat is lost. For example, the volumetric heat capacity of air at 20°C at constant pressure is ρC_p ~ 1,200 J m^{-3} K^{-1}.

It is also useful to have a macroscopic counterpart to each of the transport laws. By *macroscopic*, we mean a formula in which the concentration gradient occurs across distances larger than subcellular dimensions. When applied in this way, Fick's law can be expressed as

(2.8)
$$J_i = -D_i \frac{\Delta C}{\delta},$$

where ΔC denotes the difference in the concentration of substance i measured across a boundary layer with thickness δ (for a discussion of boundary layers, see section 2.2).

Likewise, the shear stresses that develop over the surface of a spherical object with radius r moving in air or water are given by Stokes's formula,

(2.9)
$$\tau = \frac{3}{2} v \frac{(\rho U)}{r}..$$

This equation applies to small objects moving at low speeds, more exactly at Reynolds numbers Re < 1 (for a detailed discussion of Reynolds numbers, see section 6.2).

Finally, the heat-transport formula can be expressed as

(2.10)
$$H = -\rho C_p \frac{\Delta T}{r_H},$$

where ΔT denotes the difference between ambient temperature and the temperature of the surface of interest and r_H is the resistance to heat transfer (which can be expressed in terms of the thickness of a boundary layer δ).

To be useful in most practical situations, equations (2.8)–(2.10) require us to know the values of variables such as δ and r_H, which must often be determined empirically rather than analytically. However, there are ways to manipulate the transport laws to provide considerable insights into biological phenomena that do not depend on knowing the precise numerical values for all variables. For example, Fick's law can be used to explore the effects of shape and geometry on photosynthesis for simple plant forms in meaningful ways even if we do not know the numerical value of a substance's diffusivity or its concentration gradient. Consider the effects of cell surface area A and volume V on the time it takes for the concentration of a nonelectrolyte j initially absent from a cell's interior to reach one-half the concentration of j in the external ambient medium. Denoting this half-time as $t_{0.5}$, manipulation of Fick's law and the continuity equation gives the formula

(2.11)
$$t_{0.5} - t_0 = \frac{V}{P_j A} \ln \left[\frac{(c_o - c_j)_{t_0}}{(c_o - c_j)_{t_{0.5}}} \right],$$

where P_j is the permeability coefficient of j, the expression $(c_o - c_j)_{t_0} = c_o$ is the initial difference between the external and internal concentrations of j at time zero, and $(c_o - c_j)_{t_{0.5}} = c_o / 2$ is the difference between the external and internal concentrations when the internal concentration of j reaches one-half that in the ambient medium. At first glance, equation (2.11) contains a number of variables whose magnitudes must be known before we can compute $t_{0.5}$. However, notice that $\ln \left[(c_o - c_j)_{t_0} / (c_o - c_j)_{t_{0.5}} \right] = \ln[c_o / (c_o / 2)] = \ln 2 = 0.693$ regardless of the molecular species we are considering. Thus, $t_{0.5} - t_0$ is always proportional to (V/A) such that any decrease in cell volume or any increase in cell surface area decreases $t_{0.5}$, which is adaptively beneficial to growth if j is an essential nutrient. This feature helps us to understand why palisade and spongy mesophyll cells in leaves have high A relative to their V. It may also explain why unicellular algae tend to be spheroidal in shape rather than perfect spheres (a spheroid has a larger A compared with a sphere with comparable V). Finally, it cannot escape attention that V and A can change ontogenetically as a cell, organ, or entire plant grows in size. This phenomenon, like many others, is the subject of allometric analyses (see section 10.4).

2.2 Boundary layers

We mentioned the concept of the boundary layer earlier when we discussed the macroscopic version of Fick's law of diffusion (see eq. [2.8]), and we mentioned the Reynolds number when we discussed Stokes's formula for shear stresses (eq. [2.9]). However, these concepts were not defined or explored in any detail. This is a serious omission because no discussion of mass or energy transport is complete without considering boundary layers and the resistance to molecular diffusion or heat transport that they typically provide. Nor can we properly discuss mass or energy transport without considering Reynolds numbers (see section 6.2 for a detailed description of Reynolds numbers). Indeed, we will see here that the two concepts are inextricably tied to each other.

The concept of the boundary layer was introduced by Ludwig Prandtl (1875–1953). It describes a thin envelope of relatively unmoving fluid (air or water) that surrounds every object or organism. As might be deduced from Newton's law of viscosity (see eq. [2.4] or [2.9]), the thickness of the boundary layer is affected by the velocity of the fluid U, which can vary over the surface of an object even if the ambient fluid flow remains steady. Toward the leading edge of fluid flow, the fluid sublayer within the boundary layer is generally laminar and thus dominated by shearing stresses τ, and diffusion orthogonal to the surface (i.e., at right angles) is the result of molecular motion in the laminar layer. However, diffusion becomes increasingly assisted by eddies that develop progressively along surfaces in the downflow direction as U increases (see fig. 6.5). Thus, the thickness of the boundary layer varies over a surface even when ambient U is steady.

Mathematical descriptions of how boundary layer thickness δ varies even over a relatively simple geometry can be very complex and are thus beyond the scope of this book. Fortunately, empirical studies have shown that some very simple equations describing δ for very simple geometries are often sufficient for estimating how heat or mass is transported among objects that approximate the geometries of flattened leaves, cylindrical stems, and spherical objects such as cells or fruits. For example, Schlichting (1979) gives a formula for the laminar boundary layer thickness around a flattened surface with an irregular planar outline:

(2.12a)
$$\delta = 5\left(v\,\frac{L}{U}\right)^{0.5},$$

where v is kinematic viscosity, L is the mean length, and U is the ambient flow rate. Comparable equations have also been developed empirically (Nobel 2005) for cylinders,

$$\textbf{(2.12b)} \qquad\qquad \delta = 6\left(v\frac{D}{U} \right)^{0.5},$$

and for spheres,

$$\textbf{(2.12c)} \qquad\qquad \delta = 3\left(v\frac{D}{U} \right)^{0.5} + 0.25\left(\frac{v}{U} \right),$$

where D is the diameter of the cylinder or sphere. Great care must be exercised, however, when using these or similar formulas because they are based on empirical studies that set limits on the applicability of derived physical relationships. For example, equation (2.12c) is based on a study for which the (as yet undefined concept of) Reynolds number ranged between 4×10^2 and 4×10^4. Therefore, it would be unwise to apply this formula when conditions do not fall within that range.

Each of the preceding three equations makes intuitive sense because each predicts that the boundary layer becomes thicker as the characteristic dimension of an object (i.e., L or D) increases and that it becomes thinner as an object with a constant characteristic dimension experiences increased fluid flow over its surfaces (table 2.2). However, it cannot be overemphasized that equations (2.12a–c) and others like them provide only *estimates* of δ and that they neglect a number of important biological features that are nevertheless important to mass or heat transport. For example, stems tend to oscillate and leaves tend to flutter or even curl upon themselves when exposed to increasingly higher wind speeds (Vogel 1988). These dynamic movements and transient changes in orientation

TABLE 2.2 **Estimated average boundary layer thickness δ (rounded to three significant figures; in mm) for flattened surfaces differing in average length L and experiencing different ambient wind speeds U**

L (m)	U (m/s)					
	0.1	0.2	0.5	1.0	2.0	5.0
0.01	6.12	4.33	2.74	1.94	1.37	0.87
0.05	13.7	9.68	6.12	4.33	3.06	1.94
0.1	19.4	13.7	8.66	6.12	4.33	2.74
0.5	43.3	30.6	19.4	13.7	9.68	6.12
1.0	61.2	43.3	27.4	19.4	13.7	8.66
5.0	137	96.8	61.2	43.3	30.6	19.4

Note: See equation (2.12a). For δ in water, divide each value by $(15)^{1/2} \approx 3.87$.

will alter boundary layer thicknesses locally over some surfaces and globally over entire organs. Likewise, many plant surfaces are rough, hairy, or otherwise textured in ways that can alter local fluid speed and direction. For these reasons, equations (2.12a–c) are convenient substitutes for actual wind tunnel, flume or field measurements, but they should be used only as yardsticks for the *relative* magnitude of δ.

Another important caveat is that equations (2.12a–c) are generally useful only when ambient flow rates are characterized by turbulent flow conditions (albeit laminar flow in the boundary layer). To judge whether this is the case, a simple yet invaluable tool exists with which to determine the type of fluid flow a plant organ is likely to experience. This tool is the Reynolds number, which is denoted by Re.

The formula for Re is simple and elegant:

(2.13)
$$\mathrm{Re} = \frac{\ell U}{\upsilon},$$

where ℓ is the characteristic dimension, U is the ambient velocity, and υ is the kinematic viscosity (see table 2.1). We will discuss Reynolds numbers and kinematic viscosity in much greater detail when we treat fluid mechanics in chapter 6. However, for now, it is sufficient to know that Re tells us whether fluid flow is laminar (and thus dominated by viscous forces) or turbulent (and thus dominated by inertial forces). For simple objects, such as flat plates, spheres, and cylinders with smooth surfaces, nonlaminar flow generally occurs when Re > 2,000, while turbulence in the boundary layer is observed only when Re > 200,000. This rule of thumb can be used to set limits on when to use equations (2.12a–c); that is, fluid flow conditions should fulfill the criterion $2{,}000\upsilon < \ell U < 200{,}000\upsilon$.

In passing, it is worth noting that boundary layers and mass transport are directly related to the rate at which water vapor diffuses out of stomatal pores because the thicker the boundary layer, the smaller the water vapor gradient across it, and thus the slower the rate of water vapor diffusion from the pores. Therefore, one basic feature of leaves that should be highly conserved evolutionarily across species is the relationship between maximum stomatal pore area and leaf gas exchange capacity, which is governed by the physics of diffusion through pores. The maximum area of the open stomatal pore (a_{max}) and its depth (l) can be used to define the size of stomata. Taken together, pore depth (l), the maximum area of the open stomata (a_{max}), and stomatal density (d, the number of stomata per unit area of epidermis) should determine the maximum diffusive conductance to water vapor or CO_2 ($g_{w\,max}$ or $g_{c\,max}$, respectively) according to the formula given by Farquhar and Sharkey (1982):

$$(2.14) \qquad g_{w\,max} = \frac{\left(\dfrac{d}{V_m}\right) a_{max} D}{1 + \dfrac{\pi}{2}\left(\dfrac{a_{max}}{\pi}\right)^{1/2}},$$

where D is the diffusivity of water vapor in air (with units m²/s) and V_m is the molar volume of air (with units m³/mol) (note that the maximum CO_2 conductance equals $g_{w\,max}/1.6$). It can be shown mathematically that for the same total a_{max}, smaller stomata result in higher $g_{w\,max}$ compared with larger stomata. This feature emerges because $g_{w\,max}$ is inversely proportional to the distance that gas molecules have to diffuse through the stomatal pore (l), which increases with stomatal size as the kidney bean–shaped guard cells that flank the stomatal pore inflate to become approximately circular in cross section. The physiological and mechanical aspects of stomatal opening and closing are discussed in section 3.1. An interesting treatment of how gases diffuse into the living cells of wood is presented by Nowak and Hietz (2011).

2.3 Living in water versus air

The preceding introduction to environmental biophysics provides us with many of the mathematical tools required to compare how the physical properties of water and air affect the exchange of mass and energy between a plant and its fluid environment. For simplicity, we will consider a simple hypothetical plant that is a rigid cylinder measuring 0.5 m in length and 0.01 m in diameter living in a fluid (air or water) with an ambient velocity of 2 m/s.

We will first consider the passive diffusion of carbon dioxide from the surrounding fluid using the macroscopic form of Fick's law (see eq. [2.8]) and neglecting the resistance to passive diffusion created by cell walls. Reasonable estimates (box 2.1) indicate that the uptake of dissolved CO_2 molecules is more than 1,000 times faster in air than in water. Even though these estimates are crude, it is clear that terrestrial plants have a tremendous advantage over their aquatic counterparts in terms of the passive diffusion of CO_2. It should not surprise us to learn, therefore, that many aquatic vascular plants have very filamentous, flattened, or thin leaves

BOX 2.1 **Passive diffusion of carbon dioxide in the boundary layer in air and in water**

Consider the leaves of two different vascular plants that have adapted to living on land and under water, respectively, such as a tiger lily (*Lilium*) and a water lily (*Nelumbo*). Assume that these organs have a length of 1 cm and experience a relative ambient velocity of 2 m/s. For steady state photosynthesis, a reasonable estimate of the amount of CO_2 inside a plant is 7×10^{-3} mol/m³. Numerical values for the diffusivity and concentrations of CO_2 in air and water are provided in the table below.

Using equation (2.12a), we see that in air, $\delta = 5 \, [(15 \times 10^{-6} \, m^2/s) \cdot (0.01 \, m)/ (2 \, m/s)]^{0.5} = 1.3 \times 10^{-3}$ m, whereas in water, $\delta = 5 \, [(1 \times 10^{-6} \, m^2/s) \cdot (0.01 \, m)/ (2 \, m/s)]^{0.5} = 3.5 \times 10^{-4}$ m. Inserting these values into equation (2.8), we see that in air, $J_i = -(1.47 \times 10^{-5} \, m^2/s) \, (7 \times 10^{-3} \, mol/m^3 - 1.58 \times 10^{-2} \, mol/m^3)/ 1.3 \times 10^{-3}$ m $\approx 1.0 \times 10^{-4}$ mol m⁻² s⁻¹, whereas in water, $J_i = -(1.80 \times 10^{-9} \, m^2/s) \, (7 \times 10^{-3} \, mol/m^3 - 1.17 \times 10^{-2} \, mol/m^3)/3.5 \times 10^{-4}$ m $\approx 2.4 \times 10^{-8}$ mol m⁻² s⁻¹. The foregoing, which is summarized in the table (see below), draws attention to the importance of the very large difference in the values of the diffusion constants of CO_2 in air and in water.

Property	In air	In water
Diffusivity of CO_2	1.47×10^{-5} m²/s	1.80×10^{-9} m²/s
Boundary layer thickness	1.3×10^{-3} m	3.5×10^{-4} m
Concentration of CO_2[a]	0.0158 mol/m³	0.0117 mol/m³
Concentration difference ΔC_i	-8.8×10^{-3} mol/m³	-4.7×10^{-3} mol/m³
Flux J_i	1.0×10^{-4} mol m⁻² s⁻¹	2.4×10^{-8} mol m⁻² s⁻¹

[a]The concentration of atmospheric CO_2 has increased in the past 10 years to the current level of 385 ppm.
Noting that $PV = nRT$ and that $P = 1.0$ atm = 100,000 Pa, $V = 1.0$ m³, $T = 293.15$ K, and $R = 8.3145$ Pa m³ mol⁻¹ K⁻¹, it follows that 1 cubic meter of air at 1.0 atm and 20°C has a CO_2 concentration equal to 0.0158 mol/m³.

with large surface areas for their volumes (reexamine eq. [2.11] in light of this fact). In contrast to vascular terrestrial and aquatic plants that rely on passive diffusion to acquire CO_2, most algae possess active HCO_3^- uptake physiologies (Reynolds 2006; Raven 2010; Taiz and Zeiger 2010). In spite of this physiological difference and the fact that the concentration of HCO_3^- in water is much greater than that of CO_2, many algae also have very large surface areas, which expose almost every cell to HCO_3^- dissolved in the surrounding medium. In addition, some algae, such as *Volvox*, have flagella or cilia that stir the boundary layer around them (Rushkin et al.

2010), which reduces its thickness and produces a steeper HCO_3^- gradient and thus a higher flux (see eq. [2.1]).

The difference between the transfer of heat in the aquatic environment and the transfer of heat in the terrestrial environment can be estimated from equation (2.10), expressed simply as $H = -k\Delta T/\delta$. Vogel (2009) has shown that the overheating of leaves in bright sunlight is a function of their size and shape and estimated how heat is dissipated under different wind regimes. It is clear from the numerical values for k given in table 2.1 that estimates of heat transfer in air and in water would differ significantly, albeit not as dramatically as in our preceding comparisons.

Turning our attention to the magnitudes and effects of mechanical forces, we realize that these forces are quite different in air and in water. In water, gravitational forces on plants are small, while drag forces play a dominant role (Denny 1988). For Re $\gg 1$, the drag forces generated on our hypothetical cylindrical and rigid plant can be estimated by the general equation for the drag force D_f:

(2.15) $$D_f = 0.5\,\rho\,A\,U^2\,C_D,$$

where ρ is the density of the fluid, A is the projected surface area of the plant, and C_D is known as the drag coefficient (which is a numerical factor that compensates for the effect on the drag force of the geometry and size of an obstructing object as well as for higher-order effects of fluid flow and speed). Equation (2.15) shows that the drag force increases linearly with increasing projected area A and increases as the square of fluid velocity U. It also tells us that the density of the fluid is of particular importance, since water is roughly 1,000 times denser than air (see table 2.1).

Equation (2.15) is an especially useful formula. It can be used to estimate the drag forces exerted on leaves, stems, and even entire trees. However, fluid mechanical theory provides only approximate estimates of C_D for very simple cases. Nevertheless, empirically, we know that the numerical value of C_D depends on the Reynolds number, which is easily calculated for our hypothetical plant. Specifically, because our plant is a rigid cylinder, flow at a right angle to body length (such that the characteristic dimension ℓ now becomes 0.01 m) is such that, in air, Re $= \ell U/v =$ (0.01 m) (2 m/s)/(15 × 10^{-6} m^2/s) = 1.33 × 10^3, whereas in water, Re $= \ell U/v =$ (0.01 m) (2 m/s)/(1.0 × 10^{-6} m^2/s) = 2.0 × 10^4. Surprisingly, however, this difference does not matter very much, because across the range $10^2 \le$ Re $\le 10^5$, the drag coefficient varies little numerically and holds to a value of roughly 1.0 for any smooth cylindrical object (see fig. 6.8). Inserting this value into equation (2.15) gives a drag force of approximately 0.012 kg m s^{-2} in air and roughly 10 kg m s^{-2} in water. Thus, when the flow velocity U is the same, the two drag forces can differ by nearly three orders of magnitude, simply because of the difference between the density of water and air (which differs by three orders of magnitude). These calculations illustrate once again the tremendous disparity in the mechanical forces that moving water and air can generate. Plants anchored to a substrate and living in rapidly moving water must cope with potentially very large drag forces. One solution is to bend and thus reduce projected surface areas. Another

FIGURE 2.1. Side (A) and head-on (B) views of a fig tree (*Ficus* sp.) growing in a windswept habitat where the direction of airflow is relatively uniform (along the length of the south shore of Hawaii).

is to grow in microhabitats where water moves slowly. For wave-swept environments, a third solution is to grow to a size that permits the movement of body parts to accord almost exactly with the oscillatory movements of local water currents. In this way, individual body parts are actually stationary relative to the flow of water around them (Koehl 1999).

Like moving water, wind affects plant growth and form. A well-known example is the preferential growth and survival of branches on the leeward sides of trees growing in habitats characterized by strong winds, a phenomenon sometimes referred to as "flagging" (fig. 2.1). Asymmetric growth induced by wind loading also occurs in root systems. Roots tend to grow more extensively in the direction of the maximal loads induced by the dynamic loading of the aerial portions of the tree. This phenomenon increases root anchorage. In windy environments, even without a preferred direction, trees tend to be shorter and to have thicker trunks and branches compared with those growing in sheltered environments (see section 2.7).

Indeed, wind-plant interactions cover a large range of mechanisms operating at the level of the individual plant all the way to the level of the entire forest canopy, such as tree wind-throw, long-distance dispersal of propagules (seeds, fruits, and spores), large-scale patterns of gas exchange, and thermal effects on leaves. Of particular interest from both a practical and a theoretical perspective are the effects of steady, or dynamic, wind on plants. These effects, which have been reviewed by de Langre (2008), will be discussed in greater detail in chapters 5 and 6.

2.4 Light interception and photosynthesis

With the exception of a few comparatively rare ecosystems, such as deep-sea thermal vents, the world's great food webs rely on the interception of sunlight by plants. For this reason alone, the topic of light interception is important not just to those of us interested in environmental biophysics, but to every human being. Virtually every aspect of plant life is influenced or regulated by sunlight. In particular, photosynthesis, plant growth, and many aspects of development depend on light. For these reasons, many plants have rather complex responses to even subtle changes in their light environment. Knowledge of the optical properties of different tissues is a prerequisite for understanding light absorption by chlorophyll for photosynthesis and by photoreceptors that control plant growth and development.

The physics of light interception can be approached at a variety of levels of biological organization, from the molecular and organismal levels to the structure and composition of entire communities. Here, we will review the physics of light interception at the molecular, cellular, and organ levels of organization.

We begin with Beer's law for dilute solutions, which states that the attenuation of light through a light-absorbing solution is given by the formula

(2.16)
$$A_\lambda = \log_{10}\frac{I}{I_b} = \varepsilon_\lambda\, cx,$$

where A_λ is the absorbance (also called the "optical density") of the solution, I is the incident light intensity, I_b is the attenuated light intensity measured at a distance x, and ε_λ is the extinction (also called the absorption) coefficient of a solute for wavelength λ. For dilute solutions, ε_λ is independent of the concentration c. Equation (2.16) can be generalized for any solution containing two or more light-absorbing solutes:

(2.17)
$$A_\lambda = \log_{10}\frac{I}{I_b} = \sum_j \varepsilon_{\lambda j}\, c_j x,$$

where $\varepsilon_{\lambda j}$ and c_j are the extinction coefficient and the concentration of solute j, respectively.

Both of these equations can be used to calculate the concentration of a solute, provided that its extinction coefficient is known. As an example, in box 2.2, we give a rough estimate of the chlorophyll concentration in a

---

BOX 2.2 **Absorption of light by chloroplasts**

At wavelengths around 450 nm and 675 nm, a typical dicot leaf with a transverse thickness $b = 2 \times 10^{-4}$ m absorbs about 99% of the incident sunlight passing through it. Thus, referring to equation (2.16), we see that $I_b = 0.01\, I$. The extinction coefficient $\varepsilon_{\lambda j}$ of chlorophyll in the red and blue bands of its light absorption spectrum is about 10^4 m^2/mol and less in the green band, which explains why the dorsal surfaces of most leaves look green in both transmission and reflection. Inserting these relationships into equation (2.16) (and neglecting light scattering and reflection) gives $A_\lambda = \log(1/0.01) = 2$ such that $c = 2/[(10^4$ m^2/mol$)\, (2 \times 10^{-4}$ m$)] = 1$ mol/m^3.

BOX 2.2 **(Continued)**

Naturally, chlorophyll is concentrated in chloroplasts, which in turn are found in greater numbers on the dorsal (adaxial) surfaces of most leaves. The average chlorophyll concentration in a chloroplast is roughly 30 times greater than that of a leaf—that is, $c_{\text{chloroplast}} = 30$ mol/m^3—and the average diameter of a chloroplast is 2×10^{-6} m. Therefore, $A_\lambda = (10^4$ m^2/mol$)$ $(2 \times 10^{-6}$ m$)$ $(30$ mol/m$^3) = 0.6 = \log(I/I_b)$. Thus, $I_b = I/(\text{antilog } 0.6) = 0.25 \, I$, which indicates that a chloroplast absorbs about 75% of incident red and blue sunlight.

leaf. More refined calculations have to take into account the reflection of light at external and internal surfaces, light scattering, and the so-called sieve effect. A sophisticated treatment of photon absorption in leaf tissues is provided by Fukshansky (1991) and Fukshansky et al. (1993). More broadly, plant tissue optics have been reviewed by Vogelmann (1993).

The reflection of light off the leaf and stem surface is often neglected when considering light interception. Yet as much as 5% of the light striking the upper (adaxial) surface of a leaf may be reflected away and thus photosynthetically lost. Likewise, because of the asymmetry in leaf anatomy, more than twice as much light may be reflected from the lower (abaxial) side of a leaf. In both cases, light is reflected either from the cuticle-air interface or from the numerous cell-air interfaces in the mesophyll. The spectral composition of light reflected from the interior of the leaf is strongly influenced by the absorption characteristics of the leaf pigments. In contrast, light reflected from the external surface is usually not changed spectrally. An exception is seen in the blue spruce (*Picea pungens*), whose blue coloration results from Rayleigh scattering within the waxy layer covering the cuticle. Another exception is found in some tropical plants, such as several species of *Selaginella*, whose leaves exhibit a striking blue iridescence (Lee and Lowry 1975). This optical phenomenon is due to thin-layer interference resulting from two or more thin cellulose laminae located within the outermost epidermal cell wall.

Light scattering within a tissue increases the path length of photons and therefore their probability of absorption. Since light scattering is wavelength dependent, it also has an influence on the actual absorption spectrum of a leaf, which may be different from the absorption spectrum of an isolated pigment. Therefore, comparison of an action spectrum with the absorption spectrum of a putative photoreceptor may not be straightforward.

Light scattering is also the basis of nonimaging optics in plants. Plants utilize this phenomenon in a way that is very different from its technical use (Winston 1991). A striking example is the succulent Window Plant (*Fenestraria rhopalophylla*) from Namibia and South Africa. In its native habitat, the plant body is buried in sand, except for a transparent window called the cupola. Water-filled cells in the cupola absorb infrared light and act as heat filters. Visible light enters the cupola from a wide range of angles, and it is refracted into and scattered evenly throughout the rest of the conical plant body to reach the photosynthetic cells in the periphery of the plant body, which is close to and typically buried in the cooler soil. The unusual optics of this plant can be demonstrated in reverse by illuminating the roots, whereupon the window has the appearance of a miniature floodlight (Tributsch 2001). Similar light scattering–refraction phenomena have been observed for plants with heterobaric leaves; that is, leaves with heterogeneous pigmentation resulting from a network of transparent areas that are created by extensions of the bundle sheaths around veins. This mechanism may allow leaves to grow to a greater thickness and increase their photosynthetic capacity per unit (projected) area, which would provide adaptive advantages in very dry and hot environments (see Nikolopoulos et al. 2002).

The importance of light scattering, refraction, and absorbance to plant biology cannot be overestimated. Consider a phenomenon called the *sieve effect*, which is apparent in the optical properties of suspensions, such as a suspension of chloroplasts or unicellular algae (as compared with a solution with the same amount of chlorophyll). In suspensions, the path of a photon has a finite probability of not encountering a suspended chloroplast or algal cell. Correspondingly, the absorption of light is decreased as compared with homogeneous solutions containing the same amount of pigments, in which this probability is much lower. Because the sieve effect increases with the size of the cells or plastids, the efficiency of light interception decreases with the size of the suspended particulate (which could be yet another reason why most unicellular plants are comparatively small).

For a quantitative treatment of the sieve effect, Beer's law can be recast to predict the attenuation of light per unit horizontal area as it passes through an aqueous suspension of cells:

(2.18) $$\log_{10}\frac{I}{I_b}=(k+nAa)b/\sin\beta,$$

where I_b is light intensity measured at a distance b from the surface of the suspension, I is light intensity at the surface (i.e., $b = 0$), k is the extinction coefficient of the liquid medium (water and any dissolved materials), n is the number of cells in suspension, A is their average projected area, a is the average fraction of incident light intercepted per cell, and β is the incident angle of light parallel to the surface. Notice that $\log_{10}(I/I_b) = (k + nAa)b$ when the incident angle of light is normal (at a right angle) to the surface of the suspension ($1/\sin 90° = 1$). The only variable in this formula that presents an obstacle to calculating I_b is the numerical value of a. Fortunately, for spherical cells with diameter D, Duysens (1956) has shown that

(2.19) $$a=1-\frac{2[1-(1+\varepsilon c_i D)\exp(-\varepsilon c_i D)]}{(\varepsilon c_i D)^2},$$

where ε is the absorption coefficient and c_i is the internal pigment concentration. For spherical cells (and assuming that the total volume of all cells is constant), we see that n is proportional to D^{-3} and that A is proportional to D^2. Therefore, nA is proportional to D^{-1}. Inserting this relationship into equation (2.19) reveals that nAa is a monotonically decreasing function of D; that is, the average absorption cross section decreases for any wavelength of light as D increases. The decrease in nAa is most prominent at wavelengths around 465 nm (blue) and 665 nm (red), where the extinction coefficients of chlorophyll are maximized (Kirk 1975a,b; Taiz and Zeiger 2010).

Similar calculations reveal that randomly oriented cells with prolate, oblate, or cylindrical geometries can be equally effective at capturing sunlight provided their sizes are kept below thresholds specific to each geometric class (Kirk 1976; Niklas 1994). Indeed, this general approach to estimating light interception capacity can be extended to comparatively complex geometries, some of which are relevant to terrestrial plants (box 2.3). For example, one convenient method for estimating the light-harvesting ability of any particular class of geometric objects is to calculate the area under the curve generated by plotting the quotient of the projected A_p and total surface area A of the object in question against the

BOX 2.3 **Formulas for the effective light absorption cross section of some geometric objects**

Let A_p/A be the quotient of projected area A_p and total surface area A. The area under the curve that results from plotting A_p/A against the angle θ between the solar beam and the main axis of an object provides a crude estimate of the ability of an object to capture sunlight daily. The following formulas are useful in this regard:

(2.3.1a)
$$\text{Spheres:} \frac{A_p}{A} = \frac{\pi D^2}{4\pi D^2} = \frac{1}{4}$$

(2.3.1b)
$$\text{Oblate spheroids:} \frac{A_p}{A} = \frac{ap\,|\sin(\theta+\phi)|}{2\pi a^2 + \pi b^2 \ln\left(\dfrac{1+e}{1-e}\right)}$$

(2.3.1c)
$$\text{Prolate sheroids:} \frac{A_p}{A} = \frac{1+(x-1)\cos\theta}{2x + \dfrac{2\arcsin(1-x^2)^{1/2}}{(1-x^2)^{1/2}}}$$

(2.3.1d)
$$\text{Cylinders:} \frac{A_p}{A} = \frac{\cos\theta + \dfrac{4L\sin\theta}{\pi D}}{2 + \dfrac{4L}{D}}$$

(2.3.1e)
$$\text{Cylinders with hemispherical ends:} \frac{A_p}{A} = \frac{1 + \dfrac{4L\sin\theta}{\pi D}}{4 + \dfrac{4L}{D}}$$

where a denotes the spheroid major axis, b is the spheroid minor axis, $x = b/a$, e is the eccentricity of an elliptical cross section, L is length, D is diameter, $p = (a^4 \tan^2\theta + b^4)^{1/2}/(a^2 \tan^2\theta + b^2)^{1/2}$, and $\phi = \arcsin[b^4/(a^4 \tan^2\theta + b^4)]^{1/2}$.

angle of solar radiation θ incident to a solid body as this angle changes over the course of a day.

Notice that the only geometry that cannot change A_p/A is the sphere. Obviously its light-harvesting capacity is totally insensitive to its orientation with respect to the incident angle of sunlight. There are few spherical plants, but they do exist. The green alga *Volvox* is a hollow sphere composed of a single layer of flagellated cells, which give it the ability to rotate and move through the water. In terms of other simple geometries, integrating equations (2.3.1b)–(2.3.1f) over $0° \le \theta \le 90°$ shows that the light-harvesting capacity C of oblate and prolate spheroids is very sensitive to changes in the shape of either type of solid body. For oblate spheroids, C increases as a spheroid progressively flattens, whereas for prolate spheroids, C decreases as this geometry becomes more slender. A similar approach to the one outlined in box 2.3, in which the quotient of projected surface area and actual surface area once again appears, can be used to evaluate light interception in plant canopies differing in the orientation of their leaves (box 2.4). These simple calculations help to explain why the foliage leaves of many land plants are somewhat like extremely flattened oblate spheroids with large b/a. They also help to explain why some of the earliest vascular land plants, which lacked leaves, had cylindrical photosynthetic stemlike axes, which were theoretically good at intercepting light even as they increased in length by means of apical meristems.

BOX 2.4 **Modeling light interception in canopies**

Advanced treatments of the penetration of light through a plant canopy (Ross and Marshak 1991) take into account the "sieve effect" (Fukshansky 1987; Anisimov and Fukshansky 1993) and radiation scattering, which results from multiple reflections from the morphological upper or lower sides of leaves, leading to an extended path length (Nilson 1991; Shabanov et al. 2000, 2007). Since the mathematical complexities of these approaches are beyond the scope of this book, we follow Monteith (1973) and provide only a first-order approximation of how to estimate light penetration through a canopy.

We limit our attention to direct and diffuse short-wave radiation in a hypothetical canopy consisting either of horizontally oriented or of randomly distributed leaves divided into a large number of very thin horizontal layers such that there are no overlapping leaves in any particular layer.

BOX 2.4 **(Continued)**

Under these circumstances, the radiant energy passing through one layer onto the layer below it can be expressed as

(2.4.1)

Layer	Radiant energy	$\gamma \ll 1$
0	1	
1	$1 - \gamma$	$\approx e^{-\gamma}$
2	$(1 - \gamma)^2$	$\approx e^{-2\gamma}$
n	$(1 - \gamma)^n$	$\approx e^{-n\gamma}$
N	$(1 - \gamma)^N$	$\approx e^{-N\gamma}$,

where the radiant energy in the uppermost layer is normalized to 1. For simplicity, it is assumed that the degree of shadowing γ is constant throughout the different layers. A more refined analysis must account for the exact structure of the canopy.

The fraction of the radiant energy reaching the ground level (layer N) is given by $e^{-N\gamma}$. Therefore, the fraction of the radiant energy captured in the canopy is given by

(2.4.2)
$$E_{captured} = 1 - e^{-N\gamma},$$

whereas the average radiant energy captured by the canopy is given by

(2.4.3)
$$\bar{E}_{captured} = \frac{1}{N} \int_0^{N-1} e^{-n\gamma} dn = \frac{1 - e^{-(N-1)\gamma}}{N\gamma}.$$

The degree of shadowing γ is a function of the wavelength λ of the light; that is, $\gamma = \gamma(\lambda)$. If, for particular wavelengths λ^*, the leaves are opaque, we see that $N\gamma(\lambda^*)$ is the total coverage with leaves. For other wavelengths, the values have to be corrected for the transparency of individual leaves. In addition, the orientation of leaves with respect to incident light has to be accounted for. This is done by introducing the factor (A_p/A), the quotient of the projected area A_p and the actual area A of a leaf, such that γ in the preceding equations is replaced by $\gamma(A_p/A)$, where $(A_p/A) \leq 1$.

Several hypothetical cases leading to different values of (A_p/A) and, therefore, different light capture efficiencies (eq. [2.4.2]) can be distinguished:

BOX 2.4 **(Continued)**

- Case 1: Direct sunlight from above, all leaves are oriented horizontally: $(A_p/A) = 1$.
- Case 2: Direct sunlight with an inclination angle ϕ, all leaves are oriented horizontally: $(A_p/A) = \sin\phi$. When averaged over the daylight hours, we find that

(2.4.4) $$(A_p/A)_{averaged} = \frac{2}{\pi}\sin(\phi_{max}).$$

- Case 3: Diffuse light, all leaves are oriented horizontally. Diffuse light is envisioned as light coming from an evenly illuminated half sphere. Therefore, the correction factor averaged over the half sphere is given by

(2.4.5) $$(A_p/A)_{averaged} = \int_0^{\pi/2} \sin\phi\cos\phi d\phi = 1/2.$$

- Case 4: Direct sunlight from above, the leaves are oriented randomly. Random orientation is envisioned by orienting the leaves with pitch angles $0 < \alpha < \pi$ and roll angles $0 < \beta < \pi$, each angle with the same probability. The factor is given by

$$(A_p/A)_{averaged} = \frac{1}{\pi^2}\int_0^\pi\int_0^\pi \cos(\pi/2-\alpha)\cos(\pi/2-\beta)d\beta d\alpha = \frac{4}{\pi^2}.$$

(2.4.6)

- Case 5: Direct sunlight from an inclination angle ϕ, the leaves are oriented randomly. Without loss of generality, the pitch angle may be defined in the same plane as the inclination angle. Using this definition, the morphological upper side of a leaf is lit in the range $0 \le \alpha \le \pi - \phi$ and the morphological lower side in the range $\pi - \phi \le \alpha \le \pi$. Disregarding different reflection coefficients, shadowing is the same whether the morphological upper or lower side of a leaf is lit (compare Monteith 1973). Under this premise, $(A_p/A)_{averaged} = 4/\pi^2$ for randomly oriented leaves is the same as that for case 4, independent of the inclination angle ϕ. As compared with case 2, a random orientation is advantageous if $4/\pi^2 > 2/\pi \sin(\phi_{max})$. On a yearly average, this is the case for latitudes above

BOX 2.4 **(Continued)**

$39.5°$. If leaf shedding during the winter months is taken into account, random orientation of the leaves is advantageous at even lower latitudes.

- Case 6: Diffuse light, the leaves are oriented randomly. Under the same premises as for case 5, where $(A_p/A)_{averaged}$ is independent of the inclination angle, (A_p/A) averaged over all possible orientations and all possible inclination angles is equal to $4/\pi^2$ in this case as well.

It is worth noticing that each of the formulas presented in box 2.3 contains one or more variables that describe size or shape (e.g., x, D, L, a, and b). Once again, we see that changes in plant size, shape, or geometry during ontogeny or evolution can alter the manner in which physical laws or processes affect the performance of vital biological tasks, such as the harvesting of sunlight.

No treatment of light interception is complete without an account of the fundamental physiological process that distinguishes most plants from other forms of eukaryotic life: photosynthesis, the process that allows plants to harness the energy of light and convert it into chemical energy. The basics of photosynthesis are complex, from the perspective of both physics and chemistry. Consider light energy and the manner in which it is captured by a light-absorbing pigment. Light has the properties of both waves and particles. This seeming paradox, known as wave-particle duality, stands at the basis of modern physics. In interference phenomena, as observed at the beginning of the nineteenth century by Augustin-Jean Fresnel and Thomas Young, light appears as a wave—more exactly, as electromagnetic radiation. A wave is characterized by its wavelength λ and its frequency v. The product of these two attributes gives the speed c at which the wave travels (i.e., $c = \lambda v$). In the case of light, c in a vacuum is a constant (i.e., 3.0×10^8 m/s). In other respects, as in the photoelectric effect, explained by Einstein in 1905, light appears to be composed of particles, later called photons, with a discrete quantum of energy E and a finite impulse $p = E/c$. The energy E of a photon is given by Planck's law, $E = hv$, where h is a constant ($h = 6.626 \times 10^{-34}$ J s), named after Max Planck, who first proposed that electromagnetic energy could be emitted only in quantized form.

White light is composed of photons with many different quanta. The spectrum of light that is visible to humans and many other vertebrates ranges between the wavelengths of 400 nm and 700 nm. It therefore falls between the ultraviolet and the infrared portions of the electromagnetic spectrum. Photosynthetic organisms, which include some bacteria, have evolved an assortment of pigments that can harvest the energy of different wavelengths of light, including carotenoids, bilin pigments, and phycoerythrobilin. However, with few exceptions, the main light-harvesting pigments are chlorophyll *a* and chlorophyll *b*. Both pigments have a porphyrin-like ring structure coordinated by a centrally located magnesium atom (fig. 2.2). This ring structure is attached to a hydrophobic hydrocarbon chain containing 23 carbon atoms, which anchors the chlorophyll molecule to a photosynthetic membrane. Among photosynthetic bacteria, photosynthetic membranes are part of the plasma membrane (or

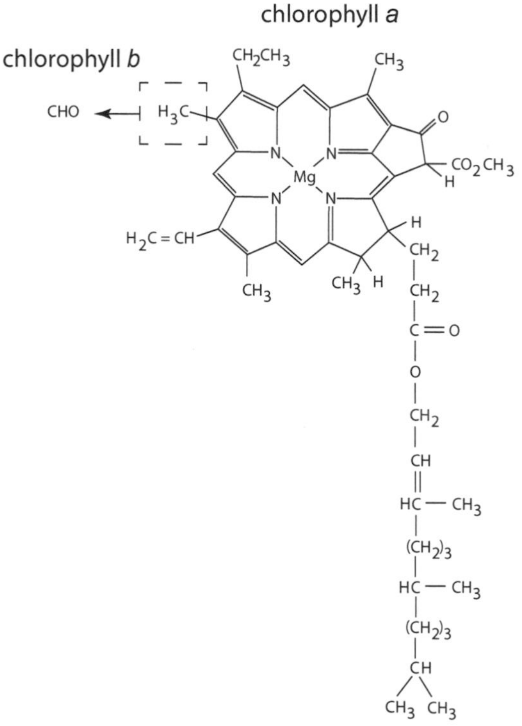

FIGURE 2.2. Molecular structure of chlorophylls *a* and *b*. Both pigments have a porphyrin-like ring structure with a central coordinated magnesium atom. The two pigments differ only in terms of whether $-CH_3$ or $-CHO$ occupies the position indicated by the dashed square.

are derived from it by invagination). Among eukaryotic photoautotrophs, the photosynthetic membranes are located in chloroplasts.

The total absorption spectrum of chlorophylls *a* and *b* falls in the range of visible light wavelengths. However, both pigments absorb predominantly in the blue and in the red portions of the visible spectrum. The wavelengths of their absorption peaks depend on the medium in which the molecules are suspended. For example, in diethyl ether, chlorophyll *a* has absorption peaks at approximately 430 nm and 662 nm, whereas chlorophyll *b* has absorption peaks at roughly 453 nm and 665 nm. Chlorophylls appear green to us because they absorb little or no energy in the middle, greenish part of the spectrum. A number of secondary light-harvesting pigments have evolved to utilize the energy that is not available to chlorophyll molecules because of their comparatively restricted absorption peaks. The carotenoids are an important category of these pigments because some of them have the capacity to transfer light energy to chlorophyll molecules when placed in an appropriate physiological context; for example, β-carotene absorbs wavelengths of light between about 380 nm and 525 nm.

When a molecule of chlorophyll *a* or *b* absorbs a photon, its energy level is elevated. In this excited state, the chlorophyll molecule is unstable and releases its excitation energy rapidly in one of four ways:

1. It can return to its ground state directly by converting its excitation energy into heat.
2. It can emit a photon with a wavelength that is longer than that of the light it absorbed and convert the remaining excess energy into heat (a process called fluorescence).
3. It can transfer its excitation energy to another molecule (a process called energy transfer).
4. It can divert its excitation energy to drive a photochemical reaction.

Photochemical reactions are among the fastest chemical reactions known (on the order of 10^{-9} s). This speed enables route 4 of energy dissipation to outcompete the other three potential routes by which chlorophyll molecules can return to their ground states.

The conversion of light energy into chemical energy involves at least 50 intermediate reaction steps that depend on the organized activity of many pigment molecules and a number of electron transfer proteins. In plants, most of the pigment molecules contribute to what are called antenna complexes. These molecules collect light energy and transfer it to chlorophyll molecules and electron transfer proteins arranged into structures called reaction centers. The chemical reactions required to store energy occur in the reaction centers. By way of a crude analogy, the antenna complex functions like a series of photovoltaic cells connected to the reaction center, which functions like a battery. The sequence of pigments in the antenna complex is not random. It is arranged so that the absorption maxima of successive molecules progressively shift toward longer and longer (red) wavelengths; for example, carotenoids are connected to chlorophyll *b* molecules, which are connected to chlorophyll *a* molecules. This arrangement of sequentially lower excitation energy levels funnels energy toward the reaction center complex.

In contrast to the large number of pigment molecules in the antenna complexes, the reaction center consists of only several hundred chlorophyll molecules that receive energy from the antenna complex. In each reaction center, specific chlorophyll molecules donate electrons to a series of molecular intermediates, collectively called the electron transport chain. An electron donor then reduces the chlorophyll molecules so that the process can be repeated. Thus, photosynthesis is a very sophisticated reduction-oxidation series of chemical reactions—indeed, it is fair to say that it may be the most sophisticated redox reaction in the world (Taiz and Zeiger 2010).

There are only a very few kinds of reaction center configurations, compared with the large number of different pigment sequences that make up antenna complexes. The conservative nature of the reaction center's makeup attests to the antiquity of its evolutionary origin and to the lethality that results if it is significantly altered by means of natural mutations. Some photosynthetic bacteria, such as the purple photosynthetic bacteria, have a single type of reaction center. In contrast, photosynthetic eukaryotes that release oxygen during photosynthesis have two kinds of reaction centers, which are used to construct photosystems designated as photosystem I and photosystem II (or simply as PSI and PSII). These two photosystems are linked by an electron transport chain of molecules into a system that is traditionally diagrammed as a zigzag or Z scheme (fig. 2.3). The chlorophyll molecules at the centers of PSI and PSII differ in their maximum absorption wavelengths. Those in PSII absorb at a maximum wavelength of 680 nm and those in PSI absorb at a maximum of 700 nm. These chlorophyll molecules are denoted by P680 and P700, respectively. In terms of their redox reactions, PSII produces a weak reductant and a

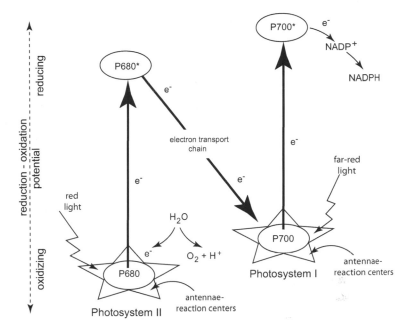

FIGURE 2.3. Schematic of the Z scheme of photosynthesis, which consists of two photosystems (PSI and PSII) connected by an electron transport chain. Red light that is absorbed by PSII produces a strong oxidant, which oxidizes water, and a weak reductant, which is an excited chlorophyll pigment (P680*); far-red light that is absorbed by PSI results in a weak oxidant and strong reductant, which reduces NADP$^+$ to form NADPH.

very strong oxidant, which oxidizes water. PSI produces a weak oxidant and a strong reductant, which reduces NADP$^+$ to NADPH via a sequence of electron transport molecules. Thus, PSII is responsible for the oxidation of water, which releases oxygen. The evolution of this photosystem in cyanobacteria-like organisms billions of years ago changed Earth's atmosphere from a reducing to an oxidizing atmosphere. This change permitted the evolution of aerobic life-forms such as ourselves. It also forced obligate anaerobic bacteria to occupy limited and, in many cases, specialized habitats.

In terms of efficiency, early experiments by R. Emerson and W. Arnold in 1932 showed that approximately 2,500 chlorophyll molecules participate to produce one molecule of O_2. Only later was it realized that most of these chlorophyll molecules serve in antenna complexes. Singsaas et al. (2001) reviewed a large number of determinations of the quantum efficiency of photosynthesis. Because of the ease with which the quantum efficiency φ can be underestimated, they rejected the hypothesis that the intrinsic efficiency of photosynthesis varies substantially among plant species or growth conditions unless plants experience stressful environments. Thus, their analysis supports the contention that the intrinsic quantum yield varies little from the mean values of 0.092 for ϕCO_2 and 0.108 for ϕO_2 in unstressed plants. This maximum quantum yield indicates that between 50 and 60 photons are required to drive the chemical reaction depicted by a simple "black box" formula for oxygenic photosynthesis; that is, $6\,CO_2 + 6\,H_2O \rightarrow 6(CH_2O) + 6\,O_2$, where $6(CH_2O)$ is the equivalent of one glucose molecule. Despite this remarkable level of efficiency, it is worth noting that chlorophyll is not an exceptionally good light-harvesting molecule. Better light-harvesting pigments are theoretically possible. Indeed, this might help to explain the slight molecular differences ("tinkerings") among the different kinds of chlorophyll molecules found in the land plants and different algal lineages. Nevertheless, despite this evolutionary "fine-tuning," the basic structure of chlorophyll has been conserved for billions of years. The reasons for this conservation may be varied, but there is an old adage that may give us insight: "if it isn't broken, don't fix it." Once a complex and physiologically important series of physical and chemical reactions such as photosynthesis evolves, dramatic changes resulting from spontaneous mutations are extremely unlikely because they are very likely to be lethal. Put differently, "if an important part of you works, don't change it because it might kill you."

2.5 Phototropism

Using the seedlings of sunflowers and other plant species, Charles Darwin (1880) was one of the first to draw attention to the bending of plants toward visible light (Iino 2001; Kutschera and Niklas 2009). This phenomenon, which is now called positive phototropism, is due to the greater expansion of cells on the shaded side than on the illuminated side of the plant axis. This differential growth is mediated by phototropins, light-sensitive proteins that preferentially absorb light at wavelengths between 350 and 500 nm (and in the UV with an absorption maximum at 270 nm) (Christie and Briggs 2001).

Curiously, however, the interpretation of classical observations (Blaauw 1909) is not always straightforward. For instance, the dose response of phototropism is not monotonic. Among vascular plants, doses of light energy between 10^{-1} and 3 J/m² lead to a positive bending response, whereas for doses between 3 and 10^3 J/m², the response is reduced. This indicates that the rate of expansion on the shaded side is decreasing with increasing doses of light, while the reduction in the rate of expansion on the side facing the light at intermediate doses (3 and 10^3 J/m²) indicates that the response is saturated. At even higher doses (between 10^3 and 3×10^4 J/m²), the response increases once again, indicating that two or more different photoreceptor systems participate in the response. Indeed, the second positive response at high light intensities, although showing the same action spectrum, has different characteristics. It is elicited by illumination of the lower parts of coleoptiles as well. Most important, the positive response at low light intensities conforms to the reciprocity law; that is, it depends only on the number of light quanta absorbed. In contrast, the second positive response does not follow the reciprocity law, but rather depends additionally on the length of exposure to light.

In *Avena* coleoptiles, even negative phototropism is observed at intermediate doses of light (Steyer 1967). This complex behavior can be explained only by inhibitory interactions between the different photoreceptor systems and the chain of events affecting elongation. Specifically, both photoreceptor systems lead to a reduction of the rate of elongation, but the high-intensity system may inhibit or even shut off the effects of the low-intensity system. Since phototropism is a differential response, this can lead to a diminished or even negative response at intermediate doses.

Positive phototropism in visible light is also observed in the sporangiophores (spore-bearing structures) of the fungus *Phycomyces* (Galland 1990). This phenomenon appears puzzling at first sight because light evokes increased growth of the sporangiophores. The explanation lies in the optical properties of these structures. With a refractive index of 1.37, the sporangiophore focuses visible light on its far side, such that the effective light intensity is higher on the far side than on the side facing the light source (Shropshire 1962). The sporangiophores of *Phycomyces* are unicellular, so their curvature toward the light must involve differential cell wall expansion within a single cell. In addition to positive phototropism in visible light, sporangiophores manifest negative phototropism when exposed to UV wavelengths. Previously, this phenomenon was attributed to the presence of gallic acid in the central vacuoles of sporangiophores, which absorb UV light and thus prevent UV focusing on the far side of each sporangiophore. It has, however, been shown that some sort of lens effect is also operating in the UV spectral range. On the basis of physiological and mutant analysis, it is therefore suggested that a second UV receptor, possibly located in the cytoplasm, is responsible for the negative phototropism under UV light (Popescu et al. 1989).

2.6 Mechanoperception

Plants perceive different static and dynamic mechanical stimuli such as gravitational forces, drag in moving fluids, wind or water currents, or touch. Braam (2005) and Telewski (2006) have reviewed the variety of ways in which plants respond to these stimuli. For example, most plant stems and leaves display negative gravitropism, while roots show positive gravitropism (see section 2.8). Nevertheless, the downward growth of roots changes when the growing tip perceives an obstacle such as a stone. In a sideways avoidance reaction, touch-stimulated roots grow away from the point of mechanical contact. This behavior, which is called thigmotropism (Θιγημα is the Greek word for "touch"), is a response not only to touch, but also to the direction of the mechanical stimulus. Another manifestation of mechanoperception is thigmomorphogenesis, the effect of mechanical perturbation on the shape of a cell, organ, or plant (see section 2.7).

One classic example of the response of plants to touch is the rapid thigmonastic movement of insectivorous plants such as the Venus flytrap (*Dionea muscipula*). This response is elicited when mechanosensitive trigger hairs on the upper leaf surface are stimulated mechanically by touch, which leads to the closure of the leaf-trap within seconds (see fig. 7.4). The movement of the tentacle-like plant hairs (called trichomes) on the leaves of the sundew *Drosera rotundifolia* caught the attention of Charles Darwin (1893), who reported that the weight of a human hair was sufficient to induce this response, whereas neither rain nor wind triggered it. Another well-known example of mechanoperception is the response of *Mimosa pudica* leaves to mechanical stimulation: a rapid folding of leaflets and the downward movement of the entire compound leaf (fig. 2.4). It has been suggested that the propagation of leaflet folding involves the transmission of electrical signals analogous to action potentials in neurons (see chap. 7). Another example of touch stimulus is seen in the behavior of the underwater traps of the carnivorous bladderwort *Utricularia*. An elegant study of these structures using finite element analysis (see section 10.5) shows that, when touched, the flexible valve of the trap buckles under

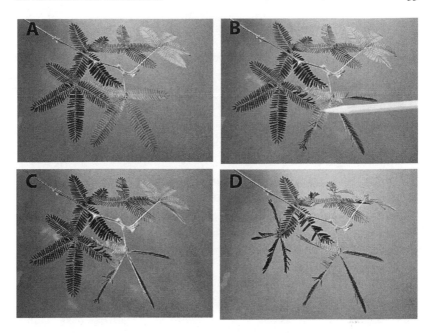

FIGURE 2.4. *Mimosa pudica* (A, unstimulated), when mechanically stimulated (with a pencil, B) shows a rapid leaf-folding response (C–D). Time intervals between B and C and between C and D are approximately 5 seconds.

the combined effects of pressure forces and the mechanical stimulation of trigger hairs (Joyeux et al. 2011).

Other, albeit less well-known, examples of mechanoperception are seen in lianas and vines. Isnard and Silk (2009) have reviewed the climbing habit and the biomechanics and hydraulics of these growth forms. For example, when the young stiff "searcher" shoots of some liana species reach a mechanically self-supporting structure (such as the trunk of a neighboring tree), stems change from producing early dense wood lacking large vessels to producing less dense wood with a high percentage of very large-diameter vessels. This developmental transition makes the wood more flexible and thus capable of tolerating the movements of the supporting tree trunk (Rowe and Speck 1996). Likewise, the stems of vines respond to mechanical stimulation by twining around vertical supports. The mechanics of the twining growth habit involves an innate tendency to form coils of smaller radii than those of a supporting pole or stem. The result is a stem placed in torsional tension as it clings to a vertical support.

More loosely coiled helices, when supported at the upper end, will wrap under tensional loads—exerted by their own weight—more closely around the supporting pole. In this case, friction between stem and supporting pole is supposed to play an important stabilizing role (Silk and Holbrook 2005).

The thigmonastic leaf tendrils of the cucurbit *Bryonia dioica* start to coil a few seconds after mechanical stimulation. This reversible reaction is called contact coiling. If the tactile stimulus persists, irreversible free coiling sets in, a process during which the whole tendril, over some 24 hours, forms a tight but highly flexible helix (Engelberth et al. 1995). In *Bryonia dioica*, 12-oxo-phytodienoic acid, which leads to IAA (indole-3-acetic acid) accumulation, is sufficient to induce tendril coiling even in the absence of a mechanical stimulus. Alamethicin (an antibiotic peptide produced by the fungus *Trichoderma viride*), which is capable of forming voltage-dependent ion channels, can also induce tendril coiling. It is suggested that alamethicin-induced membrane depolarization can trigger tendril coiling via IAA. A Ca^{2+} channel in the endoplasmic reticulum of tendrils has been identified that is believed to be operative in the signaling pathway of the touch-induced coiling response (Klüsener et al. 1997).

A groundbreaking discovery in the elucidation of the signaling pathway in mechanoperception was the discovery of four touch (*TCH*) genes in *Arabidopsis* (Braam and Davies 1990). Three of these genes (*TCH1*, *TCH2*, and *TCH3*) encode calmodulin or calmodulin-related proteins, which suggests that Ca^{2+} plays an important role in the chain of events following mechanical stimulation. This observation accords well with the subsequent work of Ding and Pickard (1993), which highlights the importance of mechanosensory Ca^{2+} channels in the plasmalemma of onion epidermal cells. The fourth touch gene (*TCH4*) encodes a cell wall–modifying enzyme. The expression of all four *TCH* genes, however, is not a specific response to mechanical stimuli, since other environmental signals, such as darkness, cold, and temperature shock, can result in their upgraded expression. Indeed, it appears that many more genes are involved in the response to mechanical stimulation, since a transcriptome analysis reveals that the expression of 760 *Arabidopsis* genes, corresponding to 2.5% of the entire genome, is affected 30 minutes after touch (Lee et al. 2005). More than half of these touch-inducible genes are also regulated by darkness, supporting the hypothesis that the transduction pathways for different signals partially overlap.

A zinc finger transcription factor has also been identified in the search for genes that play a role in the primary stages of mechanical signal transduction pathways (Leblanc-Fournier et al. 2008). This factor, which accumulates 30 minutes after the controlled bending of walnut tree (*Juglans regia*) stems, does not accumulate as a result of cold treatment, salt, or nutrient solution. The expression of a homologous transcription factor in young poplar trees (*Populus tremula* × *alba*) is linearly related to strains experienced in bending (Coutand et al. 2009). These and other recent studies indicate that the elucidation of the signal transduction pathway for mechanoperception will require intensive research efforts. Nevertheless, a provisional flow chart of the time course of events (fig. 2.5) is useful to summarize the steps following mechanoperception (Telewski 2006).

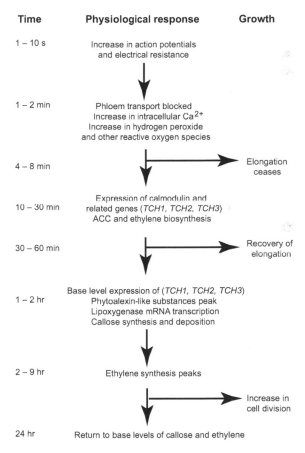

FIGURE 2.5. A provisional flow chart for the time course of physiological and growth responses to mechanical perturbation. (Adapted from Telewski 2006.)

2.7 Thigmomorphogenesis

As noted, when plants experience mechanical perturbation, their growth and development may be altered. The responses of stems to the mechanical effects of wind were first described by Knight (1803), who reported an increase in stem radial growth and a decrease in height growth. This effect was later defined by Jaffe (1973) as thigmomorphogenesis, a term that is used today in a much broader sense to describe the responses of plants not only to wind, but also to flexing or touching (Telewski 2006).

Thigmomorphogenesis can be expressed in plant organs either exhibiting or lacking secondary growth as well as in organs that have either indeterminate or determinate growth. For example, wind-induced swaying of branches or tree trunks or the stalks of inflorescences results in a reduction in organ length compared with plants growing in a protected environment. In addition, it is not atypical to observe a reduction in the elastic moduli of plant tissues in response to chronic mechanical stimulation. The reduction in the ratio of length to diameter confers a greater capacity to resist bending moments and hence a greater capacity to maintain some preferred orientation under static or dynamic mechanical loads. On the other hand, a reduction in the elastic modulus of tissues increases an organ's flexibility and thus reduces the chance that the organ will break when dynamically loaded.

Thigmomorphogenesis thus provides an adaptive, developmental response to the mechanical stresses that are typically experienced by plants. For example, Metzger and Steuceck (1974) reported differences in the thigmomorphogenic responses of two varieties of barley (*Hordeum vulgare*) that differ in their propensity for lodging (i.e., the catastrophic bending failure of stems or roots caused by excessive wind-induced drag forces). One variety, called Barsoy, is very resistant to lodging. It responds to mechanical perturbation with a significant reduction in shoot length. In contrast, another barley variety, called Penrad, which is much more susceptible to lodging, shows little variation in shoot length in response to shaking or handling. Since shoot length provides the lever arm through which the wind pressure operates, a thigmomorphogenic inhibition of shoot growth in length can be seen to provide a mechanical advantage. One of the great challenges to plant breeders is the trade-off that results from trying to maximize the yields of grain crops while simultaneously trying to reduce the probability of lodging caused by wind-induced drag on ripening seeds at the tops of stems (Foulkes et al. 2011).

Controlled bending within the linear elastic range of tomato stems (*Solanum esculentum*, formerly called *Lycopersicon esculentum*) below the actual growing zone leads to cessation of elongation within approximately 10 minutes (see fig. 2.5). This lag time reflects the time it takes for the signal to travel to elongating cells and the time required to produce a response. The minimal rate of signal propagation can be estimated at about 3 cm/min, which is in the range of the rate of propagation of hydraulic signals in the xylem. Complete cessation of elongation lasts for approximately 1 hour and the rate of elongation is reduced for up to 20 hours. An extensive biomechanical analysis of this system shows that the growth response is correlated significantly with the longitudinal *strains* integrated over the part of the stem that is bent, but not with the longitudinal *stresses* (Coutand and Moulia 2000). The analysis of these experiments did not allow for the precise identification of the tissues responsive to the thigmomorphogenic signal. However, it seems likely that only living cells are competent and that the outermost cells (e.g., the epidermis and outer cortex) are involved in signal perception.

2.8 Gravitropism

Gravity is an important environmental stimulus that plants can use to optimize resource exploitation and light interception. Two hypotheses exist concerning the mechanisms of gravity sensing. One hypothesis posits that gravity sensing involves the sedimentation of intracellular particles (called statoliths) in the gravitational field and their interaction with susceptible cellular components, which creates a physiological signal. The second hypothesis posits that gravity sensing involves the perception of small pressure differences between the plasma membrane and the extracellular matrix at the upper and the lower sides of cells by means of mechanosensitive ion channels (Hemmersbach and Braun 2006). These two concepts are not mutually exclusive. Pickard and Thimann (1966) have shown that wheat coleoptiles depleted of statoliths display gravitropic responses similar to intact controls.

A model organism for the study of the early processes in gravity sensing is the multicellular green alga *Chara* (fig. 2.6). This alga produces rhizoidlike structures that grow downward and anchor its thallus to the sediment, while protonemata grow upward against the direction of the gravity vector. In both cases, actin microfilaments regulate the positioning of $BaSO_4$-filled vesicles, which serve as statoliths. Upon gravitational stimulation, the actin-myosin system directs sedimenting statoliths to the gravity-sensitive region of the plasma membrane. Experiments in a

FIGURE 2.6. Vertical portions of the multicellular (filamentous) thallus of the green alga *Chara*, showing the whorls of branched filaments (bf) separated by long internodal cells (inc).

microgravity environment reveal a 0.1 *g* threshold for the displacement of these statoliths. What remains to be shown is whether the physiological response is elicited by the contact of statoliths with gravity sensors or by the pressure exerted on them (Braun and Limbach 2006).

The role of statoliths remains additionally problematic because these structures are absent in several gravity-responsive cells, such as the internodal cells of *Chara*. They are also lacking in the flagellated cells of the unicellular alga *Euglena*. Stretch-activated, asymmetrically localized ion channels are proposed to exist in the anterior cell membrane of *Euglena*. Conceivably, these ion channels are closed when cells swim upward but open when cells swim downward. Channel opening causes a depolarization of the plasma membrane, which triggers a chain of intracellular events leading to a reorientation behavior. In small cells, like those of *Euglena*, the pressure differences between the anterior and the posterior side are likely to be extremely small. Energetic considerations show that the gating energy for mechanosensitive channels is on the order of thermal noise (2×10^{-21} J). It seems likely that amplifying components such as microfilaments must be involved to achieve sensitivity with a threshold down to 0.1 *g*.

Among land plants that have the capacity for secondary growth, the gravitropic response includes the formation of reaction wood (see section 8.9): tension wood in angiosperms and compression wood in gymnosperms. Tension wood is formed on the upper side of a displaced stem. Compared with other wood, this wood has fibers with a lower lignin content, and the cellulose microfibrils in the S2 layer of its secondary cell walls are oriented more in the longitudinal direction. Compression wood is formed along the lower side of a displaced stem. It is characterized by tracheids with a high lignin content, thickened secondary cell walls, and cellulose microfibrils in the S2 layer oriented at $45°$ to $60°$ with respect to the longitudinal axis of the stem. The maturation of xylem cells requires alterations in cell length and diameter. Changes in cell shape participate in the restitution of a stem's vertical posture. When cells elongate within the relatively rigid structure of a stem, they develop compressive stresses, and when cells shrink, they induce tensile stresses in the stem. These internal growth stresses overcompensate for the stresses due to an externally imposed displacement from the vertical direction and lead to a realignment of the stem to an upright position.

However, a kinematic analysis of growth and curvature along the stems of young poplar tree hybrids (*Populus deltoides × Populus nigra*), initially tilted by $35°$ against the vertical, reveals that the gravitropic response is far more complex. After a lag phase, stem growth involves an upward curving and an autotrophic decurving that sets in before the upper parts of the stem reach a vertical orientation. During upward curving, tension wood is produced along the upper side of the stem. However, concomitant with decurving, in the distal parts of the stem, tension wood is also produced on what initially was the lower side. Modeling shows that an upward curving alone would not normally lead to a straight vertical orientation in the upper part of the stem, and that decurving is necessary to reach this orientation (Coutand et. al. 2007). In this sense, gravitropism is analogous to a feedback system in which the difference of any current geometric configuration from a straight vertical orientation is sensed and used to influence where tension wood is formed.

2.9 Root growth, root anchorage, and soil properties

The mechanical properties of soils are important to plant survival and growth because they influence the ease with which roots penetrate their substrate, the establishment of seedlings (parts of which must penetrate the soil to reach the light), and the force required to dislodge roots and overturn individual plants. Therefore, we will consider soil mechanics in this chapter, although many of the engineering concepts that apply to the behavior of soils are taken up in chapter 4.

The mechanical behavior of any particular type of soil depends on its water content. As the water content increases, the consistency or state of a particular soil can change from that of a solid to that of a plastic or, in extreme cases, a non-Newtonian fluid. For example, soils with a high silty clay or organic content behave like plastics or liquids when they have a moderate water content. When they behave as solids, soils can fracture. When they behave as plastics or liquids, they will flow once a yield or threshold stress has been achieved. In large part, these phenomena explain how and why trees dislodge from their soils when their trunks are subjected to high wind pressures. The trunk of a tree will experience a bending moment when the tree canopy is exposed to wind. The maximum bending moment occurs at the base of the trunk. When high winds are accompanied by heavy rains, the mechanical properties of the soil may be altered from those of a solid to those of a very plastic material. By the same token, the bending and torquing of a tree trunk can displace soil as roots are placed alternately under tension and compression. This displacement reduces the strength of the soil and increases the ability of rain to infiltrate deeper. As the rain continues to modify the consistency of the soil, further bending and torquing of the trunk can produce stresses sufficient to cause the soil to yield completely, even under modest wind pressures. Liquefaction of soil can have devastating consequences, particularly on steep slopes, where its consistency can change to that of a non-Newtonian fluid, resulting in rapid flow rates when yield stresses develop from the soil's own weight.

The extent to which a given soil behaves as a solid, plastic, or liquid can be experimentally determined in the field or under laboratory conditions by measuring the water content at its lower (drier) and upper (wetter) limits of plasticity. Typically, soil samples are sieved in the laboratory to remove particles larger than 2 mm in diameter; the remaining particulates are then molded into a paste by adding water (Hillel 1980; American Society for Testing and Materials 1984). Needless to say, sieving a soil sample alters its material properties from what plants usually experience in the field.

Traditionally, the lower (drier) limit of a soil's plastic behavior is defined as the water content at which the soil crumbles when it is rolled into a cylinder 3 mm in diameter (Atterberg 1911). The upper (wetter) limit of plasticity is determined by a penetrometer. Water is added to the sample until the cone of the penetrometer penetrates to a depth of 30 mm. The lower and upper limits of plasticity are known as the plastic and liquid limits, while the difference between the two is called the plasticity index (Marshall and Holmes 1988). The plastic and liquid limits of soils typically increase as their clay and organic matter contents increase, and as might be expected, they are highly correlated over a broad range of soil types. In general, soils with a low plasticity index relative to their liquid limits (soils with a high silty clay or organic content) are the least desirable for engineering purposes, since they flow easily and behave like a liquid when placed under compression.

From the perspective of root growth mechanics, however, soils that behave as a plastic with moderate water content might be considered desirable, since the growing tips of roots would encounter low resistance to the compressional stresses they generate as they advance through the soil. Since reversible elastic strains are limited in most soils, large permanent (plastic) deformations and the yield stresses at which they occur dominate the mechanical environment attending root growth. These deformations can result from soil fracture or plastic flow. Since the strength of a soil generally decreases as its water content increases, our initial expectation that wet soils will be more mechanically desirable for root growth than dry soils appears reasonable.

This expectation can be mathematically formalized by means of the Mohr-Coulomb equation, which provides a way to calculate the shear strength $S\tau$ of a soil based on a constant c expressing the cohesiveness of the soil, the stress τ normal (perpendicular) to the plane of shearing, and the $\tan\theta$, which is the coefficient of internal friction (see Marshall and Holmes 1988):

(2.20) $$S\tau = c + \tau\tan\theta.$$

In standard engineering practice, a soil sample is examined over a range of values for τ, and plots of $S\tau$ versus τ are used to calculate c and $\tan\theta$. The soil sample's cohesiveness c is the y-intercept, and $\tan\theta$ the slope, of the linear plot of $S\tau$ versus τ.

According to the Mohr-Coulomb equation, the shear yield stress will equal the cohesiveness of the soil when there is no stress component operating normal to the shear plane. In fact, there is always some normal stress component, since the weight of the soil above the plane of shearing exerts some force, but at shallow depths this force is negligible. Not surprisingly, the Mohr-Coulomb equation indicates that for any given soil cohesiveness, the shear stresses required to pull a root from the ground increase as a function of soil depth. Accordingly, if root systems are designed to operate as tensile members, then for any given soil type, the depth of growth is the most important factor. Further, soil cohesiveness tends to increase as clay content increases. For example, c is zero for sand and increases to an upper limit of about 3×10^4 N/m^2 for clay. Therefore, all other things being equal, root systems growing in clayey soils require extremely high stresses to dislodge them. There are clear trade-offs, however, when ease of root penetration and the requirement for roots to grip the soil are considered together. The Mohr-Coulomb equation indicates that either low normal stresses or low cohesiveness, or both, will aid root penetration. Shallow growth (low τ) in sandy soils (with low to zero c) facilitates root penetration, while deep growth (high τ) in clayey soils (high c values) ensures an efficient means of gripping the soil.

From first principles, we would expect organs to exert compressive stresses as they grow through a soil. (If roots are well supplied with oxygen, they can exert tip pressures up to about 1 MPa.) If a soil is wetter than its plastic limit, the compressive stresses generated by root growth will cause the soil to deform plastically as the root advances, progressively reducing the pore sizes among soil aggregates in front of the root tip. Thus, the loading caused by growth is progressively resisted by the larger contact area among soil particles as soil aggregates flatten against one another. How tightly aggregates in a soil are compressed can be expressed by the soil void ratio e, which is the instantaneous volume of pores within the soil divided by the volume of solid particulates. The initial void ratio e_0 depends on the initial compressional stress τ_0 to which the soil is subjected. It decreases as further compressional stress τ is applied. This phenomenon can be mathematically expressed by the formula

(2.21)
$$IC = \frac{e - e_0}{\log(\tau / \tau_0)},$$

where IC is the compressional index, which is experimentally determined by taking a soil sample and measuring e at different values of τ. IC is the slope of the semilog plot of e versus τ. For any specified water content, IC is a constant for any sample. The data from such an experiment can be used to estimate the compressional stresses that a subterranean organ must exert to progress through a soil type as a function of the resulting changes in the void ratio of the soil. Intuitively, we can see that as growth proceeds, these stresses ought to increase. But roots typically grow around large soil aggregates and follow channels made by earthworms and former (decayed) roots, which offer comparatively little resistance to penetration by young roots. Additionally, roots expand in girth at some distance from their growing tips. This lateral expansion can propagate soil fractures, opening up low-resistance avenues for subsequent growth in length.

Measurements of soil strength by means of penetrometers invariably overestimate the actual resistance the soil offers to root growth, since the penetration tip of a penetrometer cannot avoid large soil aggregates as root tips can. Nonetheless, the rate of root elongation correlates well with resistance measurements made with the use of a penetrometer. Taylor and Ratliff (1969) measured the rate of root elongation of peanut plants grown in loamy sand at three water contents (0.07, 0.055, and 0.038 g H_2O/ cm^3) and reported that the rate of elongation dramatically declines as soil strength increases from 0 to 6 MPa. Regression analysis of the data provided by these authors shows that, regardless of water content, the rate of elongation r_e declines exponentially with penetrometer resistance R according to the formula $r_e = 3.32 \times 10^{-0.208R}$ ($n = 18$, $r = 0.899$), where r_e is expressed in units of millimeters per hour and R in MPa. Clearly, other factors, such as the availability of oxygen to roots and leaf transpiration, influence root growth.

Even a brief review of the physical and material properties of soils demonstrates that the mechanics of underground growth is very complex. Nonetheless, a few equations and some experimentation can provide considerable insight into the obstacles that confront the delicate growing tips of underground roots and stems. Fortunately, the Poisson's ratios (see section 4.6) of young, growing tissues are high, making them essentially incompressible materials that operate mechanically as hydrostats. As such, root and stem tips can exert tremendous compressive stresses on soil. Another feature of growing roots is that they can alter the chemistry of soils by excreting compounds. They can also lubricate and slough off their surfaces, thereby reducing simple and pure shear stresses. The capacity of roots to chemically alter the soil they grow through can have surprising consequences. Palm trees that survived a hurricane that struck the Hawaiian Islands in 1979 were found to have remained anchored to their growth site by means of massive concretions of soil formed around their root systems (fig. 2.7).

FIGURE 2.7. The fibrous root system of this coconut palm (growing on the island of Kauai) was exposed by the erosion of beach sand in a hurricane in 1979, which washed away the foundations of large hotels (a portion of one hotel foundation is indicated by the downward-pointing arrow).

Literature Cited

American Society for Testing and Materials. 1984. Annual book of ASTM standards. Vol. 4. Philadelphia: ASTM.

Anisimov, O., and L. Fukshansky. 1993. Light-vegetation interaction: A new stochastic approach for description and classification. *Agr. For. Meteorol.* 66: 93–110.

Atterberg, A. 1911. Die Plastizität der Tone. *Int. Mitt. Bodenk.* 1:10–43.

Blaauw, A. H. 1909. Die Perzeption des Lichtes. *Rec. Trav. Bot. Neerl.* 5:209–372.

Braam, J. 2005. In touch: Plant responses to mechanical stimuli. *New Phytol.* 165:373–89.

Braam, J., and R. W. Davies. 1990. Rain-, wind-, and touch-induced expression of calmodulin and calmodulin-related genes in *Arabidopsis. Cell* 60:357–64.

Braun, M., and C. Limbach. 2006. Rhizoids and protonemata of characean algae: Model cells for research on polarized growth and plant gravity sensing. *Protoplasma* 229:133–42.

Coutand, C., M. Fournier, and B. Moulia. 2007. The gravitropic response of poplar trunks: Key roles of prestressed wood regulation and the relative kinetics of cambial growth versus wood maturation. *Plant Physiol.* 144:1166–80.

Coutand, C., L. Martin, N. N. Leblanc-Fournier, M. Decourteix, J.-L. Julien, and B. Moulia. 2009. Strain mechanosensing quantitatively controls diameter growth and *PtaZFP2* gene expression in poplar. *Plant Physiol.* 151:223–32.

Coutand, C., and B. Moulia. 2000. Biomechanical study of the effect of a controlled bending on tomato stem elongation: Local strain sensing and spatial integration of the signal. *J. Exp. Bot.* 51:1825–42.

Christie, J. M., and W. R. Briggs. 2001. Blue light sensing in higher plants. *J. Biol. Chem.* 276:11457–60.

Darwin, C. 1880. *The power of movement in plants.* London: John Murray.

———. 1893. *Insectivorous plants.* London: John Murray.

De Langre, E. 2008. Effects of wind on plants. *Annu. Rev. Fluid Mech.* 40:141–68.

Denny, M. W. 1988. *Biology and the mechanics of the wave-swept environment.* Princeton, NJ: Princeton University Press.

Ding, J. P., and B. Pickard. 1993. Mechanosensory calcium-selective cation channels in epidermal cells. *Plant J.* 3:83–110.

Duysens, L. N. M. 1956. The flattening of the absorption spectrum of suspensions as compared to that of solutions. *Biochim. Biophys. Acta* 19:1–12.

Emerson, R., and W. Arnold. 1932. The photochemical reaction in photosynthesis. *J. Gen. Physiol.* 16:191–205.

Engelberth, J., G. Wanner, B. Groth, and E. Weiler. 1995. Functional anatomy of the receptor cells in tendrils of *Bryonia dioica* Jacq. *Planta* 196:539–50.

Farquhar, G. D., and T. D. Sharkey. 1982. Stomatal conductance and photosynthesis. *Annu. Rev. Plant Physiol.* 33:317–45.

Foulkes, M. J., G. A. Slafer, W. J. Davies, P. M. Berry, R. Sylvester-Bradley, P. Martre, D. F. Calderini, S. Griffiths, and M. P. Reynolds. 2011. Raising yield potential of wheat. III. Optimizing partitioning to grain while maintaining lodging resistance. *J. Exp. Bot.* 62:469–86.

Fukshansky, L. 1987. Absorption statistics in turbid media. *J. Quant. Spectros. Ra.* 38:389–406.

———. 1991. Photon transport in leaf tissue: Application in plant physiology. In *Photon-vegetation interactions*, edited by R. B. Myneni and J. Ross. Berlin: Springer Verlag.

Fukshansky, L., A. Martinez V. Remisowsky, J. McClendon, A. Ritterbusch, T. Richter, and H. Mohr. 1993. Absorption spectra of leaves corrected for scattering and distributional error: A radiative transfer and absorption statistics treatment. *Photochem. Photobiol.* 57:538–55.

Galland, P. 1990. Phototropism of the *Phycomyces* sporangiophore: A comparison with higher plants. *Photobiology* 52:233–48.

Hemmersbach, R., and M. Braun. 2006. Gravity-sensing and gravity-related signaling pathways in unicellular model systems of protists and plants. *Signal Transduct.* 6:432–42.

Hillel, D. 1980. *Fundamentals of soil physics.* New York: Academic Press.

Iino, M. 2001. Phototropism in higher plants. In *Photomovement*, edited by P.-D. Häder and M. Lebert. Amsterdam: Elsevier Science.

Isnard, S., and W. K. Silk. 2009. Moving with climbing plants from Charles Darwin's time into the 21st century. *Amer. J. Bot.* 96:1205–21.

Jaffe, M. J. 1973. Thigmomorphogenesis: The response of plant growth and development to mechanical stimulation. *Planta* 114:588–94.

Joyeux, M., O. Vincent, and P. Marmottant. 2011. Mechanical model of the ultrafast underwater trap of *Utricularia*. *Phys. Rev. E* 83, doi:10.1103/PhysRevE.83.021911.

Kirk, J. T. O. 1975a. A theoretical analysis of the contribution of algal cells to the attenuation of light within natural waters. I. A general treatment of suspension of pigmented cells. *New Phytol.* 75:11–20.

———. 1975b. A theoretical analysis of the contribution of algal cells to the attenuation of light within natural waters. II. Spherical cells. *New Phytol.* 75:21–36.

———. 1976. A theoretical analysis of the contribution of algal cells to the attenuation of light within natural waters. III. Cylindrical and spheroidal cells. *New Phytol.* 77:341–58.

Klüsener, B., G. Boheim, and E. Weiler. 1997. Modulation of the ER Ca^{2+} channel BCC1 from tendrils of *Bryonia dioica* by divalent cations, protons and H_2O_2. *FEBS Lett.* 407:230–34.

Knight, T. A. 1803. Account of some experiments on the descent of sap in trees. *Phil. Trans. Roy. Soc. London* 96:277–89.

Koehl, M. A. R. 1999. Ecological biomechanics of benthic organisms: Life history, mechanical design and temporal patterns of mechanical stress. *J. Exp. Biol.* 202:3469–76.

Kutschera, U., and K. J. Niklas. 2009. Evolutionary plant physiology: Charles Darwin's forgotten synthesis. *Naturwissenschaften* 96:1339–54.

Leblanc-Fournier, N., C. Coutand, J. Crouzet, N. Brunel, C. Lenne, B. Moulia, and J.-L. Julien. 2008. *Jr-ZFP2*, encoding a Cys2/His2-type transcription factor, is involved in the early stages of the mechano-perception pathway and specifically expressed in mechanically stimulated tissues in woody plants. *Plant Cell Env.* 31:715–26.

Lee, D. W., and J. B. Lowry. 1975. Physical basis and ecological significance of iridescence in blue plants. *Nature* 254:50–51.

Lee, D., D. H. Polisensky, and J. Braam. 2005. Genome-wide identification of touch- and darkness-regulated *Arabidopsis* genes: A focus on calmodulin and XTH genes. *New Phytol.* 165:429–44.

Marshall, T. J., and J. W. Holmes. 1988. *Soil physics.* 2nd ed. Cambridge: Cambridge University Press.

Metzger, J., and G. L. Steuceck. 1974. Response of barley (*Hordeum vulgare*) seedlings to mechanical stress. *Proc. Pa. Acad. Sci.* 48:114–16.

Monteith, J. L. 1973. *Principles of environmental physics.* New York: Elsevier.

Niklas, K. J. 1994. *Plant allometry: The scaling of form and process.* Chicago: University of Chicago Press.

Nikolopoulos, D., G. Liakopoulos, I. Drossopoulos, and G. Karabourniotis. 2002. The relationship between anatomy and photosynthetic performance of heterobaric leaves. *Amer. J. Bot.* 129:235–43.

Nilson, T. 1991. Approximate analytical methods for calculating the reflection functions of leaf canopies in remote sensing applications. In *Photon-vegetation interactions*, edited by R. B. Myneni and J. Ross, 161–90. Berlin: Springer Verlag.

Nobel, P. S. 2005. *Physicochemical and environmental plant physiology.* 3rd ed. Amsterdam: Elsevier.

Nowak, W. G., and P. Hietz. 2011. An improved model for the diffusion of oxygen into respiring wood. *J. Biol. Syst.* 19:101–12.

Pickard, B. G., and K. V. Thimann. 1966. Geotropic response of wheat coleoptiles in absence of amyloplast starch. *J. Gen. Physiol.* 49:1065–85.

Popescu, T., A. Roessler, and L. Fukshansky. 1989. A novel effect in *Phycomyces* phototropism. *Plant Physiol.* 91:1586–93.

Raven, J. A. 2010. Inorganic carbon acquisition by eukaryotic algae: Four current questions. *Photosynth. Res.* 106:123–34.

Reynolds, C. 2006. *Ecology of phytoplankton.* Cambridge: Cambridge University Press.

Ross, J., and A. Marshak. 1991. Monte Carlo methods. In *Photon-vegetation interactions*, edited by R. B. Myneni and J. Ross, 441–67. Berlin: Springer Verlag.

Rowe, N. P., and T. Speck. 1996. Biomechanical characteristics of the ontogeny and growth habit of the tropical liana *Condylocarpon guianense* (Apocynaceae). *Int. J. Plant Sci.* 157:406–17.

Rushkin, I., V. Kantsler, and R. E. Goldstein. 2010. Fluid velocity fluctuations in a suspension of swimming protists. *Phys. Rev. Lett.* 105, doi:10.1103/PhysRevLett.105.188101.

Shabanov, N. V., D. Huang, Y. Knyazikhin, R. E. Dickinson, and R. B. Myneni. 2007. Stochastic radiative transfer model for mixture of discontinuous vegetation canopies. *J. Quant. Spectros. Ra.* 107:236–62.

Shabanov, N. V., Y. Knyazikhin, F. Baret, and R. B. Myneni. 2000. Stochastic modeling of radiation regime in discontinuous vegetation canopies. *Remote Sens. Environ.* 74:125–44.

Schlichting, H. 1979. *Boundary layer theory.* New York: McGraw-Hill.

Shropshire, W. Jr. 1962. The lens effect and phototropism in *Phycomyces. J. Gen. Physiol.* 45:949–58.

Silk, W. K., and N. M. Holbrook. 2005. The importance of frictional interactions in maintaining the stability of the twining habit. *Amer. J. Bot.* 92:1820–26.

Singsaas, E. L., D. R. Ort, and E. H. DeLucia. 2001. Variation in measured values of photosynthetic quantum yield in ecophysiological studies. *Oecologia* 128:15–23.

Steyer, B. 1967. Die Dosis-Wirkungsrelationen bei geotroper und phototroper Reizung. Vergleich von Mono- mit Dicotyledonen. *Planta* 77:277–86.

Taiz, L., and E. Zeiger. 2010. *Plant physiology.* 5th ed. Sunderland, MA: Sinauer Associates.

Taylor, H. M., and L. F. Ratliff. 1969. Root elongation rates of cotton and peanuts as a function of soil strength and soil water content. *Soil Sci.* 108:113–19.

Telewski, F. W. 2006. A unified hypothesis of mechanoperception in plants. *Amer. J. Bot.* 93:1466–76.

Tributsch, H. 2001. Bionische Vorbilder für eine solare Energietechnik. In *Bionik, Ökologische Technik nach dem Vorbild der Natur,* 2nd ed., edited by Armin von Gleich. Stuttgart: Teubner Verlag.

Vogel, S. 1988. *Life's devices: The physical world of animals and plants.* Princeton, NJ: Princeton University Press.

———. 2009. Leaves in the lowest and highest winds: Temperature, force and shape. *New Phytol.* 183:13–26.

Vogelmann, T. C. 1993. Plant tissue optics. *Annu. Rev. Plant Physiol. Plant Mol. Biol.* 44:231–51.

Winston, R. 1991. Nonimaging optics. *Sci. Amer.* 264:76–81.

CHAPTER THREE

Plant Water Relations

One of the nice things about water plants is that they never need watering.
—Christopher Lloyd, *The Well-Tempered Garden*

This chapter examines how the flow of water and nutrients through the multicellular plant body is achieved and maintained. This topic is treated before the mechanical attributes of cells, tissues, organs, or the whole plant body are discussed because the availability of water and nutrients influences every aspect of the survival, growth, reproduction, and evolution of all plant life-forms—algae and embryophytes, past and present.

Before proceeding with the topic of water relations, a brief word about units of measure is appropriate. All of the relationships discussed in this chapter deal with pressure or can be expressed in terms of pressure. The older literature dealing with plant water relations expressed pressure in units of either atmospheres (abbreviated atm) or bars, which are practically identical for most purposes (they differ by about 1.3%). The popularity of these units may be attributed to the fact that 1 atmosphere equals 1 kg/cm^2, which has a force that can be visualized easily. However, the SI unit for pressure is the Pascal (abbreviated Pa). One Pascal is the force per square meter exerted by one kg at an acceleration of 1 m/s^2. It is equivalent to one Newton per square meter (abbreviated as N/m^2).

3.1 The roles of water acquisition and conservation

The survival of terrestrial plants, and even that of their individual organs, in large part depends on a supply of water from the substrate to which they are anchored. The functional lifetime of a plant and of its individual leaves depend on a positive average net photosynthesis and the maintenance of nonlethal temperatures, which in turn require access to a sufficient quantity of water. In the absence of photosynthesis, even for a comparatively short time, leaves typically senesce because there is no known mechanism to import nutrients such as sugars into mature leaves. Provided light is not limiting in the environment, net photosynthesis depends on the water balance of the entire plant and its photosynthetic organs because water deprivation limits the capacity of plant tissues to exchange gases with the external atmosphere. When gas exchange occurs, water vapor is inevitably lost, thereby cooling plant tissues that might otherwise achieve physiologically deleterious temperatures.

Plant growth and the expansion of cells also depend on a supply of water. The influx of water and the reduction of the yield stress of the cell wall result in cell expansion to accommodate an increase in water content (see section 8.3). An adequate supply of water is also necessary to maintain the stiffness of mature cells with thin and elastic walls. Fully differentiated tissues composed of thin-walled cells operate mechanically as hydrostatic devices; that is, their apparent stiffness increases as their cell walls are placed under tensile stress by their water-filled contents (see section 8.5).

Although land plants acquire water principally from their substrates, dew or fog that has precipitated on aerial organs can also be absorbed, providing a supplementary source of water, or more rarely, as in the case of *Welwitschia mirabilis*, the bulk of water needed for growth. Epiphytes and parasitic plants can be exceptions to these generalizations. For example, the twining epiphyte *Dischida rafflesiana* produces pitcher-shaped leaves with very thick cuticles. These leaves operate like cisterns that trap water vapor lost through transpiration from their inner (adaxial) surfaces. Occasionally, they also collect rain runoff. The inner surface of these leaves typically has twice the number of stomata found on the outer leaf surface. As a result, little water is lost during photosynthesis. The little that is lost is replaced by adventitious roots attached to the trees on which this species lives.

Nevertheless, the vast majority of vascular land plants absorb water from substrates by means of a root system and lose water by transpiration attending the exchange of carbon dioxide and oxygen with the atmosphere, which is essential for photosynthesis. Thus, hydraulic continuity must be maintained between the root system and transpiring leaves. Among vascular plants, this continuity is achieved by the xylem, a complex tissue composed of a variety of cell types that include xylary elements (tracheids

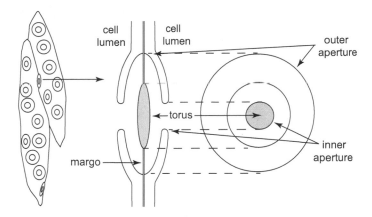

FIGURE 3.1. Schematic of the bordered pits of tracheids. Tracheids are fusiform (spindle-shaped) in geometry and have cell walls that are pitted to allow water to pass from one cell into another (*as shown at left*). Some tracheids have bordered pits, so called because they have an inner and an outer aperture composed of cell wall material (*shown in the middle and at the right*). The middle lamella that is suspended across the outer aperture is modified in each pit. It has a thickened, padlike torus, suspended by a thinner, trampoline-like margo. If the water pressure in a tracheid drops precipitously, the torus in each of its pits is pushed against the cell's inner apertures, thereby preventing the water pressure from dropping in neighboring cells (see also fig. 3.4). For a morphospace of tracheid physiology see Wilson and Knoll (2010).

and vessel members). Xylary elements are dead cells that provide low resistance to the bulk transport of water through their lumens. Vessels are composed of individual cells (vessel members) stacked end to end. The end walls of each vessel member are perforated, which allows water to flow with little or no resistance from one vessel member into another.

Angiosperms that inhabit environments in which water is abundant typically have vessels that are large in diameter, whereas angiosperms living in dry habitats have narrow vessels, in part because the probability of cavitation increases with increasing vessel diameter (Sperry et al. 2008). Most gymnosperm species produce wood composed of tracheids. Like vessel members, tracheids lack protoplasts when they become fully functional. Unlike vessel members, tracheids do not have perforated end walls. Water flows among tracheids through small perforations that are formed over the cell wall surface. Among conifers, these perforations take the form of bordered pits (fig. 3.1). These pits possess a remarkable set of structures called the torus and margo. The torus is a swollen region of the middle lamella that is suspended by the margo, which is a thinner region of the middle lamella. Provided that the water pressure between adjacent cells is at or near equilibrium, the torus-margo structure permits water to flow through the bordered pit. If, however, one of the adjacent cells experiences a pressure drop, as, for example, one caused by the formation of a cavitation, the margo flexes and presses the torus against the inner aperture of the pit, thereby sealing the cell and preventing the passage of a vapor bubble (Pittermann et al. 2005).

The evolutionary acquisition of xylem contributed to the ability of vascular plants to achieve heights that are not possible when the internal transport of water relies predominantly on passive diffusion through cell walls, as it does in many nonvascular plant species. However, the ability to form xylem is not essential for the survival of plants on land or, more precisely, for their survival in air, provided that the path lengths between the external supply of water and transpiring plant surfaces are kept short. The nonvascular land plants (mosses, liverworts, and hornworts) are ecologically very successful and taxonomically diverse. Most of these plants rely on the transport of water over their external surfaces by means of molecular adhesion and capillarity, although the presence of an external layer of water can impede the diffusion of gases through the epidermis (Proctor 2005). Some nonvascular plants, such as the giant moss *Dawsonia superba*, can grow to 60 cm in height, thus rivaling the vertical stature of many herbaceous vascular species. *Dawsonia*, like many other types of moss, has internal cells called hydroids that conduct water internally (or, at the least, provide an internal water reserve). Hydroids converge on tracheids in their appearance and are part of a complex tissue system that contains additional cell types that function like phloem cell types.

The conservation of water within the vascular plant body relies on the presence of the cuticle and the hydrostatic behavior of stomata. The cuticle is chemically and ultrastructurally very complex (see section 8.1). In addition to serving as a partial barrier against water loss, it functions as a UV filter, as a deterrent against microbial attack and herbivory, and as a mechanically strong tensile "skin." It also serves as a barrier that prevents the anomalous fusion of adjoining organs as they develop and mature (Weng et al. 2010). The capacity of the cuticle to reduce water loss depends in part on the extent to which it is externally hydrated. When wet, the cuticle can become moderately permeable to the diffusion of water molecules, thereby permitting the absorption of dew or fog that has precipitated on leaf or stem surfaces or aerial roots. When dry, however, the hydrophobic outermost external waxy layer that is typical of most cuticles provides substantial resistance to the rapid loss of water.

Although it helps to conserve water, the cuticle is also a barrier to the diffusion of O_2 and CO_2, both of which are essential for the survival of aerobic photoautotrophic organisms such as plants. This property of the cuticle introduces a "conflict in design," which was resolved by an ancillary adaptation: a perforated epidermis that permits the passage of atmospheric gases into and out of the multicellular plant body. With very few exceptions, the epidermis of land plants, both nonvascular and vascular, has small openings, or stomata. For example, the dorsal surface of the thalloid liverwort *Marchantia* has numerous pores that open into chambers lined with filaments of cells that are rich in chloroplasts. These invaginations provide large surface areas through which O_2 and CO_2 can be exchanged with the external atmosphere by passive diffusion. The cells surrounding these pores cannot change the pore diameter appreciably, which confines *Marchantia* and other nonvascular plants like it to moist microhabitats. In contrast, mosses and vascular plants have a pair of guard cells flanking each stoma. These guard cells actively regulate pore diameters hydraulically. When supplied with water by other epidermal cells, they expand and thus increase stomatal diameters. When they lose water, they contract and close the stomata.

Stomatal opening and closing involves two components: (1) a complex physiological feedback mechanism that drives changes in turgor pressure within the guard cells, and (2) mechanically anisotropic cell walls in the guard cells that respond to these changes by expanding or contracting, thereby opening and closing the stoma.

As early as 1856, H. von Mohl proposed that changes in the turgor pressure within guard cells provided the mechanical force required to open and close the opening between adjoining guard cells. In 1908, F. E. Lloyd suggested that the turgor of guard cells was regulated by osmotic changes resulting from the interconversion of sugar and starch. According to this hypothesis, the hydrolysis of starch to produce glucose would increase the solute concentration within the cytoplasm of the guard cells, which in turn would cause water to enter the guard cells from adjoining cells. The polymerization of glucose to form starch would reverse the process. More recent work has shown that the sugar-starch interconversion hypothesis is only one of at least four metabolic pathways that can regulate the concentrations of osmotically active solutes in guard cells (fig. 3.2). These pathways are

1. The blue light–stimulated uptake of K^+ and Cl^- and the biosynthesis of malate^{2+} within guard cells
2. The production of sucrose via starch hydrolysis
3. The production of sucrose by photosynthetic carbon fixation
4. The influx/efflux of sucrose synthesized in cells adjoining guard cells

These pathways are not redundant in the sense that if one fails to operate, another is available to take its place. It is more likely that these pathways are adaptations to different sets of environmental conditions that increase or reduce water stress.

For example, stomata typically open when hydrated photosynthetic organs are exposed to light. Research has shown that this response is induced by blue light, which provides the environmental cue. The mechanism driving stomatal opening is blue light stimulation of proton extrusion from guard cells, a process that requires ATP. The efflux of protons generates a more negative electric potential difference across the cell membranes of the guard cells. In turn, this difference is a driving force for the passive influx of K^+ by means of voltage-regulated potassium channels (see chap. 7). The influx of very large concentrations of K^+ results in a decrease in the water potential within the guard cell protoplast, which induces an influx of water from neighboring cells. Indeed, depending on the species and environmental conditions, the increase in K^+ concentration in guard cells can be dramatic—from 100 mM (when stomata are closed) to 400–800 mM (when stomata open). The increase in the concentration of K^+ is electrically balanced by the influx of Cl^- (and/or the synthesis of malate^{2-} anions), which is concentrated in guard cell vacuoles (see fig. 3.2). Malate anions are synthesized in the cytoplasm of guard cells by means of a metabolic pathway involving the accumulation of sucrose from the hydrolysis of starch, which occurs in guard cell chloroplasts. The influx of K^+ and Cl^- ions involves secondary transport mechanisms driven by the gradient of electrochemical potential for H^+ generated by a proton pump.

The sucrose concentration within guard cells increases slowly during the morning, but sucrose becomes the dominant osmotically active solute when K^+ efflux occurs. This osmoregulatory pattern indicates that stomatal opening and closing is primarily associated with K^+ transport. However, as noted, there exist at least three other mechanisms by which stomatal opening and closing is regulated by changes in the sucrose concentration within guard cells (see fig. 3.2). This is evident from the fact that the starch concentration in guard cell chloroplasts decreases as a result

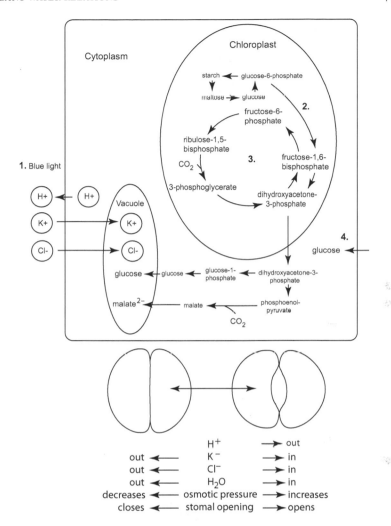

FIGURE 3.2. *Top:* Schematic of the four metabolic pathways that regulate stomatal opening and closing (each pathway is indicated by a number in boldface). *Bottom:* Schematic showing the effects of pathway 1 on the appearance of the two guard cells.

of starch hydrolysis. This change increases the sucrose concentration and thus increases the osmotic pressure of the cytoplasm, which leads to the opening of stomata. Sucrose can also be synthesized directly by guard cell chloroplasts. In addition, sucrose that is synthesized by neighboring cells can pass through the cell walls of guard cells.

Other factors influence the opening and closing of stomata. For example, when leaves are illuminated and photosynthesizing, the concentration of CO_2 in their intracellular spaces decreases. This change results in a decrease in the CO_2 concentration within guard cells, which triggers stomatal opening. Conversely, when leaves are left in the dark, their internal CO_2 concentrations increase, and stomata close. Experiments also indicate that stomatal closure is affected by abscisic acid (ABA). For example, microinjection of a physiologically inactivated ("caged") form of ABA and subsequent activation of the hormone by UV irradiation results in rapid stomatal closure. The mechanism for this response appears to involve inhibition by ABA of K^+ channels, which are required for stomatal opening.

The mechanical operation of guard cells has received considerable attention. It is still not clear, however, how the cell walls of guard cells deform and achieve their different shapes when strained by increasing internal osmotic pressure, in part because of the mechanical diversity of stomata (Franks and Farquhar 2007). Research as well as theory has shown that the mechanical response of a guard cell requires a nonrandom orientation of cellulose microfibrils in the unstrained state of the cell wall. This nonrandom network of microfibrils results in walls that become increasingly elastically anisotropic as the cytoplasm inflates. Observations suggest that guard cell walls operate nearly as an isotropic polymer when subjected to small or moderate strains. As the volume of the cytoplasm increases beyond some threshold, however, the microfibrillar network becomes strained to the point that it enters a second, anisotropic phase of mechanical behavior. The shear modulus of guard cell walls has been estimated to be 2 MPa (Wu and Sharpe 1979), which is within the limits of values reported for many other plant tissues.

From a mechanical as well as a physiological perspective, it must be recognized that changes in the volume and shape of guard cells do not occur in isolation. The water entering a guard cell when stomata open must be supplied by neighboring cells, and these cells must be, to some extent, mechanically affected as guard cells expand or contract. The generalized picture of the mechanical anisotropy of guard cell walls, therefore, belies the large range observed in guard cell morphology and the extent to which guard cells interact physiologically and mechanically with subsidiary cells. Among angiosperms, stomata develop from asymmetric divisions of specific protodermal cells. These divisions result in a guard cell "mother cell," which divides to form the guard cells and either two or four subsidiary cells that lack chloroplasts. The stomata of lycopods lack subsidiary cells (i.e., they are anomocytic) and have large and comparatively broad guard cells, which undergo minimal swelling or lateral movement during stomatal opening. The stomata of ferns, such as *Nephrolepis exaltata*, are also typically anomocytic. However, their guard cells can achieve wider apertures than those of lycopods, which assists with higher rates of gas exchange. The stomata of angiosperms such as *Tradescantia virginiana* and *Triticum aestivum* differ from those of lycopods and ferns in that they have one or two subsidiary cells parallel to each guard cell. Perhaps as a consequence, angiosperm guard cells are capable of substantial lateral movement. They interact physically with subsidiary cells during stomatal opening in ways that can result in wide stomatal pores, which facilitate high rates of photosynthetic gas exchange. Despite this similarity, the stomata of grasses such as *T. aestivum* have characteristic dumbbell-shaped guard cells, which may explain why grasses are capable of very rapid stomatal opening and closure.

3.2 Some physical properties of water

Among the most abundant substances on earth, water has the largest number of anomalous physical properties. As compared with other chemicals of comparable molecular weight, which are gases at room temperature, water is a liquid between 0°C and 100°C at 1 atm. Water owes its remarkable properties (including a very high specific heat and the highest known heat of vaporization) to the distribution of its positive and negative electrical charges (its dipole moment) and intermolecular hydrogen bonds.

The hydrogen bonds in water, which have a length of 0.177 nm, can bind water molecules into a highly symmetrical lattice that has unusually wide intermolecular spacing. This property, which is most evident in ice crystals, accounts for the lower density of the solid phase (ice) than the liquid (water) phase, which explains why ice floats, and thus why earth's ocean floors are not covered with icebergs. As ice melts, only some of its hydrogen bonds break, which nevertheless leads to a collapse of the low-density ice lattice and a dynamic structure of aggregates of water molecules. Correspondingly, the density of water increases as the temperature rises from 0°C to 4°C.

Very large amounts of energy must be supplied to break water's hydrogen bonds, and two bonds must be broken for one molecule to evaporate.

These properties explain why water has the highest heat of vaporization and the highest specific heat of any naturally occurring substance on earth. They also explain why large bodies of water stabilize the temperatures of islands, peninsulas, and coastal areas. And they help us to understand why the large, fleshy, water-packed stems of cacti may confer an advantage: they take a long time to cool down in deserts, where nighttime temperatures can drop below freezing, and conversely, a long time to heat up to potentially dangerous temperatures during the daytime. By the same token, water's high specific heat of vaporization helps to cool transpiring leaves, stems, and aerial roots as water is lost by transpiration as CO_2 and O_2 are exchanged with the external atmosphere.

Water has an extremely high surface tension due to its large intermolecular cohesive forces, which become dramatically unbalanced at the liquid-gaseous interface. These large cohesive forces confer great tensile strength, which allows water to be drawn up vertically under negative pressures over considerable heights without breaking under its own weight-force in a narrow conduit (see section 3.8). This high tensile strength helps to explain how some trees, such as the General Sherman tree, can achieve heights in excess of 80 m. Hydrogen bonding also substantially increases the ability of molecules to resist rearrangements and accounts for water's unusually high viscosity. Finally, water has a very high dielectric constant, which accounts for the high solubility of ions in water—a feature that allows roots to absorb minerals dissolved in water that are essential elements for growth.

Many other physical properties of water influence the biology and ecology of plants. For example, pure water is not optically transparent. It attenuates all wavelengths of visible light and thus all wavelengths of photosynthetically active radiation (400–700 nm) (abbreviated PAR). This attenuation, however, is not uniform across all PAR wavelengths. Water's absorption of light begins to rise noticeably as wavelengths increase above 550 nm and increases significantly at the red end of the spectrum, particularly in the near-infrared. Therefore, although water appears colorless, it is actually a blue liquid. One consequence of this property that is beneficial to plants is that water provides an excellent heat-absorption filter; for example, a water column 1 m deep absorbs about 35% of light with wavelengths of more than 680 nm (Kirk 1983). On the other hand, a negative consequence is its reduction in the intensity of PAR, since most of the light available for photosynthesis falls between 400 and 500 nm and between 650 and 700 nm; for example, the absorption peaks of chlorophyll *a* are 430 nm and 662 nm. This reduction can be exacerbated by substances dissolved in water (collectively called *Gelbstoff*) that absorb or scatter sunlight. As might be expected, given the existence of aquatic plants capable of surviving at considerable depths below the air-water interface, the adaptive evolution of aquatic plants has compensated for this physical limitation. The in vivo absorption spectra of carotenoids and other secondary light harvesting pigments in various algae increase their light-harvesting capacity over the entire PAR spectrum by shunting additional irradiant energy into chlorophyll molecules (see Kirk 1983; Owens 1988). Nevertheless, it is very possible that the colonization of land by plant life was propelled by the benefits of exchanging a blue and often turbid liquid for an essentially transparent mixture of gases.

3.3 Vapor pressure and Raoult's law

When substances or very small particulates are dissolved in water, they have the ability to bind to water molecules, thereby reducing the free energy of the solvent-solute system. This phenomenon is vital to the establishment of water gradients in plant tissues and organ systems, which provide a mechanism for directing the passive diffusion of water through neighboring cells and the entire multicellular plant body.

In order to explore this phenomenon, we first turn to the relationship between the vapor pressure of pure water, v_w, and that of a solution of water, v_s, which is expressed by what has become known as Raoult's law:

$$(3.1) \qquad v_s = v_w \left(\frac{n_w}{n_w + n_s} \right) \approx v_w \left(1 - \frac{n_s}{n_w} \right),$$

where n_w is the number of moles of the solvent (in this case, water) and n_s is the number of moles of the solute. (One mole is the amount of a substance containing one Avogadro's number of molecules of the substance; Avogadro's number is equal to about 6.0225×10^{23} entities.) Raoult's law, which holds true only for dilute solutions, shows that the vapor pressure of water in solution is proportional to the mole fraction of the solute. This proportional relationship helps to explain why the addition of solutes to water raises the boiling point and lowers the freezing point of the solution. The addition of solutes to water raises the boiling point because water boils when its vapor pressure is raised to that of the surrounding atmosphere and because the addition of solutes lowers the vapor pressure of water. Thus, water containing solutes must be heated to a higher temperature in order to boil. Adding solutes to water depresses the freezing point because it decreases the solution's vapor pressure (but not that of ice, which essentially excludes all solutes). This change lowers the equilibrium temperature at which the liquid and solid phases coexist.

The influence of solutes on the freezing point of water has meaningful consequences for plants. As early as 1912, N. A. Maximov suggested that the primary cause of freezing injury in plants was the disruption of the plasma membrane by the formation of ice crystals within the cytoplasm. Raoult's law shows that the freezing point of water can be depressed by the addition of solutes. Therefore, an increase in the cytoplasm's solute concentration can lower the temperature at which ice crystals form. Indeed, intracellular solute concentrations typically increase as temperatures gradually decline with the advent of winter. Likewise, the cytoplasmic solute concentrations of comparable organs differ among conspecifics growing at different elevations. Intracellular solute concentrations are typically changed metabolically by the catalysis of large organic polymers; for example, by the conversion of starch to sugar or proteins to amino acids. Solute concentrations can also be changed by extrusion of water. The mechanism for this process is not yet known. At subfreezing temperatures, ice may begin to form on external plant surfaces, such as those of leaves, stems, or bud scales. The loss of water from cells results in higher intracellular solute concentrations, which can significantly depress the freezing point of cytoplasm. This phenomenon of extracellular freezing can occur on internal as well as external surfaces. The horsetail *Equisetum hyemale* has stems with hollow internodes whose chambers frequently contain large "icicles" during cold winter months in high-latitude localities (Niklas 1989). These masses of ice melt during spring thaws and may provide aerial shoots with a source of liquid water before the ground thaws and water becomes available to roots. In addition to their ability to store water in the form of ice, the various chambers running through the stems of some horsetails facilitate pressurized airflow through aboveground and belowground stems (Armstrong and Armstrong 2009).

3.4 Chemical potential and osmotic pressure

An important concept in thermodynamics that was developed in the nineteenth century independently by Gibbs and von Helmholtz is that of free energy. For nonelectrolytes, the change in free energy ΔG equals the product of the volume V and the change in pressure ΔP minus the product of the entropy S and the change in temperature ΔT in the system plus the sum of the chemical potentials μ_i times the change in the number of moles Δn_i of components i. Effects of surface tension like those in narrow capillaries can be taken into account by an additional term $\sigma \Delta A$, where σ is the surface tension and ΔA the change in the surface area. Thus, the change in free energy is given by the formula

$$(3.2) \qquad \Delta G = V \Delta P - S \Delta T + \sigma \Delta A + \sum \mu_i \Delta n_i,$$

wherein the chemical potential μ_i is the free energy per mole of substance i, or more exactly,

$$(3.3) \qquad \mu_i = \left(\frac{\partial G}{\partial n_i} \right)_{T,P,n_{j \neq i}}$$

at constant temperature, pressure, and number of moles $n_{j \neq i}$ for all other substances.

The chemical potential of a substance depends on a number of factors, among which the concentration of the substance is very important. As the concentration of a substance decreases, so does its ability to do work. Thus, solute gradients within tissues, organs, or the entire plant body establish gradients of water's chemical potential. In particular, a gradient of chemical potential is established as water enters the plant body in roots and exits the plant body through leaves and as photosynthates, which contribute to the intracellular solute concentration, are produced in leaves and exported into roots.

The extent to which the chemical potential of water is decreased by the presence of one or more solutes is given by the formula

$$(3.4) \qquad \mu_{ws} - \mu_{pw} = RT \ln \left(\frac{a_{ws}}{a_{pw}} \right) = RT \ln \left(\frac{v_{ws}}{v_{pw}} \right),$$

where μ_{ws} is the chemical potential of an aqueous solution, μ_{pw} is the chemical potential of pure water at the same temperature and pressure, R is the gas constant (8.314 J mol^{-1} K^{-1}), and T is the temperature (in K). The quotient a_{ws}/a_{pw} gives the relative activity of water defined over the relative vapor pressure (v_{ws}/v_{pw}), which for dilute solutions, as Raoult's law shows, equals $n_w/(n_w + n_s)$. Notice that $\ln(v_{ws}/v_{pw}) = 0$ when the "solution" is pure water. Also notice that the natural logarithm of the relative vapor pressure is negative for any solution because the relative vapor pressure of any solution must be less than that of pure water.

Thermodynamic equilibrium is attained if the free energy reaches a minimum; that is, if ΔG equals zero. It follows that at a constant temperature, water molecules diffuse through a permeable membrane separating pure water from an aqueous solution until a pressure difference is reached to satisfy equation (3.2) in its equilibrium form (i.e., $V\Delta P + \mu_w \Delta n_w = 0$). Put differently, water moves from a region of high chemical potential to a region of low chemical potential until equilibrium is reached. This phenomenon is readily apparent when cell protoplasts are stripped of their cell walls and submerged in pure water. They swell and burst when the tensile strength of their plasma membranes is exceeded by the hydrostatic pressure that develops in them as water molecules enter the cytoplasm. Much greater internal pressures are required to burst cells with even thin primary walls because those walls have a tensile strength on the order of 10^2–10^3 MPa (Iraki et al. 1989). The primary cell wall, therefore, provides a mechanical constraint on the volumetric expansion of its protoplast and thus prevents the influx of water molecules that would otherwise destroy the protoplast.

The pressure (i.e., the force per unit area of cell membrane or wall) that must be applied by an expanding protoplast against its cell wall to prevent additional water molecules from moving across the cell membrane is called the osmotic pressure or osmotic potential, which is traditionally symbolized by Π. For dilute solutions of nondissociating solutes, this pressure is given by van't Hoff's law:

(3.5)
$$\Pi = \left(\frac{n_s}{V_w}\right)RT = cRT,$$

where V_w is the volume of water in solution (in liters) and n_s/V_w is the solute concentration, c. Van't Hoff's equation indicates that, for any given temperature and cell volume, the osmotic pressure Π increases linearly with the number of moles of the solute n_s. When c is very high, or when electrolytic ionization or molecular dissociation occurs, large deviations may result between predicted and observed osmotic pressures. For example, equation (3.5) predicts that the osmotic pressure of a molar solution of NaCl should be 2.27 MPa, whereas it is actually 4.32 MPa. (In passing, it is worth noting that this observation [which dates back to van't Hoff's time] strongly

supported Arrhenius's proposition of a dissociation of NaCl in water into
1 mol of Na^+ and 1 mol of Cl^- ions.) At high solute concentrations, van't
Hoff's equation predicts higher values than those observed when nondissociating molecules become hydrated because those molecules bind water to
them and thus reduce the activity (= effective concentration) of water in solution (e.g., a single sucrose molecule can bind six water molecules). Thus,
all dissociating solutes, most nondissociating solutes, and all water-binding
surfaces reduce the chemical potential of water in solution.

In a general form, the chemical potential of substance i can be formulated as

$$(3.6) \quad \mu_i = \mu_i^0 + RT \ln a_i + (P - P^0)\overline{V}_i - g(h - h^0)m_i - \sigma \left(\frac{\partial A}{\partial V} \right) \overline{V}_i,$$

where μ_i^0 is the chemical potential under standard conditions ($a = 1$ mol/
L, atmospheric pressure $= P^0$, $T = 293.15$ K, and an operationally defined
reference height, e.g., ground level or $h^0 = 0$ m), a_i is the activity (for dilute solutions, it is equal to the concentration in mol/L), $\overline{V}_i$ is the partial
molar volume (considered constant for dilute solutions), m_i is the mass
of substance i, and ghm_i is the gravitational term. The term $\sigma \left(\frac{\partial A}{\partial V} \right) \overline{V}_i$ accounts for the effects of surface tension. Water-binding surfaces, as, for
example, in the soil, reduce the activity of water. The contribution of the
water-binding surfaces is therefore implicitly included in the second term
of equation (3.6). It should be noted that the constant μ^0 is a reference
level that by convention is set to zero for pure water. Since for all applications only the difference of chemical potentials is considered, only relative
values are of importance (see Nobel 2005).

3.5 Water potential

We are now in a position to consider the ability of water's chemical potential to do work. This ability can be quantified in a different way when both
sides of equation (3.6) are divided by the partial molar volume of water
$\overline{V}_w$ (in units of L/mol). This manipulation allows us to define a quantity
that is called the water potential ψ_w in units of energy per unit volume:

$$(3.7) \quad \psi_w = \left(\frac{\mu_w - \mu_w^0}{\overline{V}_w} \right) = \frac{RT}{\overline{V}_w} \ln a_w + (P - P^0) - \rho_w g(h - h^0) - \sigma \left(\frac{\partial A}{\partial V} \right).$$

Comparing equations (3.6) and (3.7) shows that $\psi_w = 0$ for pure water under standard conditions. Because the activity of water a_w will be reduced when solutes are added to pure water, ψ_w becomes more negative as more and more solutes are added to pure water. The addition of water-binding surfaces (generically referred to as matrices) will also reduce water's activity, which results in more negative ψ_w values. Equation (3.7) also shows the effects of temperature, pressure, gravity or surface tension on water potential. An increase in temperature above the standard conditions will increase ψ_w. Positive pressure also increases ψ_w; negative pressure decreases ψ_w; and any surface tension reduces ψ_w. These relationships are summarized by a simplified formula that is frequently presented in plant physiology textbooks:

(3.8) $$\psi_w = \psi_s + \psi_m + \psi_p,$$

where ψ_s is the solute potential (which is always expressed as a negative number), ψ_m is the matric potential (which is always a negative number), and ψ_p is the pressure potential (which may be either positive or negative). In this equation, ψ_s describes numerically the decrease in the free energy of water due to the addition of solutes. The solute potential is related to the osmotic pressure, which for very dilute solutions, such as xylem water, is $\psi_s = -\Pi = -cRT$. The term ψ_m summarizes the decrease in free energy resulting from surface tension effects and from the addition of water-binding surfaces (e.g., colloids, particulates, and cell surfaces). The matric potential ψ_m may be thought of as a "bookkeeping term" that explicitly accounts for interfacial molecular interactions and capillary effects, which for small capillaries with an inner radius r is given by

(3.9) $$P_{capillary} = \frac{2\sigma}{r}.$$

It is difficult at times to separate conceptually (and sometimes even technologically) the difference between "interfacial molecular interactions" and "capillary effects," particularly in the context of the role played by the cell wall, which is chemically complex and porous.

Finally, ψ_p describes the effects of all internal or external fluid pressures in the absence of internal fluid flow (see section 3.8). When dealing with isolated organs or tissue samples, external fluid pressures can be neglected. However, the effect of gravity on tall columns of water, such as those in the stems of trees or vines, can significantly affect the pressure potential's contribution to ψ_w. Thus, when gravity "pulls down" on a column of water, the pressure potential will be $\psi_p = (P - P^0) - \rho_w gh$, where ρ_w is the density of water (998.2 kg/m³ at 20°C), g is the acceleration due to gravity (9.807 m/s² at sea level and 45° latitude), and h is the height of the water column. For a water column measuring 10 m in height, we see that the water potential at the top of the column is decreased by $-\rho_w gh \approx -0.10$ MPa.

A number of techniques are available to measure ψ_w directly from isolated plant tissues or organs such as leaves. For many kinds of leaves and short stems bearing a few leaves, water potential can be measured by the pressure chamber technique (Scholander et al. 1965). This technique capitalizes on the fact that when a leaf or branch is cut off a plant, the xylem water within it is placed at atmospheric pressure and the sample as a whole is at some negative water potential (fig. 3.3). Thus, the xylem water rapidly moves into the cells of the sample. The water potential is then determined by measuring the magnitude of the external pressure uniformly applied over the sample's surface, which drives water from the living cells into the xylem tissue (see section 9.4). This negative balancing pressure has been called *xylem pressure*. However, xylem pressure is not necessarily the pressure *in the xylem* (see section 3.8). If, for example, as in the spectacular experiments by Koch et al. (2004), the balancing pressure in a branch high up in a tree is determined in the laboratory, the actual pressure at any point in

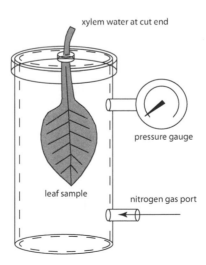

xylem water at cut end

pressure gauge

leaf sample

nitrogen gas port

FIGURE 3.3. Schematic of a Scholander pressure chamber (see also fig. 9.9).

the xylem below has to be corrected for the effect of gravity on the column of water between this point and the height of the origin of the branch. In addition, if transpiration occurs, an additional pressure difference results from the resistance to flow in narrow conduits (see section 3.7).

The actual pressure in the xylem can be measured directly with a pressure probe (see section 9.5). This technique, introduced by Green (1968), uses a micromanometer connected to a fine capillary tube inserted into a plant cell or a water-conducting vessel. At equilibrium, the pressure of the liquid within the device is equal to the pressure within the cell or vessel. Difficulties arise in the measurement of large negative pressures, since under these circumstances the insertion of the capillary tube can induce cavitation, which compromises the interpretation of the measurements (for a discussion of both the technique and its pitfalls, see Wei et al. 2001).

3.6 Turgor pressure and the volumetric elastic modulus

A cell is said to be fully turgid when the positive pressure exerted by its protoplast against the inner surfaces of the cell wall is sufficient to exclude the net entry of additional water molecules. Under these circumstances, the value of the pressure difference between the inside and the outside of the cell is called turgor pressure ψ_{TP}. Inspecting equation (3.8) shows that turgor pressure is the difference between the water potential outside the cell and the sum of the solute potential and matric potential within the cell (i.e., $\psi_{TP} = \psi_w - \psi_s - \psi_m$). In many cases, the matric potential can be neglected. Thus, under standard conditions where $\psi_w = 0$, if the turgor pressure of a cell is +3 MPa, the solute potential must equal −3 MPa. As a cell loses water, its turgor pressure decreases and the solute potential becomes more negative. Accordingly, its water potential also decreases. When viewed under a microscope, the protoplast of a cell that has lost a significant amount of water will be seen to have pulled away from its cell wall and undergone plasmolysis. Indeed, observing how cells (or tissue samples) change in volume when submerged in solutions differing in solute potential provides a way to determine their water potential.

Turgor pressure has important effects on the mechanical behavior of plants because it influences the apparent stiffness of cells, tissues, and entire organs (compare box 8.1), and thus the ability of plants to resist bending or twisting under their own weight or an externally applied force. The influence of turgor pressure is particularly pronounced on cells with thin walls (e.g., parenchyma and collenchyma) and much less apparent on cells with very thick walls (e.g., sclerenchyma and secondary xylem). When the protoplast of a cell is fully turgid and pressed against its wall, the materials in the wall are placed in hydrostatic tension and appear to stiffen, much as a bicycle tire or balloon becomes stretched and stiffens when the air pressure within it increases. The primary walls of plant cells are well suited to deal with large tensile stresses by virtue of their cellulose microfibrils, which have very high breaking tensile stresses. However, when cells are deprived of water and their turgor pressure decreases, their walls become progressively free to bend, buckle, or twist. This phenomenon, called wilting, is visible at the organ or whole plant level.

Plant physiologists and agronomists have long known that some species wilt less rapidly in response to water deprivation than others. These species maintain turgor pressure by virtue of osmotic adjustment. Water loss from cells invariably results in an increased solute concentration, which can result in the influx of water molecules if water becomes available again. However, osmotic adjustment is a physiologically controlled process that results in a *net* increase in the concentration of solutes within cells, tissues, or organs. This process allows cell enlargement and growth to be maintained for a limited time under near-drought conditions. The solutes involved in osmotic adjustment may be inorganic ions (e.g., Na^+, K^+, and Cl^-), simple carbohydrates, a variety of organic acids, or substances that would be normally used for the construction of primary cell walls. One side effect of using cell wall materials for this purpose is a reduction in the tensile strength of cell walls, which is correlated with a reduction in the proportions of crystalline cellulose and the amino acid hydroxyproline in the insoluble cell wall protein fraction (see Iraki et al. 1989). Hydroxyproline is an important component of extensin, a protein that influences the ability of cell walls to stretch.

The biomechanics of plant growth has been reviewed recently (Schopfer 2006). Osmotic adjustment and alterations in cell wall stiffness are required for cell enlargement and growth. The former process establishes a water potential gradient that allows for the influx of water and thus an increase in volume. However, this "growth" in size is not possible unless the cell wall stretches and deforms, nor is it permanent unless the rigidity of the cell wall is restored physiologically, which results in a permanent increase in cell volume. The process of reducing and then increasing the rigidity of cell walls must be accompanied by the production of new wall materials if a constant cell wall thickness is to be maintained. Cell growth in size, therefore, is determined by the maintenance of positive turgor pressure and the regulation of the physical properties of cell walls.

Chapter 8 treats cell growth in terms of the physical properties of plant cell walls. Here, we only wish to briefly explore a property called the volumetric (or bulk) elastic modulus E_V and its relationship to changes in the turgor pressure ΔP_T and the volume ΔV of a cell with an original volume of V_0 as water enters or leaves it. The relationships among these parameters are given by the formula

$$(3.10) \qquad E_V = \Delta P_T \left(\frac{V_0}{\Delta V} \right).$$

Numerical values for E_V can be experimentally determined by placing a tissue or organ in a pressure chamber (see section 9.4) and collecting the liquids expressed as the pressure within the chamber is slowly increased (Scholander 1965). The slope of the linear portion of the applied pressure versus the cumulative expressed liquid volume equals the change in the applied pressure ΔP_A divided by the change in the expressed liquid volume ΔV_E, and V_0 is the total amount of water in the sample at full turgor. Experiments such as these have shown that the magnitude of E_V depends on the thickness of cell walls and can range between 1 and 50 MPa, indicating that cell volume can change from as little as 0.2% to as much as 10% in response to an increase of 0.1 MPa internal pressure.

As might be expected, the range over which E_V can vary influences the time t_e it takes for a cell to expand, as does a cell's osmotic potential, volume, and surface area. The interrelationships among these parameters are given by the formula

$$(3.11) \qquad t_e = \frac{V_0}{A} \frac{1}{\Omega(E_V + \Pi)},$$

where t_e is the time required to complete all but $1/e$ of the change in cell volume ($\sim$37%), A is cell surface area, and Ω is the permeability of the cell to water flow in or out. The numerical values of Π tend to be on the order of 0.5% to 2.0% that of E_V. Therefore, $t_e \sim V_0/(A\Omega E_V)$. Biologically reasonable values for Ω are on the order of 10^{-12} m s^{-1} Pa^{-1}. Therefore, a spherical cell with a surface area-to-volume quotient equal to 1×10^{-2} μm^{-1} and a volumetric elastic modulus of 20 MPa requires about 5 s to change its volume. Clearly, if either the cell's permeability to water flow or the volumetric elastic modulus is small, the time it takes to alter its volume increases significantly. For example, if $\Omega = 10^{-13}$ m s^{-1} Pa^{-1}, the time constant for our hypothetical cell is about 50 s.

3.7 Flow through tubes and the Hagen-Poiseuille equation

Water molecules can move through the plant body either by passing from one living cell into another or by means of bulk flow through the lumens of dead cells specialized to conduct water. For small, short terrestrial plants, the former is sufficient to meet the demands of transpiration. However, as the distance between transpiring surfaces and the substrate supplying water increases, bulk flow of water becomes increasingly important.

Engineers and physicists have long appreciated that tubes provide an excellent means of transporting water rapidly over long distances and that the rate of transport increases with tube diameter and the gradient of hydraulic pressure. Indeed, Gotthilf Hagen (in 1839) and Jan Poiseuille (in 1840) independently showed that the volume of water moving per unit time $\Delta V/\Delta t$ through a cylindrical tube with radius r scales as the fourth power of r and linearly with the hydrostatic pressure gradient $\partial P/\partial x$. In 1856, Hans Wiedemann formalized these relationships with the equation

$$(3.12) \qquad \frac{\Delta V}{\Delta t} = -\frac{\pi}{8\mu}\left(\frac{\partial P}{\partial x}\right)r^4,$$

where $\Delta V/\Delta t$ is flux, μ is the dynamic viscosity of the solution passing through a tube, and the negative sign indicates that flow is in the direction of decreasing hydrostatic pressure. Equation (3.12) is called the Hagen-Poiseuille equation, although Wiedemann deserves equal credit, particularly because he provided its mathematical formality (see box 6.2). Physiological studies often find it useful to consider the flow speed φ, which is the flux per unit area. Because the average flow speed is the flux divided by the cross-sectional area of a tube, it follows that

$$(3.13) \qquad \varphi = -\frac{1}{8\mu}\left(\frac{\partial P}{\partial x}\right)r^2.$$

Either way, the Hagen-Poiseuille equation assumes that the flow regime is laminar and that the flow speed profile across the tube's longitudinal profile is parabolic. These assumptions account for the fourth power of r in equation (3.12) and should make it clear that $\Delta V/\Delta t$ and φ are averaged values.

Equation (3.13) helps to show that the pressure gradients required to promote water flow through cell walls are exceptionally large compared with those that promote flow through narrow tubes. To illustrate this point, consider water flow through a primary cell wall with interfibrillar spaces that, on average, measure 10 nm (= 100 Å) in diameter (i.e., $r = 5$ nm). Given the dynamic viscosity of water (i.e., 1.003×10^{-3} Pa s at 20°C; see table 2.1), equation (3.13) indicates that a flow speed of water equal to 1 mm/s requires a pressure gradient of $-[8\ (1.003 \times 10^{-3}$ Pa s$)\ (1 \times 10^{-3}$ m/s$)]/(5 \times 10^{-9}$ m$)^2 \approx -3 \times 10^5$ MPa/m. In contrast, the same equation indicates that a pressure gradient of $\partial P/\partial x \approx -0.08$ MPa/m is required to achieve the same flow speed through a tube measuring 20 μm in diameter. In reality, this comparison is unrealistic because adjoining, living cells are typically interconnected by plasmodesmata, which have diameters of 30 to 60 nm, through which water can flow symplastically.

The Hagen-Poiseuille equation is frequently used to estimate the hydraulic capacity of xylary elements (tracheids and vessel members) in stems or roots. The stipulation that laminar flow occurs through a tracheid or a vessel member is not a concern. Using the heat pulse method, which allows the flow rates of water through stems to be measured, Huber and Schmidt (1936, 1937) report peak water flow velocities of 14–45 m/hr through wood with vessels measuring 200–400 μm in diameter. Taking 400 μm as the characteristic dimension of a vessel's diameter, 45 m/hr (≈ 0.013 m/s) as the maximum flow speed, and 1.0×10^{-6} m²/s as the kinematic viscosity of water (at 20°C), we see that the Reynolds number under these conditions is Re $= (400 \times 10^{-6}$ m$)\ (0.013$ m/s$)/(1.0 \times 10^{-6}$ m²/s$) \approx 5$, which is well below the Reynolds number at which the transition from laminar to nonlaminar flow occurs (for a detailed discussion of Reynolds numbers, see section 6.2).

In contrast, other assumptions that are required when using the Hagen-Poiseuille equation are undoubtedly violated (fig. 3.4). Tracheids and vessel members typically have thickened secondary walls with surface irregularities, and their cross-sectional geometries are more elliptical or polygonal than circular in outline. The flow speed profiles within these cells are therefore likely not to be paraboloid, but a good deal more complex. In addition, as noted earlier, although they have perforated walls, tracheids are tapered and closed at their ends, whereas the end walls of vessel mem-

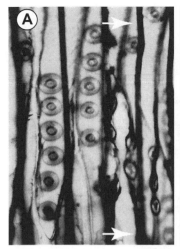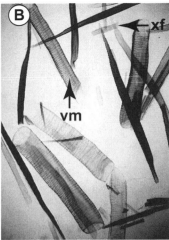

FIGURE 3.4. Xylary cells participating in the bulk flow transport of water through the roots, stems, and leaves. (A) Close-up of a group of tracheids from pine (*Pinus*) wood showing the tapered end walls of a tracheid (indicated by pair of arrows) with six bordered pits along one of its lateral sides (for a diagram of bordered pits, see fig. 3.1). (B) Close-up of a maceration of oak (*Quercus*) wood showing dissociated xylem ray cells (xf) and vessel members with opened end walls (vm).

bers are perforated with different and varying degrees of closure. Likewise, individual vessels do not span the entire length of the plant body, and very large resistances to flow can occur when water passes out of one vessel and enters another. These and other biological features undoubtedly result in resistances to the flow of water, which can make estimates of flow rates problematic unless the Hagen-Poiseuille equation is adjusted to compensate for them. Under any circumstances, it is important to remember that the use of this formula when dealing with real water-conducting cells provides nothing more than an approximation.

Given these caveats, it is not surprising that experimentally measured flow rates of water through plants agree only roughly with those predicted by the Hagen-Poiseuille equation. For example, Tyree and Zimmermann (1971) report that the maximum (peak) flow rates of water through the trunks of *Acer rubrum* (red maple) are between 33% and 67% of those calculated using the Hagen-Poiseuille equation, whereas Petty (1978, 1981) reports values between 34% and 38% for *Betula pubscens* (birch) and 38% for *Acer pseudoplatanus* (sycamore). Similar percentage values are reported for conifer species, although the water-conducting cells in the wood of these species are tracheids, cells that deviate from being capillary tubes even more than vessel members. For example, Münch (1943) reports values ranging between 26% and 43% for *Abies alba* (white spruce), while Ewart (1905) reports a value of 43% for *Taxus* (yew). In contrast, the hydraulics of some vine species appears to be remarkably well approximated using the Hagen-Poiseuille equation. For example, Zimmermann and Brown (1971) report values of 100% for vines, indicating that the vessels in these species behave like ideal capillary tubes!

3.8 The cohesion-tension theory and the ascent of water

The most ancient land plants lacked tissues specialized for water conduction. Much like many modern nonvascular land plants, some probably relied on their external surfaces to wick up water from their substrates and carry it by capillarity over their external surfaces. Others may have hugged their substrates, thereby keeping the distance between an external supply of water and their uppermost tissues to a minimum. In either case, the extent to which these plants could grow vertically was limited by the path lengths over which water had to travel to reach aerial tissues. The evolution of tissue systems specialized to conduct water and sap and which simultaneously provide mechanical support for the weight of aerial organs permitted plants to achieve great vertical stature. By the end of the Devonian, vascular plants reached heights achieved by modern trees. The tallest and largest organisms (measured by weight) on earth are trees.

As plant stature increased, however, the path lengths between water-absorbing organs and transpiring photosynthetic surfaces increased, as did the daily demand for water. For example, over 4 L of water per day are required for the growth of a typical corn plant with a total leaf area of 2 m^2; over 200 L per day are required for the growth of a moderately sized oak tree. With a xylem cross-sectional area of 0.2 cm^2 and a transpiration rate of 200 mL/hr, the average speed of flow at the base of a corn plant is about 10 m/hr, whereas the peak flow speeds in trees reach values four to five times this rate. It is not surprising that the maximum diameter of water-conducting cells reported for fossil plants increased throughout the Devonian, the period when plant height increased at its fastest evolutionary pace. Likewise, the high negative pressures required for these flow rates help to explain why the secondary walls of vessel members and tracheids are secondarily thickened: these thickenings provide mechanical reinforcements that prevent the cell walls from imploding (see fig. 3.4B).

The mechanism by which water is transported through the xylem to the tops of the tallest trees has been the object of intense research and considerable debate. By the end of the nineteenth century, however, it was generally agreed that the ascent of water in xylary elements did not require living cells. In 1891, Strasburger demonstrated that water could rise through stems that had been killed by heat, which led him to conclude that transpiration is sufficient to carry water through tracheids or vessels. This demonstration foreshadowed what has become known as the cohesion-tension theory for the ascent of water, which is based on three observations:

1. The high cohesive force of water molecules
2. The continuity of water within the plant body as a whole and in the xylem in particular
3. The existence of a water potential gradient that, on average, becomes more negative along the path length separating roots from leaves and other transpiring organs

According to this theory, water transport in plants occurs along a gradient of negative pressure (tension) in the dead, tubelike cells of the xylem. Transpiration, water cohesion (and adhesion to the inner surfaces of cell walls), osmotic pressures, and surface tension (in menisci formed in leaf aerenchyma, etc.) provide the factors necessary to lift water against the pull of gravity (for an excellent review, see Zimmermann 1983).

According to equation (3.4), water potential is related to water vapor pressure. It should be noted here that ψ_w is the water potential at the interface between liquid and gaseous phase at the aerenchymatous surfaces vented by stomata. Usually the water vapor pressure within the aerenchymatous cavities will be higher than the water vapor pressure in the atmosphere, often expressed as relative humidity. The difference depends on a number of factors, particularly on the thickness of the boundary layer through which the water vapor has to diffuse and thus on the wind speed. Most important, the difference can be regulated by the degree to which stomata are closed or open.

Transpiration is the main driving force for xylem flow. Using the simplification that the radius r of the xylem vessels is constant over the entire length of the tree, equation (3.13) allows us to calculate the pressure gradient required for a given flow speed. For angiosperms with $r = 20$ μm (Sperry et al. 2006), a flow speed of 1 mm/s requires a pressure gradient of 0.02 MPa/m. For a tree 100 m high, this amounts to negative pressures on the order of −2 MPa, in addition to −1 MPa due to the effect of gravity on the water column. It is important to remember that tree morphology determines the distribution and speed of xylem flow. These high negative pressures explain the observation that trees shrink in girth under conditions of extensive transpiration. Indeed, tree trunk diameters may decrease by as much as 6% during an afternoon as the result of the negative pressures necessary for the rapid water transport toward leaves.

Cavitation under these large negative pressures is a major obstacle to xylem flow, as it leads to embolisms in vessels (Sperry et al. 2006). The probability of cavitation decreases with vessel diameter; that is, vessels with a smaller diameter have a higher factor of safety. Here we see yet another example of a design constraint. The unfortunate reality for many plants is that there is an inverse relationship between the efficiency of water conduction (as gauged by the Hagen-Poiseuille equation) and the danger of embolism caused by cavitation. The probability of cavitation of water is difficult to determine precisely because it depends on the geometry and chemical composition of the surfaces with which water molecules make contact as well as on the solutes or dissolved gases that the water contains. Thus, estimates of the tensile strength of water vary. However, Briggs (1950) showed that columns of pure water contained in glass capillaries could withstand centrifugal forces in excess of 25 MPa before they "snapped" when their tensile strength was exceeded. This tensile strength is remarkably high, even compared with some metals, and much larger than the forces that are required to pull water to the tops of trees.

Of particular importance is the fact that tracheids and vessels, once embolized, can refill. This refilling can occur at night, when rates of photosynthesis and transpiration are low, and even during the daytime if the soil moisture content is sufficiently high. Embolisms can also be dissolved by surface tension effects, but only at pressures $P > -2\sigma/r$ (compare eq. [3.9]), which for $r = 20$ μm is about −7 kPa. Putative mechanisms of embolism repair under high tension, as in tall trees, are still debated (Tyree and Zimmermann 2002; Clearwater and Goldstein 2005). It has been suggested that xylem and phloem form an interactive "countercurrent" system (Holbrook and Zwieniecki 2005). Understanding their controlled interaction may be the key to understanding embolism repair at high negative pressures.

These considerations do not address another important point. Although water potential gradients and the tensile properties of water are sufficient to lift water up the trunks of very tall trees, the rate at which water is supplied to growing cells in the uppermost canopies of those trees may be insufficient to sustain turgor pressures at levels that permit cell expansion and growth. In theory, this would result in a progressive reduction in the rate of growth in height (Koch et al. 2004; Ning et al. 2008), ultimately resulting in a maximum height that depends on the hydraulic architecture of the species and local environmental conditions, rather than on strictly mechanical considerations such as the strength of wood. Likewise, the rates at which water is transported meet increasing hydraulic path length resistance (which is a function of height) such that water columns in the xylem can be pulled to their breaking point and cavitate (by the formation of water vapor embolisms). Koch et al. (2004) examined *Sequoia semper-virens* (redwood) trees in the wet temperate forests of Northern California and showed that a number of leaf functional traits changed along the heights of individual trees. Based on regression analyses, these authors found that leaf water potential declined with increasing height, as did the negative balancing pressure (from -1 MPa at 30 m height to -2 MPa at 115 m height) and light-saturated photosynthetic rates per unit leaf mass. In addition, leaf mass per surface area increased, and overall leaf size declined, with height above ground, suggesting that leaves become progressively more water-stressed toward the tops of very tall trees.

Indeed, hydraulics, rather than mechanics, seems to limit growth. Niklas and Spatz (2004) have shown that the proportional relationship between the height and basal stem diameter of a tree emerges directly from a consideration of plant growth and hydraulics and requires no assumption regarding mechanical stability (*sensu* McMahon 1973). Our analysis shows that the size-dependent relationships for small and intermediate-sized plants differ from those observed for large trees, for which height is generally proportional to the 2/3 power of basal stem diameter. Therefore, no single allometric scaling rule holds true across the entire size range of self-supporting plants. The predictions from the hydraulic model agree with data gathered from a broad taxonomic spectrum of vascular plants with self-supporting stems differing in height.

3.9 Phloem and phloem loading

The evolution of xylem permitted terrestrial plants to grow to great heights. However, this anatomical innovation required parallel evolutionary innovations involving the translocation of photosynthates. Among the vascular plants, the most highly specialized tissue for this purpose is called phloem. This tissue achieves its greatest anatomical complexity in the angiosperms. In this section, we briefly describe the anatomy of angiosperm phloem and discuss a theory, called the pressure-flow model, that helps to explain how metabolites produced in mature photosynthetic leaves are transported to other parts of the plant.

The phloem is generally close to the xylem, and it is typically external to the xylem. The phloem consists of a number of morphologically and anatomically different cell types. Some cells provide mechanical support (i.e., phloem fibers and sclereids). Others are involved with storage (e.g., parenchyma and laticifers). In the angiosperms, the cells in the phloem that conduct photosynthates and other solutes are called sieve tube members (the comparable cells in gymnosperms are called sieve cells). These cells, which are arranged end to end, function as long-distance conducting structures called sieve tubes. The sieve tube members are characterized by sieve plates, containing numerous sieve plate pores, at their adjoining end walls. They also have numerous lateral sieve areas along their sides. The sieve plate pores and the lateral sieve areas are regions in which numerous plasmodesmata are aggregated. The plasmodesmata in the sieve plate pores provide for symplastic (= cytoplasmic) continuity among all the sieve tube members that constitute a sieve tube. The lateral sieve areas provide symplastic continuity between each sieve tube member and the adjacent living cells. Each sieve tube member is the product of the asymmetric division of a procambial cell, which also produces one or more companion cells. When fully differentiated, sieve tube members lack nuclei. Their metabolism is regulated by the nuclei of their companion cells. As we will soon see, the close physiological link between a sieve tube member and its companion cell has profoundly important consequences for the long-distance transport of photosynthates and other molecules, including water.

There are three kinds of companion cells: ordinary companion cells, transfer cells, and intermediary cells. The kind that develops depends on where the phloem is located in the plant body. All three kinds have abundant mitochondria, which attest to their comparatively high metabolic rates. Ordinary companion cells are distinguished by having few plasmodesmata connecting to cells other than their corresponding sieve tube cells. Transfer cells have fingerlike extensions on their inner cell walls that greatly increase the surface area of their cell membrane. Both of these cell types are thought to be involved with the transport of molecules via the apoplast (i.e., through cell walls). In contrast, intermediary cells have numerous plasmodesmata that symplastically connect them to their neighboring cells, particularly the bundle sheath cells in leaves. Intermediary cells are thought to be involved in symplastic transport.

The transport of molecules through the apoplast or the symplast of the phloem follows a source-to-sink pattern. That is, transport proceeds from a location where solutes are produced (as, for example, in mature, photosynthetic leaves) to a location where they are metabolized or stored (as, for example, in roots or tubers). This transport cannot be the result of passive diffusion within the phloem because rates of phloem translocation are much faster than those predicted for passive diffusion (i.e., on the order of 1 m/hr). There are numerous theories about how phloem transports solutes. The most widely accepted theory, called the pressure-flow model, was first proposed by Ernst Münch (1930). According to this model, solutes move through the phloem as the result of an osmotically generated pressure gradient ($\Delta \psi_p$) between a source and a sink (see eq. [3.8]). The first step is generally referred to as *phloem loading*: the export of carbohydrates (typically sucrose and sugar alcohols) from mesophyll (source) cells into the intermediary companion cells of the minor vein phloem in leaves (Turgeon and Ayre 2005). There are two possible routes for this to occur, either through plasmodesmata (symplastic transport) or through the cell wall (apoplastic transport). Symplastic transport is driven by the diffusion of carbohydrates down a concentration gradient. This results in a low (negative) solute potential ($\Delta \psi_s$), which in turn lowers the water potential ($\Delta \psi_w$), in the sieve tube member cells into which the carbohydrates enter (fig. 3.5). Apoplastic transport is driven by the active transport of carbohydrates through cell walls into the phloem. In either case, as a result of the increased solute concentrations in the sieve tube members and their companion cells, water from the neighboring xylem enters the sieve tube members, increasing their turgor pressure ($\Delta \psi_p$). Phloem unloading involves the sequestration or metabolism of the transported carbohydrates, which results in a lower solute concentration and a higher (more positive) solute potential in the sieve tube members in the sink tissue. Water leaves the phloem in response to the water potential gradient, resulting in a drop in turgor pressure in the phloem (fig. 3.5).

The role played by the sieve plates in this scenario is critical. The pressure difference between the source and the sink would rapidly vanish if the transport pathway operated in a completely open system. The sieve plates provide resistance along the transport pathway that maintains the

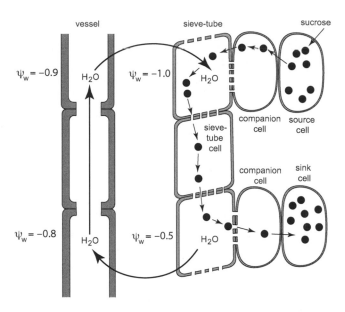

FIGURE 3.5. Schematic of the pressure-flow model of phloem translocation first proposed by Münch (1930). The model hypothesizes that carbohydrates such as sucrose synthesized in source (mesophyll) cells are either actively or passively transported into the companion cells of minor vein phloem and that a pressure gradient between source and sink cells drives the flow of the solution. Water flows from the neighboring xylem into the phloem near the source cells. The flow of water along this pathway is driven by a pressure gradient (rather than the water potential gradient; see representative values for the water potential at four locations).

pressure gradient in the sieve tubes. The sieve plates are also important because water and solute molecules are not passing across cell membranes in the sieve tubes. Instead, these molecules are being transported within a continuous symplast. Notice that the values of the water potentials along the sink-to-source pathway are such that water appears to be flowing against the water potential gradient (see fig. 3.5). These values therefore appear to violate the laws of thermodynamics. However, the water in this system is not moving by osmosis. It is moving in bulk flow at the same rate as the transported solutes. Consequently, the solute potential does not contribute to the driving force for the movement of water. The movement of water is driven by the pressure gradient, which is generated by active transport mechanisms.

Phloem loading strategies differ among species. These strategies are characterized on the basis of plasmodesmata connectivities, the presence of a concentration difference between leaf mesophyll and minor phloem veins, and the type of carbohydrates that are transported. Several general trends have emerged (Turgeon et al. 2001). Currently, all species examined that have low numbers of plasmodesmata connecting the mesophyll to minor vein companion cells have apoplastic phloem loading. Species with high plasmodesmatal connectivities between the mesophyll and companion cells tend to rely on polymer trapping (to prevent imported carbohydrates from passively diffusing back into the mesophyll). Species that load actively, either apoplastically or by polymer trapping, tend to have lower total soluble carbohydrate concentrations than species that rely on passive diffusion to carry carbohydrates into their phloem.

Literature Cited

Armstrong, J., and W. Armstrong. 2009. Record rates of pressurized gas-flow in the great horsetail, *Equisetum telmateia*: Were Carboniferous Calamites similarly aerated? *New Phytol.* 184:202–15.

Briggs, L. J. 1950. Limiting negative pressure of water. *J. Appl. Phys.* 21:721–22.

Clearwater, M. J., and G. Goldstein. 2005. Embolism repair and long distance water transport. In *Vascular transport in plants*, edited by M. N. Holbrook and M. A. Zwieniecki. Amsterdam: Elsevier.

Ewart, A. J. 1905–1906. The ascent of water in trees. *Phil. Trans. Roy. Soc. London*, ser. B, 198:41–45.

Franks, P. J., and G. D. Farquhar. 2007. The mechanical diversity of stomata and its significance in gas-exchange control. *Plant Physiol.* 143:78–87.

Green, P. M. 1968. Growth physics in *Nitella*: A method for continuous in vivo analysis of extensibility based on micro-manometer techniques for turgor pressure. *Plant Physiol.* 43:1169–84.

Holbrook, M. N. and M. A. Zwieniecki. 2005. Integration of long distance transport systems in plants: Perspectives and prospects for future research. In *Vascular transport in plants*, edited by M. N. Holbrook and M. A. Zwieniecki. Amsterdam: Elsevier.

Huber, B., and E. Schmidt. 1936. Weitere thermo-elektrische Untersuchungen über den Transpirationsstrom der Bäume. *Tharandt. Forst. Jb.* 87:369–412.

———. 1937. Eine Kompensationsmethode zur thermo-elektrischen Messung langsamer Saftströme. *Ber. Deutsch. Bot. Ges.* 55:514–29.

Iraki, N. M., R. A. Bressan, P. M. Hasegawa, and N. C. Carpita. 1989. Alteration of the physical and chemical structure of the primary cell wall of growth-limited plant cells adapted to osmotic stress. *Plant Physiol.* 91:39–47.

Kirk, J. T. O. 1983. *Light and photosynthesis in aquatic ecosystems.* Cambridge: Cambridge University Press.

Koch, G. W., S. C. Sillet, G. M. Jennings, and S. D. Davis. 2004. The limits to tree height. *Nature* 428:851–54.

McMahon, T. A. 1973. The mechanical design of trees. *Science* 233:92–102.

Münch, E. 1930. *Die Stoffbewegungen in der Pflanze.* Jena, Germany: Gustav Fischer.

———. 1943. Durchlässigkeit der Siebröhren für Druckströmungen. *Flora* 136:223–62.

Niklas, K. J. 1989. Extracellular freezing in *Equisetum hyemale. Amer. J. Bot.* 76:627–31.

Niklas, K. J., and H.-C. Spatz. 2004. Growth and hydraulic (not mechanical) constraints govern the scaling of tree height and mass. *Proc. Nat. Acad. Sci. USA* 101:15661–63.

Ning, D., F. Jinto, C. Shuo, and L. Yang. 2008. A hydraulic-photosynthetic model based on extended HLH and its application to coast Redwoods (*Sequoia sempervirens*). *J. Theor. Biol.* 253:393–400.

Nobel, P. S. 2005. *Physicochemical and environmental plant physiology.* 3rd ed. Amsterdam: Elsevier.

Owens, T. G. 1988. Light-harvesting antenna systems in chlorophyll *a*/*c*-containing algae. In *Light-energy transduction in photosynthesis: Higher plant and bacterial models,* edited by S. E. Stevens Jr. and D. A. Bryant, 122–36. Rockville, MD: American Society of Plant Physiologists.

Petty, J. A. 1978. Fluid flow through the vessels of birch wood. *J. Exp. Bot.* 29:1463–69.

———. 1981. Fluid flow through the vessels and intervascular pits of sycamore wood. *Holzforschung* 35:213–16.

Pittermann, J., J. S. Sperry, U. G. Hacke, J. K. Wheeler, and E. H. Sikkema. 2005. Torus-margo pits help conifers compete with angiosperms. *Science* 310:1924.

Proctor, M. C. F. 2005. Why do Polytrichaceae have lamellae? *J. Bryol.* 27:221–29.

Scholander, P. F., E. D. Bradstreet, E. A. Hemmingsen, and H. T. Hammel. 1965. Sap pressure in vascular plants. *Science* 148:339–46.

Schopfer, P. 2006. Biomechanics of plant growth. *Amer. J. Bot.* 93:1415–25.

Sperry, J. S., U. G. Hacke, and J. Pittermann. 2006. Size and function in conifer tracheids and angiosperm vessels. *Amer. J. Bot.* 93:1490–1500.

Sperry, J. S., F. C. Meinzer, and K. A. McCulloh. 2008. Safety and efficiency conflicts in hydraulic architecture: scaling from tissues to trees. *Plant Cell Env.* 31:632–45.

Turgeon, R., and B. G. Ayre. 2005. Pathways and mechanisms of phloem loading. In *Vascular transport in plants,* edited by N. M. Holbrook and M. A. Zwieniecki, 45–68. Amsterdam: Elsevier.

Turgeon, R., R. Medville, and K. C. Nixon. 2001. The evolution of minor vein phloem and phloem loading. *Amer. J. Bot.* 88:1331–39.

Tyree, M. T., and M. H. Zimmermann. 1971. The theory and practice of measuring transport coefficients and sap flow in the xylem of red maple (*Acer rubrum*). *J. Exp. Bot.* 22:1–18.

———. 2002. *Xylem structure and the ascent of sap.* 2nd ed. Berlin: Springer Verlag.

Wei, C., E. Steudle, M. T. Tyree, and P. M. Lintilhac. 2001. The essentials of direct xylem pressure measurements. *Plant Cell Env.* 24:549–55.

Weng, H., i. Molina, J. Shockey, and J. Browse. 2010. Organ fusion and defective cuticle function in a *lacs1 lacs2* double mutant of *Arabidopsis. Planta* 231:1089–1100.

Wilson, J. P., and A. H. Knoll. 2010. A physiologically explicit morphospace for tracheid-based transport in modern and extinct seed plants. *Paleobiology* 36:335–55.

Wu, H.-I., and P. J. H. Sharpe. 1979. Stomatal mechanics. 2. Material properties of guard cell walls. *Plant Cell Env.* 2:235–44.

Zimmermann, M. H. 1983. *Xylem structure and the ascent of sap.* Berlin: Springer Verlag.

Zimmermann, M. H., and C. L. Brown. 1971. *Trees: Structure and function.* Berlin: Springer Verlag.

CHAPTER FOUR

The Mechanical Behavior
of Materials

For all particularity in natural science reduces to the discovery of definite magnitudes and relations of magnitudes.—Ernst Cassirer, *The Concepts of Natural Science*

The major premise of this book is that organisms cannot violate the fundamental laws of physics. A corollary to this premise is that organisms have evolved and adapted to mechanical forces in a manner consistent with the limits set by the mechanical properties of the materials out of which they are constructed (e.g., Skotheim and Mahadevan 2005). We see no better expression of these assertions than when we examine how the physical properties of different plant materials influence the mechanical behavior of plants. Nor do we find any better evidence for the evolution and adaptation of plants to mechanical forces than when we compare their mechanical attributes with those of fabricated materials and observe that many plant materials mechanically outperform some of the most common materials used by engineers and architects. For instance, the nutshell of the macadamia, an Australian evergreen (*Macadamia ternifolia*), is as hard as annealed, commercial-grade aluminum, resists twice the force necessary to cause fracture in aluminum and many other metals, and is stronger than silicate glasses, concrete, porcelain, or domestic brick. Yet the nutshell is less than half as dense as many of these materials, which is an advantage in that it contributes disproportionately less to the overall weight a plant must sustain.

Strength is only one of the physical attributes that must be considered when assembling a structure like a nutshell, tree trunk, or flower stalk, since vertical construction carries with it the design constraint of self-loading (see chap. 5). In addition to being strong, materials should be light. In fact, plants have evolved a number of materials that are extremely strong for their unit weight (density). For example, cellulose, for its density, is the strongest material known when placed in tension. The trade-off between strength and density within a nutshell illustrates a much more general principle: the mechanical behavior of any single material is defined by a number of material properties, and not all of these properties can be maximized simultaneously. Each material must be used according to its particular qualities and the types and magnitudes of the mechanical forces it must sustain.

In addition to their low density and comparatively great strength, biological materials have other advantages over many engineered counterparts. First, biological materials are versatile. Their material properties can change as a function of their age or immediate physiological condition; for example, young plant cell walls are ductile, while older cell walls tend to be much more elastic and resilient. Second, the material properties of plant substances and organs can change, through their capacity for growth, in response to the magnitudes of the forces they are subjected to. Finally, by virtue of growth, many biological materials can be repaired when damaged.

Indeed, the reason for differences in the experimental results reported by early workers was that many failed to appreciate the importance of comparing plant materials isolated from plants of equivalent age and physiological condition. Another reason for acrimony among early plant biomechanicists is that they failed to normalize the mechanical forces applied to organs in terms of the transverse areas through which these forces operated. As we will see in this chapter, this normalization procedure gives us a quantity called stress (which is force/area). Clearly, the same magnitude of force applied to structures constructed out of the same materials but differing in their dimensions will have very different mechanical consequences. Unfortunately, the transverse areas of plant organs or parts of organs are difficult to measure accurately, even with present-day instrumentation. Testing small biological samples, therefore, presents many challenges (see chap. 9).

The capacity of biological materials to change their material properties through growth and development confers a spatial and temporal heterogeneity on the mechanical behavior of the plant body and its constituent parts. This capacity for change sets biological materials and the structures built with those materials apart from all engineered artifacts. Therefore, an important goal is to understand not merely how biological materials can be studied in terms of engineering practice and theory, but how engineering theory and practice can be extended and enriched by what we learn about plants, which is the subject of the relatively new discipline called biomimetics (see section 9.8).

Nonetheless, because they *can* change over time, the mechanical properties of most plant materials are difficult to quantify. And even at a particular developmental instant, most plant materials exhibit properties that conform to those of neither ideal solids nor ideal fluids. Fortunately, with care and a full appreciation of the many limits involved, we can approximate the behavior of many plant materials as if they were elastic solids or ideal fluids. Indeed, the fabricated materials used in everyday engineering practice are anything but ideal materials, yet their analyses can be approximated as such with considerable success. For example, almost all solids manifest to some degree the property known as elasticity—that is, they deform when subjected to an applied force and restore these deformations when the force is removed. The elementary theory of elasticity, developed by engineers, assumes that materials are ideal elastic materials—that they completely and instantly restore their deformations when the magnitude of applied forces drops to zero. Under certain boundary conditions, most metals behave very nearly as ideal elastic solids and perform within the parameters set by the elementary theory of elasticity. By the same token, it is not unusual for an engineer to treat a real fluid as if it were an ideal fluid. And even though the behavior of the real and ideal fluids may differ, the extent to which theory can predict the behavior of a real fluid is typically satisfactory for most practical situations.

For pedagogical reasons, we shall use the same tactic followed by the majority of engineers and treat plant materials initially as if they were ideal solids or ideal fluids. In this way, we can explore some fundamental concepts even though we know that biological reality is far more complex. For example, when an external force is applied to a material, internal forces are produced. These internal forces, normalized by the area through which they operate, are stresses that produce deformations, which if normalized by the undeformed dimensions of a material, are called strains. Hooke's law ("*ut tensio sic vis*") states that, for small deformations, the strains are proportional to the stresses. Within this range of proportionality, stresses and strains are related to one another by numerically discrete material moduli that can be used to distinguish among materials under certain conditions of loading, such as tension, compression, shearing, torsion, and bending. Once treated, these fundamental concepts will be expanded to treat nonlinear mechanical behavior.

Another physical attribute of most plant materials is viscoelastic behavior, which will be examined in detail in section 4.11. Viscoelastic materials are those that show both elastic and viscous components in their behavior. Thus, in a crude sense, they are hybrid materials exhibiting the properties of both solids and fluids. Unfortunately, we will see that our ability to treat viscoelastic materials is limited to phenomenological descriptions of their behavior. This inadequacy stems from the fact that most biological materials manifest nonlinear viscoelasticity, for which no adequate mechanistic theory has (as yet) been developed. Thus, it is always beneficial to retain an awareness of when and how theory and reality fail to coincide.

4.1 Types of forces and their components

Regardless of its apparent complexity, any externally applied force or combination of forces can be divided into two fundamental force components, distinguishable in terms of the direction in which they operate with respect to a surface of interest. These components are called the normal force component, which operates perpendicular to the surface of interest, and the tangential force component, which operates parallel to the surface of interest. The normal force component results in either tension or compression, while the tangential force component results in shear. Some external forces can be applied in such a fashion that one or the other of the two force components predominates. With reference to a material, an external force or forces can be directed inward or outward (resulting predominantly in compression or tension, respectively), or the external force can operate tangentially to one or two surfaces of a material (resulting in shear or torsion). Bending is the deformational manifestation of the simultaneous operation of normal and tangential force components (which will be treated in chap. 5). Which of these two force components is more important from the perspective of subsequent mechanical performance or mechanical failure depends on the properties of the material, the geometry of the object, and the nature of the physical environment in which the material operates.

There are two general categories of externally applied forces that invariably operate on any material or structure: surface forces and body forces. A surface force is any force distributed over the external boundaries of a material; the most pervasive is hydrostatic pressure because all organisms are surrounded by some form of fluid that exerts a pressure. In contrast, forces distributed within a material are called body forces. Among terrestrial organisms, the most pervasive of these forces is the gravitational force. Clearly, all aquatic and terrestrial plants experience surface and body forces, but depending on their particular habitat, how much these forces influence form-function relationships can vary widely. For example, aquatic plants typically experience lower body (gravitational) forces because their tissues are buoyed by water, thereby compensating for, or at least lessening, the influence of gravity, whereas they typically experience high surface (hydrostatic) forces due to the osmotic potential of their cells (see chap. 3). In contrast, the body forces generated by gravity and the surface forces generated by episodically high wind pressures have dictated many of the structural features typical among terrestrial plants, particularly those that can grow to great heights.

4.2 Strains

The extent of the deformations that result when a material is subjected to a force is called strain. Strain is a dimensionless quantity that reflects the quotient of the magnitude of a deformed dimension and that of the undeformed dimension. Strains can be expressed as pure numbers or as percentages (e.g., 0.05 or 5%). Since there are three normal stress components and three shear stress components (see section 4.8), there are six components of strain. Three of these are the unit elongations measured along the three Cartesian axes of the material. These unit elongations are the deformations resulting from the three normal stress components. (A contraction in a dimension can be thought of as a negative elongation.) These strain components, symbolized by ε_x, ε_y, and ε_z, are referred to as the normal strain components. The remaining three components of strain are the unit shear strains, symbolized by γ_x, γ_y, and γ_z, which are called the shear strain components.

Normal strains can be calculated in one of three ways. Each method can yield a different absolute value of strain (even for the same material experiencing the same stress level), depending on the magnitude of the strains that develop. Therefore, it is vital to know (and to report) precisely how strains are calculated. The first method of calculating strains gives what are called Cauchy strains (or engineering strains or conventional strains—all three terms are synonymous). These strains are calculated by

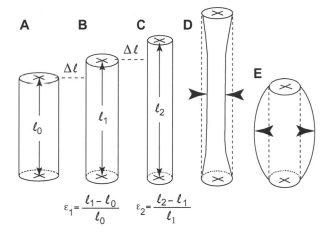

FIGURE 4.1. Illustration of Cauchy strains (A–C) and necking and barreling (D–E) in a cylindrical sample of material subjected to a longitudinal force. As the force is increased, the sample either stretches (deforms) in the longitudinal direction and contracts transversely, or compresses in the longitudinal direction and expands transversely. (A–C) Cauchy strains (measured in the longitudinal direction) are calculated by taking the instantaneous change in sample length ($\Delta \ell$) divided by the instantaneous length. (D–E) Necking in tension and barreling in compression are respectively characterized by a nonuniform contraction or expansion in the cross sections of samples. The maximum transverse contraction or expansion typically occurs at the midpoint of a specimen, as shown by the two sets of arrows.

dividing the difference between any deformed dimension, denoted here by l, and the original dimension, denoted by l_0 (fig. 4.1A–C):

(4.1a)
$$\varepsilon = \frac{l - l_0}{l_0} = \frac{\Delta l}{l_0}.$$

In a related way, normal strains can be calculated as a stretch ratio, symbolized here by λ, which is simply the extension ratio of the material:

(4.1b)
$$\lambda = \frac{l}{l_0} = \frac{l_0 + \Delta l}{l_0} = 1 + \varepsilon.$$

Normal strains of rubbery materials are typically (but not exclusively) reported in terms of the stretch ratio.

Tensile or compressive strains may have gradients of deformation, as, for example, when necking or barreling occurs (fig. 4.1D–E). Necking is the regional diminution in cross-sectional area resulting from the application of tensile forces; barreling is the regional expansion in cross-sectional area resulting from compression. Under these circumstances, the strain is not constant throughout the specimen. Rather, it is a function of where the strain is measured along the axis of the specimen. In cases like this, Cauchy strains should be given in the differential form,

(4.1c)
$$\varepsilon(l') = \frac{d\Delta l'}{dl'} \quad 0 \le l' \le l.$$

A different method of calculating normal strains gives what are called Henchy strains or true strains or natural strains (all three are synonymous), denoted by ε_{true}. This type of strain is calculated as the integral of the change in the reference dimension, denoted here by l, over the limits of the original dimension to the altered size:

(4.2)
$$\varepsilon_{true} = \int_{l_0}^{l} \left(\frac{dl}{l}\right) = \ln\left(\frac{l}{l_0}\right).$$

True strains should be measured for any material exhibiting significant deformation ($\ge 5\%$). Clearly, true and Cauchy strains are mathematically related, $\varepsilon_{true} = \ln(1 + \varepsilon)$, and yield essentially identical values for small strains ($<5\%$). However, it is not wise to mix different types of stresses and strains.

Typically, true strains provide a more realistic picture of mechanical phenomena than do Cauchy strains, particularly when materials undergo large elastic or plastic deformations. Consider a rubber band placed in uniaxial tension. If its original length is 10 cm and it is extended to a length of 11 cm, then the true strain equals $\ln(11/10) = 0.095$, whereas the Cauchy strain equals 0.10. If the same rubber band is further extended up to 25 cm, then the true strain is $\ln(25/11) = 0.82$, whereas the Cauchy strain is 1.5 (where 10 cm is used as the original dimension). Clearly, the two strains produced by equivalent extensions are not equal, and deformation is better dealt with in terms of the magnitude of the dimension immediately preceding each incremental deformation.

Shear strains are slightly more complicated. A shear strain equals the deformation of an element divided by the width of the element. This is typically formalized mathematically by the formula

(4.3a)
$$\gamma = \Delta x/z = \tan\Theta,$$

where Δx is the deformation in the axis that parallels the direction of the shearing force and z is the distance over which the forces act. This is il-

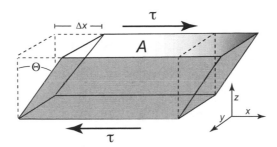

FIGURE 4.2. Diagrammatic representation of a rectangular solid (original shape indicated by dashed lines) rigidly attached at its base deformed by a shearing force (direction indicated by opposing arrows) into a parallelepiped (showed by shaded surfaces). The shear strain γ equals the tangent of the deformation angle Θ, which equals $\Delta x/z$.

lustrated in figure 4.2 for a rectangular element of material deformed into a parallelepiped.

However, shear strain components in materials subjected to simple or pure shear are more appropriately calculated in terms of the substantial gradient of deformation that typically occurs within materials. The shear strain is thus given in its differential form,

(4.3b)
$$\gamma = \frac{dx}{dz}.$$

4.3 Different responses to applied forces

We will treat the concepts of stress and strain by first examining the relatively simple uniaxial stresses that result when a cylindrical bar is subjected to a tensile or a compressive load; that is, a "pulling" or a "compressing" force F, respectively. The intensity of internal forces is uniformly distributed within the principal surface of interest, which is the plane perpendicular to the axis of length; that is, the tensile or the compressive stress is calculated by dividing F by the transverse area of the bar. These stresses, which operate normal (perpendicular) to each cross section through the bar (and are called normal stresses), are denoted by σ and are given by the formula

(4.4)
$$\sigma = F/A.$$

The bar reacts to these stresses by deforming in length Δl. When this deformation is normalized with respect to the bar's initial length l_0, we obtain the corresponding Cauchy strain ε (see eq. [4.1]) for the one-dimensional case.

The relationship between normal stress and the resulting strain (eqs. [4.4] and [4.1a]) can now be used to distinguish among the five major types of materials that could be used to construct our hypothetical bar:

1. The material can store the strain energy supplied by the applied force and use it to return instantaneously to its original shape when the force is removed (fig. 4.3). Materials of this sort exhibit elastic behavior and usually a linear stress versus strain relationship. In this case, the modulus of elasticity E is defined as the slope of the stress versus strain relationship, which is mathematically expressed as

(4.5a)
$$E = \frac{\partial \sigma}{\partial \varepsilon} = \frac{\sigma}{\varepsilon}.$$

   If the material is homogeneous, this slope is called Young's modulus (in honor of Thomas Young, 1773–1829). Young's modulus is a characteristic quantity for any linear elastic material. Its values measured in tension and in compression are identical. It is important to note, however, that this generality holds true for linear elastic *materials*, but not necessarily for *structures* made of these materials (see chap. 5).

2. Some materials undergo large changes in shape and internal structure that increase over time and can be partially or totally recovered slowly once the external force diminishes or drops to zero. This type of behavior is called viscoelastic behavior (see fig. 4.3). Depending on the time given for recovery, the energy stored in the material is only partially recovered.

3. Beyond the linear range, the material may dissipate the strain energy by deforming permanently. This response is called plastic behavior. Typically, the stress versus strain relationship is no longer linear over the entire range; that is, the modulus of elasticity E is a function of the strain:

(4.5b)
$$E = \frac{\partial \sigma}{\partial \varepsilon} \neq \frac{\sigma}{\varepsilon}.$$

   The strain at which the response ceases to be linear is called the yield strain, and the corresponding stress is called the yield stress. In the engineering sciences, the yield stress is the critical stress beyond which a material may be weakened. In biology, because self-repair processes are an essential part of life's devices, the yield stress may be less relevant.

4. Usually, beyond the linear range, the material may fail by tearing apart under tensile loads or collapsing under compressive loads (see fig. 4.3). It may fail either suddenly, as in brittle materials, or gradually, as in ductile materials.

5. Some materials have no elastic component; they rapidly dissipate all the energy supplied by external forces and permanently deform by irreversibly flowing in the direction of the applied force. This response is called fluid behavior (which is reviewed in chap. 6).

In general, the strain energy, denoted by U, can be obtained using the formula

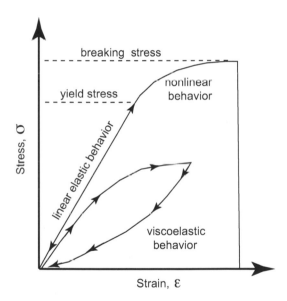

FIGURE 4.3. Schematic of different stress (σ)-strain (ε) responses of a material placed in tension. An ideal elastic material exhibits a perfectly elastic response to a tensile force provided that the force does not exceed its elastic limit. If this limit is reached (if the yield stress is reached), the elastic material may exhibit nonlinear stress-strain behavior, or it may break immediately. In this diagram, the elastic material continues to deform until it reaches a breaking stress that exceeds the yield stress. An alternative σ versus ε behavior is viscoelastic behavior, in which the material deforms nonlinearly with increasing or decreasing stress levels. A hysteresis loop results when a material manifests different stress-strain pathways when it is loaded and unloaded (as shown here for a viscoelastic material).

$$\textbf{(4.6)} \qquad U = \int\limits_0^{l_{max}} F\, dl = \int\limits_0^{\varepsilon_{max}} A\sigma l d\varepsilon = \int\limits_0^{\varepsilon_{max}} V\sigma d\varepsilon,$$

where A is the area and l the length of a cylinder or prismatic bar and V is its volume (i.e., $V = Al$). If the volume is constant during straining, equation (4.6) can be simplified to

$$\textbf{(4.7)} \qquad U = V \int\limits_0^{\varepsilon_{max}} \sigma d\varepsilon.$$

In the linear elastic range, where σ is proportional to ε, equation (4.7) can be simplified further by introducing the modulus of elasticity (see eq. [4.5a]) to give the formula

$$\textbf{(4.8)} \qquad U = EV \int\limits_0^{\varepsilon_{max}} \varepsilon d\varepsilon = \frac{1}{2} EV\varepsilon^2.$$

Clearly, some materials can exhibit more than one of the foregoing five responses, as, for example, when an elastic material undergoes plastic deformation. Nonetheless, when most biological materials are tested at a particular age and in a particular metabolic condition, they typically exhibit one of these types of behavior. When different materials are used to assemble a structure, however, the mechanical behavior of the structure can differ significantly from the mechanical behavior of any of its constituent materials. What makes organisms so interesting from a mechanical perspective is that even the simplest unicellular organism consists of a variety of materials, and thus its collective mechanical attributes often differ radically from those of its constituent materials. These material composites we call organisms are sophisticated in their material properties and possess mechanical versatility and complexity that permit them to survive in habitats where many engineered artifacts do not.

4.4 A note of caution about normal stresses and strains

The area through which normal axial stresses operate is easily measured for prepared specimens of metals or plastics. By contrast, measuring the

transverse areas of intact biological structures, whose cross-sectional geometries naturally vary, can pose difficulties. Usually, geometric simplifications of irregular transverse geometries are sufficient for first-order
approximations when calculating stresses. In many cases, however, we
need greater precision and accuracy. With the aid of computer systems
and appropriate software, the transverse areas of even complex and irregular geometries can be measured empirically.

However, a serious problem arises if a specimen is composed of several
materials that have different mechanical properties. A stress versus strain
curve of such a specimen gives only an average modulus of elasticity,

(4.9)
$$\bar{E} = \frac{\sum_i A_i E_i}{\sum_i A_i},$$

where A_i is the cross-sectional area of material i with a modulus of elasticity E_i. In some cases it may be possible to determine the elastic modulus of
each constituent material by subtracting the mechanical contributions of
each of the other components. If that is not possible, the average modulus
of elasticity determined in testing should be referred to as the structural
modulus of elasticity (compare Niklas 1989).

The definition of stresses becomes difficult when the transverse area of
a specimen deforms significantly under an applied load (see fig. 4.1D–E).
These deformations can take the form of either necking under tension or
barreling under compression. In either case, the cross-sectional area exhibits a temporal variation in absolute dimension as a specimen deforms
under a constant force. Therefore, instantaneous stresses can differ in
magnitude from those that immediately precede or follow them. For some
fabricated materials that are very inextensible, such as steel, the changes
in transverse area may be small over a very large range of loadings. For
these types of materials, stresses may be adequately calculated based on
the original cross-sectional area of the unloaded material. This manner of
calculating stress yields what engineers call engineering stress or nominal
stress, symbolized by σ_n.

Calculations of nominal stresses should be avoided when the conditions of loading change markedly over relatively short spans of time or
when the deformations resulting from loadings are large (~5%). Since
many plant materials typically deform substantially under low stress levels,

and since virtually all plant materials can deform more than 5% before critical stress levels are reached, stress should be calculated by dividing the applied force by the *instantaneous* transverse area (see fig. 4.1). This procedure gives a stress called true stress, symbolized by σ_{true}.

4.5 Extension to three dimensions

Thus far, we have considered some very simple loading conditions and some very idealized materials. However, most biological materials are far more complex. Part of this complexity comes from the fact that most exhibit anisotropic behavior; that is, they do not show the same mechanical behavior if tested in their three major orthogonal directions. For these materials, it is necessary to consider the three-dimensional case.

When considered in this way, stresses and strains are represented by three-dimensional vectors such that the modulus of elasticity E takes the form of a 3×3 tensor. In a Cartesian coordinate system, this situation reads mathematically as

(4.10)
$$\begin{pmatrix} \sigma_x \\ \sigma_y \\ \sigma_z \end{pmatrix} = \begin{pmatrix} E_{xx} & E_{xy} & E_{xz} \\ E_{yx} & E_{yy} & E_{yz} \\ E_{zx} & E_{zy} & E_{zz} \end{pmatrix} \begin{pmatrix} \varepsilon_x \\ \varepsilon_y \\ \varepsilon_z \end{pmatrix}.$$

For simplicity, we will restrict our discussion of equation (4.10) to elastic behavior, although it is not difficult to apply the same formalism to nonlinear mechanical behavior. The entries in the diagonal of the tensor (eq. [4.10]) are those described by equation (4.5a). The off-diagonal entries in the tensor describe the interactions among the changes in size in different directions. This can be intuited by considering a rubber band that has a rectangular cross section when relaxed but which, when pulled and extended, takes on a square cross section (fig. 4.4). The change from a rectangular to a square cross section shows that the two lateral strains are not equivalent even though the tensile stress is uniform throughout each cross section. This anisotropic mechanical behavior typically indicates that a material has a molecular composition or microstructure with preferred orientations. Rubber is a cross-linked polymer whose molecular structure reorients under tension. At small strains, it behaves as an isotropic material; at larger strains, it becomes anisotropic. Most plant tissues conferring

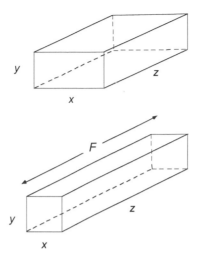

FIGURE 4.4. Diagrammatic representation of a short length of an anisotropic rubber band that has a rectangular cross section that is wider than high (i.e., $y \ll x$) when relaxed, but when subjected to a tensile force (F) along its z-axis, contracts in width and height to produce a square cross section (i.e., $y = x$).

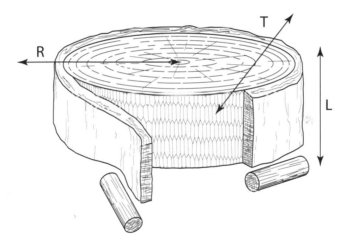

FIGURE 4.5. Diagram through a section of the trunk of the cork oak, showing the orientations of the three major planes of reference: radial (R), tangential (T), and longitudinal (L). The bark is shown as if peeled away to reveal the vascular cambium (shown by the surface composed of hexagonally shaped cells). The orientations of two wine bottle corks removed from the outer bark are also shown. Note that the length of each cork is oriented along the tangential plane of reference.

mechanical support, such as wood and sclerenchyma, show varying degrees of anisotropy as a result of their cellular heterogeneity, the preferred orientation of their constituent cells, or both.

Anisotropic materials include axisymmetric materials, which have two mutually perpendicular directions of symmetry; that is, they have equivalent material properties when measured in two of their three Cartesian dimensions. Bottle cork (the outer bark of the oak *Quercus suber*) is an example (fig. 4.5). The cells in this tissue have more or less equivalent radial and tangential dimensions but are prismatic in the longitudinal direction. Thus, when they are pulled or compressed, they exhibit similar material properties in two of their three dimensions. Upon compression in the longitudinal direction, cork expands only very little in the other two directions, such that it can be pressed uniformly into the necks of wine bottles. This orientation is not only mechanically desirable, but also precludes wine from flowing through the tubular channels within cork, called lenticels.

4.6 Poisson's ratios

Virtually every common material undergoes lateral contraction when it is stretched and lateral expansion when it is compressed. The relative magnitudes of these deformations are governed by a material property known as Poisson's ratio, named in honor of the French mathematician and mechanicist Siméon-Denis Poisson (1781–1840). Poisson's ratio, symbolized by v, is defined as the negative transverse strain divided by the axial strain in the direction of the externally applied force; that is,

(4.11)
$$v_{xy} = -\frac{\varepsilon_y}{\varepsilon_x} \text{ or } v_{xz} = -\frac{\varepsilon_z}{\varepsilon_x}.$$

The first letter in the subscript of each Poisson's ratio identifies the axis parallel to the direction of the applied force (in this example, the x-axis is the axis of the force's application), whereas the second letter indicates the dimension in which the transverse strain has been measured (the y- or z-axis). The Poisson's ratio provides an important material property because it measures the ability of a material to change in volume versus its ability to change in shape.

Anisotropic materials have six Poisson's ratios for each coordinate system used to describe the material. Within the linear elastic range, these ratios can be expressed by a 3×3 tensor relating strain ε and stress σ (and thus the elastic modulus E) in a three-dimensional Cartesian coordinate system:

$$\textbf{(4.12)} \qquad \begin{pmatrix} \varepsilon_x \\ \varepsilon_y \\ \varepsilon_z \end{pmatrix} = \begin{pmatrix} \dfrac{1}{E_x} & -\dfrac{v_{xy}}{E_x} & -\dfrac{v_{xz}}{E_x} \\[2mm] -\dfrac{v_{yx}}{E_y} & \dfrac{1}{E_y} & -\dfrac{v_{yz}}{E_y} \\[2mm] -\dfrac{v_{zx}}{E_z} & -\dfrac{v_{zy}}{E_z} & \dfrac{1}{E_z} \end{pmatrix} \begin{pmatrix} \sigma_x \\ \sigma_y \\ \sigma_z \end{pmatrix}.$$

For most commonly occurring materials, we find that $0 \leq v \leq 0.5$, although negative Poisson's ratios are thermodynamically possible (see below). As the magnitude of v approaches a value of 0.5, the material will increasingly resist a change in its volume, but will increasingly respond to stress by changing its shape. Fluidlike materials have Poisson's ratios approaching 0.5; an ideal fluid has a Poisson's ratio of 0.5 (box 4.1). In contrast, the Poisson's ratios of crystalline solids are smaller. The Poisson's ratio of cellulose measured in tension and transversely along its polymeric chain length is 0.10 (Mark 1967), indicating that this carbohydrate polymer changes its transverse dimensions little in response to tension. Homogeneous biological materials have Poisson's ratios that fall between these two extremes. However, these values can be very different for composite biological materials such as most tissues, for which Poisson's ratios greater than 1.0 can be found. A calculation presented in box 4.2 shows that in a sclerenchyma cell, in which practically inextensible cellulose microfibers provide the strengthening material in the cell wall, the Poisson's ratio strongly depends on the microfibrillar angle; that is, the angle between the fibers and the longitudinal axis of the cell.

For isotropic materials, in which the elastic material properties are the same regardless of the direction in which they are measured, all six Poisson's ratios are equivalent in magnitude, and only one Poisson's ratio v is required to describe the relationship between the two transverse strains. For such materials, Poisson, on the basis of his analytical investigation of the molecular theory of the structure of materials, found $v = 0.25$. Isotropy is a consequence of extreme molecular or infrastructural

BOX 4.1 **Poisson's ratio for an incompressible fluid**

Consider a cylindrical cell with a compliant cell wall filled with a nearly in-compressible fluid such as water. The volume of the cell is therefore $V = \pi L R^2$, where length L and radius R denote the dimensions of the fluid column enclosed by the cell wall. Upon uniaxial tension or compression, both L and R will change to $L + \Delta L$ and $R + \Delta R$. Since the fluid is regarded as incompressible, the volume remains unchanged:

(4.1.1) $$V = \rho(L + \Delta L)(R + \Delta R)^2 = \rho L R^2.$$

Expanding this equation and neglecting the higher-order terms such as $L(\Delta R)^2$ and $\Delta L \Delta R$, which is justified for small volumetric changes, we obtain

(4.1.2) $$\rho(L R^2 + 2 L R \Delta R + \Delta L R^2) = \rho L R^2,$$

or

(4.1.3) $$2 L R \Delta R + \Delta L R^2 = 0.$$

Dividing this equation by $(L R^2)$ yields

(4.1.4) $$2 \Delta R / R = -\Delta L / L.$$

The Poisson's ratio is therefore

(4.1.5) $$v = -\frac{\Delta R / R}{\Delta L / L} = 0.5.$$

This last relationship is equally applicable to any cellular geometry.

---

homogeneity within a material. Most engineering materials evidence iso-tropic mechanical behavior and tend to have Poisson's ratios within the range $0.25 \leq v \leq 0.30$.

Although many of the cell types in wood are axisymmetric in their indi-vidual mechanical behavior, the anatomical juxtaposition of these cell types

BOX 4.2 **Poisson's ratio for a cell**

Consider a cylindrical plant cell (with length L and radius R) containing nearly inextensible cellulose fibers wound in the form of a spiral placed under uniaxial tension or compression. The microfibrillar angle between the fiber and the longitudinal axis is α (fig. 4.2.1). If the number of turns (or the fraction of turns) of the spiral is $\Omega > 0$, the length D of the fiber is given by

(4.2.1) $D = L / \cos \alpha = 2\pi\Omega R / \sin \alpha.$

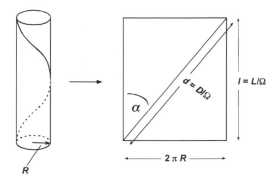

FIGURE 4.2.1. Schematic drawing of one gyration of a fiber around a portion of a cylindrical cell (with circumference $2\pi R$ and length ℓ) in three dimensions and projected on a flat surface.

It follows that

(4.2.2) $2\pi\Omega R / L = \sin \alpha / \cos \alpha = \tan \alpha.$

It also follows from this geometry that the length of the fiber is given by

(4.2.3a) $D = [L^2 + (2\pi\Omega R)^2]^{1/2}.$

If the cell is under uniaxial tension or compression, L changes to $L + \Delta L$ and R changes to $R + \Delta R$. If no twisting of the fiber is permitted, Ω is constant. The unchanged length D of the inextensible fiber can therefore be obtained by

(4.2.3b) $D = \{(L + \Delta L)^2 + [2\pi\Omega(R + \Delta R)]^2\}^{1/2}.$

BOX 4.2 **(Continued)**

Combining equations (4.2.3a) and (4.2.3b) yields

(4.2.4) $L^2 + (2\pi\Omega R)^2 = (L + \Delta L)^2 + [2\pi\Omega(R + \Delta R)]^2.$

Expanding equation (4.2.4) and neglecting the higher-order terms such as $(\Delta R)^2$ and $(\Delta L)^2$, which is justified for small changes, we obtain

(4.2.5) $L\Delta L = -4\pi^2\Omega^2 R\Delta R.$

Inserting the trigonometric relationship given by equation (4.2.2) into equation (4.2.5) gives

(4.2.6) $L\Delta L = -(L/R)^2 \tan^2\alpha R\Delta R$

and

(4.2.7) $\tan^2\alpha = -(\Delta L/L)/(\Delta R/R).$

The Poisson's ratio is therefore given by the formula

(4.2.8) $v = -(\Delta R/R)/(\Delta L/L) = 1/\tan^2\alpha.$

Similar calculations for different cross sections yield the same relationship.

At $\tan^2\alpha = 2$ (i.e., a microfibrillar angle α of 54.7°), the Poisson's ratio equals 0.5, which implies that the volume of the cell is constant for small changes of the microfibrillar angle. This answer can be derived in a different way: $\alpha = 54.7°$ is the angle at which the volume of the cell is maximized (compare Vogel 1998, 150). Using equation (4.2.1) to calculate the volume V as a function of the microfibrillar angle α, we obtain

(4.2.9) $V = \pi R^2 L = [D^3/(2\pi\Omega)]\cos\alpha\sin^2\alpha.$

Setting the derivative $dV/d\alpha$ to zero yields the condition $\tan^2\alpha = 2$ for constant length D of the fiber, or $\alpha = 54.7°$ for the microfibrillar angle at maximal volume.

results in three principal directions of symmetry: the longitudinal direction along the grain (denoted by L) and the radial (R) and tangential (T) directions to the grain (see fig. 4.5). In this coordinate system, wood has six Poisson's ratios, denoted by v_{LR}, v_{RL}, v_{LT}, v_{TL}, v_{RT}, and v_{TR}. As in equation (4.11), the first letter in each of these subscripts indicates the direction of the applied force, and the second letter indicates the direction in which the transverse strain is measured. For balsa wood, the values for the six Poisson's ratios are 0.229, 0.488, 0.665, 0.217, 0.011, and 0.007; for yellow birch, the ratios are 0.426, 0.451, 0.697, 0.447, 0.033, and 0.023. Although the absolute values of the six Poisson's ratios vary among species, v_{LT} is typically the largest of the six. Most important, an average Poisson's ratio for a material like wood is a meaningless quantity. As mentioned before, cork is an example of an axisymmetric material with a Poisson's ratio that approaches zero; that is, there is virtually no radial expansion upon longitudinal compression.

All of the foregoing assumes that the Poisson's ratio is a constant. For many types of nonbiological materials this assumption is valid. For most living plant materials, however, the Poisson's ratio can change, and great caution must be exercised in using any single Poisson's ratio to infer mechanical behavior under all loading conditions. For example, Chappell and Hamann (1968) report that the initial values of the Poisson's ratio of apple flesh (a parenchymatous tissue with a Poisson's ratio within the range 0.21–0.34) decreased over time under a constant stress level. Under a given loading regime, the greatest decrease in v (about 16%) was observed during the first 30 seconds of testing. But higher stress levels resulted in a more rapid decline in v; that is, the Poisson's ratio of apple flesh depends on both the duration and the magnitude of loading. A decrease in v indicates a decrease in the resistance of a material to a change in volume and an increase in its ability to resist shear. Thus, over time and under a constant stress level, apple flesh appears to have the capacity to change its state from one tending to behave more like a liquid to one behaving more like a solid. This change in state appears to accelerate as the level of stress is increased. From a functional perspective, the alterations in the Poisson's ratio reported for apple flesh could confer a mechanical benefit, since the tissue can resist shearing over relatively short intervals. More important, the behavior of apple flesh highlights the desirability of considering the Poisson's ratios of organic materials as potential variables rather than as constants.

Before leaving the general topic of Poisson's ratio, it is worth noting that *negative* ratios are theoretically permissible for materials. That is, a material can expand orthogonally to the direction of an applied tensile load or can contract orthogonally to the direction of an applied compressive load. Negative Poisson's ratios have been reported for some synthetic polymers whose polymeric units separate laterally as they are extended in length. Negative Poisson's ratios have also been reported for some foams: commercially fabricated cellular solids with a beamlike or strutted infrastructure (Lakes 1987). The subject of cellular solids is treated in section 8.7. It is also treated extensively by Gibson et al. (2010). These materials have numerous commercial applications, since they can inflate when they are pulled. They may also have botanical analogues. For example, some aerenchyma (a spongy tissue consisting of numerous strutlike interconnected cells) is anatomically similar to commercially fabricated foams. To our knowledge, no one has measured the Poisson's ratios of aerenchyma or plant organs that characteristically have a spongy infrastructure consisting of many beams and struts attached to an external wall. However, since aerenchyma is attached to the walls of many otherwise hollow stems and leaves, and since these leaves and stems experience ovalization of the cross section when they are bent (see section 5.11), negative Poisson's ratios would confer a mechanical advantage, since they would permit aerenchyma cells to "push outward" and thus resist the ovalization of the leaf or stem. It would be particularly interesting if some of the tissues found in roots had a negative Poisson's ratio, since that would produce an expansion of the root when it is placed in tension, anchoring it more firmly in its substrate.

4.7 Isotropic and anisotropic materials

Although most biological materials are anisotropic, a few simplifications for isotropic materials are worth discussing. These materials are characterized by only one modulus of elasticity E, one shear modulus G, and one Poisson's ratio v. An object subjected to stresses σ_x, σ_y, and σ_z will manifest strains given by the formula

(4.13a)
$$\varepsilon_x = \frac{1}{E}[\sigma_x - v(\sigma_y + \sigma_z)]$$

(4.13b)
$$\varepsilon_y = \frac{1}{E}[\sigma_y - v(\sigma_x + \sigma_z)]$$

(4.13c) $$\varepsilon_z = \frac{1}{E}[\sigma_z - v(\sigma_x + \sigma_y)].$$

At uniform pressure, it follows that, $P = \sigma_{volume} = \sigma_x = \sigma_y = \sigma_z$, such that equations (4.13a–c) sum as

(4.13d) $$\varepsilon_x + \varepsilon_y + \varepsilon_z = \frac{1}{E}[3\,\sigma_{volume}(1-2v)].$$

With $\sigma_{volume} = K\Delta V/V$, where K is the volumetric elastic modulus and the volumetric change is $\Delta V/V = \varepsilon_x + \varepsilon_y + \varepsilon_z$, we obtain the following formula:

(4.14) $$E = 3K(1-2v).$$

Note that for isotropic materials, v cannot be larger than 0.5. Therefore, E will always take on a value greater than zero.

When hydrostatic pressure is applied to the exterior surface of a compressible material, the volume of the material will decrease and its density will increase. The reciprocal of K is called the compressibility of a material. Clearly, materials with a high Poisson's ratio, which resist a change in volume, have very high bulk moduli and very low compressibility. Conversely, materials with low bulk moduli have a high compressibility and densify relatively easily when subjected to modest or high hydrostatic pressure. Indeed, the bulk moduli of liquids are on the order of GPa (compare eq. [9.2b]), whereas the bulk moduli of aerenchymatous plant tissues, which have many large air spaces within them, can approach zero as the volume fraction of air spaces increases.

Many parenchymatous plant tissues are reported to be isotropic, or nearly so (see, however, Vincent 1989, who shows that parenchyma can exhibit anisotropy because of differences in cell size and shape as well as tissue density, even in the same organ). The isotropic behavior of these tissues is consistent with the geometry of their constituent cells. Parenchymatous tissues are often composed of nearly isodiametric cells that have little or no preferred orientation with respect to that of the tissue or organ that contains them. The Poisson's ratio of some parenchyma approaches or equals 0.5, indicating that these tissues behave very much like an incompressible fluid.

Some biological materials and many fabricated materials, such as metals, can be treated as isotropic elastic materials, or nearly so; therefore, v

and E alone can be used to predict their mechanical behavior. For aniso-tropic materials, however, the relationship between stresses and strains and the material moduli must be empirically determined. For these ma-terials, the moduli must be reviewed in greater detail, starting with the elastic range at which stress and strain are proportionally related to one another (for linear elastic materials) and then progressing to a treatment of the range at which stresses and strains are not proportionally related (for nonlinear elastic materials).

Unfortunately, the literature rarely provides the elastic modulus for each of the various directions in which forces can act on an anisotropic plant material (or the Poisson's ratios from which some of the elastic moduli could be calculated). Nonetheless, these elastic moduli are essential. For instance, the elastic modulus of wood submitted to uniaxial compression along the direction of the grain, symbolized by E_L, can differ by one or two orders of magnitude from the elastic moduli measured in the tangential and radial directions to the grain (denoted by E_T and E_R; see fig. 4.5). For balsa, E_L = 3.12 GN/m^2, E_R = 0.144 GN/m^2, and E_T = 0.0468 GN/m^2; thus, E_T/E_L = 0.015, and E_R/E_L = 0.046, indicating that balsa can sustain smaller tangential and radial loadings than loadings along the grain. Unfortu-nately, data for E_R and E_T from other wood species are not extensive. Those that are available indicate that when a tangential elastic modulus is required but empirically unavailable, a reasonable approximation is given by E_T = 0.06 E_L. Axisymmetric materials are characterized by $E_L \neq E_R = E_T$ or $E_L = E_R \neq E_T$ or $E_L = E_T \neq E_R$.

4.8 Shear stresses and strains

In a two- or three-dimensional world, externally applied forces can have tangential force components that produce internal forces operating par-allel to the plane of each cross section. The stresses that result from the tangential force component are called shear stresses, symbolized by τ.

Like normal stresses, a shear stress is calculated by dividing the applied shear force by the area through which it operates. For example, when two planks of wood are glued together over a portion of their surfaces and then pulled in directions parallel to their lengths, the shear stress is calculated by dividing the applied force F by the interface surface area A between the two joined planks, as shown in fig. 4.6A:

(4.15) $\tau = F/A.$

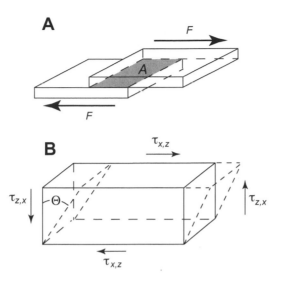

FIGURE 4.6. Illustration of direct and simple shear. (A) Direct shear occurs when two objects (here represented by rectangular solids) are forced to slide past each other (as indicated by the directions of the two arrows). The shear stress τ is calculated by dividing the applied shear force F by the area of interface A (shaded area) ($\tau = F/A$). (B) Simple shear results when a solid experiences tensile and compressive strains as a result of direct shear (in the diagram, elements within the solid are deformed such that the entire solid is distorted into a parallelepiped). The shear stress components ($\tau_{x,z}$ and $\tau_{z,x}$) have corresponding shear strains γ and can be calculated from the translation angle (i.e., $\gamma = \tan\Theta$).

When dealing with shear stress, however, it is important to recognize that a solid can shear in any of three ways: direct shear, simple shear, and pure shear: Direct shear occurs at the interface between two objects that are forced to slide past each other (see fig. 4.6A), as when you rub your hand over a tabletop or when two branches rub against each other. In direct shear, a material element moves relative to the surface of another material element so that the materials deform little or not at all, except as a consequence of friction and the resulting abrasion of surfaces. Simple shear occurs when elements of material within a solid slide past one another and simultaneously experience tensile and compressive deformation in the direction of shearing (fig. 4.6B). This is illustrated when a rectangular piece of gelatin is deformed into a parallelepiped by pushing its upper surface parallel to the surface of a supporting plate. The parallelism between simple shear and direct shear can be demonstrated in a crude fashion when a deck of playing cards is deformed into a parallelepiped by pushing it across a tabletop with one's palm. Although direct shear occurs between any two cards, each card can be taken as an analogue to a single

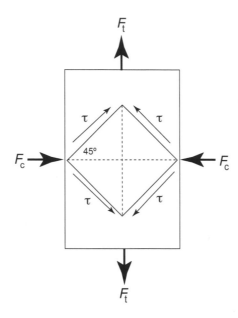

FIGURE 4.7. A schematic drawing illustrating the condition of pure shear, which can be understood as a result of the simultaneous operation of a tensile force F_t in one direction and an equivalent compressive force F_c in the perpendicular direction.

rectangular material element whose surface moves parallel to the plane of shearing.

Pure shear is equivalent to the state of stress produced within a material that has been submitted to tension in one direction and equal compression in the perpendicular direction (fig. 4.7). Here, the adjective *pure* refers to the fact that only the shear force component has magnitude. Pure shear may appear to be an abstraction, in that the normal stress components within the peripheral volume of a sheared solid typically never equal zero. However, at the very center of a large object undergoing simple shear, the net magnitudes of the normal tensile and compressive stress components approach zero, and the condition of pure shear is realized. In this sense, pure shear is actually a special case of simple shear. Indeed, the behavior of a material in pure shear is often examined by placing a large specimen of the material in simple shear. Likewise, the condition of pure shear can be realized by an element of a material on the surface of a circular tube subject to torsion resulting from a very small rotation of one end of the tube with respect to the other. We shall return to a consideration of simple and pure shear in the next chapter, where we will consider torsion in greater detail.

The quotient of shear stress τ and shear strain γ (see eqs. [4.15] and [4.3]) is called the shear modulus G:

(4.16) $$G = \tau/\gamma.$$

And in a fashion analogous to equations (4.7) and (4.8), the shear energy U_{shear} is given by

(4.17) $$U_{shear} = V \int_0^{\gamma_{max}} \tau \, dy.$$

In the linear elastic range, the equation for shear energy takes the form

(4.18) $$U_{shear} = 1/2 G V \gamma^2.$$

For isotropic materials, the relationship between the shear modulus and the modulus of elasticity E is given by

(4.19) $$G = \frac{E}{2(1+v)}$$

(see Timoshenko and Goodier 1970), while the shear modulus is related to the volumetric elastic modulus by

(4.20) $$G = \frac{3K(1-2v)}{2(1+v)}.$$

If we consider an anisotropic material, normal and shear stresses operate within each of the three Cartesian dimensions, as shown in figure 4.8 for a cubic material element whose dimensions of depth, width, and length are designated by the Cartesian coordinates x, y, and z, respectively. As shown in this figure, three symbols (σ_x, σ_y, and σ_z) are needed to describe the normal stresses, while six symbols (τ_{xy}, τ_{yx}, τ_{xz}, τ_{zx}, τ_{yz}, and τ_{zy}) are required to describe the shear stresses. These six can be reduced to three by realizing that the equilibrium of moments requires that $\tau_{xy} = \tau_{yx}$, $\tau_{xz} = \tau_{zx}$, and $\tau_{yz} = \tau_{zy}$ (Timoshenko and Goodier 1970). Therefore, only six stresses,

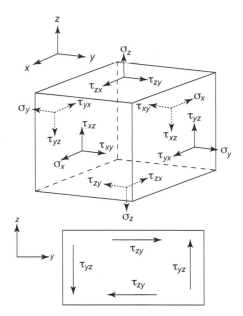

FIGURE 4.8. Normal and shear stresses within a cubical element of a material with depth x, width y, and length z. Three normal stresses (σ_x, σ_y, and σ_z) act perpendicular to the three axes of the element. Parallel sides have equivalent magnitudes of normal stress. Six shear stresses (τ_{xy}, τ_{yx}, τ_{xz}, τ_{zx}, τ_{yz}, and τ_{zy}) act parallel to the surfaces of the element. The shear stresses acting perpendicular to the line of intersection of the perpendicular sides of element are equivalent ($\tau_{xy} = \tau_{yx}$, $\tau_{xz} = \tau_{zx}$, and $\tau_{yz} = \tau_{zy}$). Therefore, only six stress components are required to specify the internal forces generated within the element subjected to a shearing force (σ_x, σ_y, σ_z, $\tau_{xy} = \tau_{yx}$, $\tau_{xz} = \tau_{zx}$, and $\tau_{yz} = \tau_{zy}$).

called the components of stress, are needed to describe all the stresses operating within the material element (σ_x, σ_y, σ_z, $\tau_{xy} = \tau_{yx}$, $\tau_{xz} = \tau_{zx}$, and $\tau_{yz} = \tau_{zy}$). In pure shear, the relationship between shear stress and shear strain is given by the following formula (which is analogous to eq. [4.10]):

$$(4.21) \qquad \begin{pmatrix} \tau_{xz} \\ \tau_{yx} \\ \tau_{zy} \end{pmatrix} = \begin{pmatrix} G_{xx} & G_{xy} & G_{xz} \\ G_{yx} & G_{yy} & G_{yz} \\ G_{zx} & G_{zy} & G_{zz} \end{pmatrix} \begin{pmatrix} \gamma_x \\ \gamma_y \\ \gamma_z \end{pmatrix}.$$

Shear is not an uncommon biological phenomenon. It can be observed in many kinds of leaves but nowhere more prominently than in the simple leaves of the banana (*Musa acuminata*) or the traveler's palm (*Ravenala madagascariensis*). These leaves shear between the parallel vascular bun-

FIGURE 4.9. Mechanical damage to the leaves of a traveler's palm (*Ravenala madagascariensis*) resulting from shearing stresses generated by wind. The youngest leaves are located at the top; the oldest leaves are farthest from the top of the plant. These stresses act in such a manner that the softer tissues between the stronger parallel vascular bundles shear and break.

dles (which characterize the leaves of many monocots) where wind pressure pulls the softer, nonvascular tissues apart (fig. 4.9). Young leaves or the older leaves of plants sheltered from the wind suffer little damage, but the leaves of unsheltered plants can be shredded to such a degree that they cease to function prematurely. By contrast, the leaves of most dicots, which have a reticulated vascular network, show little evidence of destructive shearing by wind, in large part because the softer tissues within these leaves are held together by an interweaving fabric of stiffer, stronger tissues.

Recall that wood has three principal planes of reference—*LR*, *LT*, and *RT*—where *L* represents the direction parallel to the grain and *R* and *T* represent the radial and tangential directions to the grain (see fig. 4.5). Using this notation, the corresponding shear moduli for wood are designated by G_{LR}, G_{LT}, and G_{RT}. These shear moduli can differ, often substantially, in their magnitudes. For example, in the case of balsa wood, $G_{LR} = 0.169$ GN/m^2, $G_{LT} = 0.115$ GN/m^2, and $G_{RT} = 0.0156$ GN/m^2. Thus, $G_{LT}/G_{LR} = 0.685$ and $G_{RT}/G_{LR} = 0.111$, indicating that balsa resists shear best in the

LR plane and most poorly in the *RT* plane, as is generally true for the majority of wood species. In addition, shear moduli are always less than the corresponding elastic moduli, so that shearing failure is likely to occur in bending or torsion (see chap. 5). Turning again to balsa wood, we see that $G_{LR}/E_L = 0.054$, $G_{LT}/E_L = 0.037$, $G_{RT}/E_R = 0.11$, and $G_{RT}/E_T = 0.33$. However, great caution should be exercised in treating the mechanical properties of botanical materials such as wood as constants, since they vary with the age and relative moisture content of a sample. The same caveat applies to treating the elastic moduli of plant tissues as constants (see chap. 8).

4.9 Interrelation between normal stresses and shear stresses

At equilibrium, the sum of all forces and the sum of all moments in every element of an object must be zero. Converted into stresses and expressed in differential form, this physical stipulation leads to

(4.22a)
$$\frac{\partial \sigma_x}{\partial x} + \frac{\partial \tau_{xy}}{\partial y} + \frac{\partial \tau_{xz}}{\partial z} + F_x = 0,$$

(4.22b)
$$\frac{\partial \sigma_y}{\partial y} + \frac{\partial \tau_{yx}}{\partial x} + \frac{\partial \tau_{yz}}{\partial z} + F_y = 0,$$

(4.22c)
$$\frac{\partial \sigma_z}{\partial z} + \frac{\partial \tau_{xz}}{\partial x} + \frac{\partial \tau_{yz}}{\partial y} + F_z = 0,$$

where F_x, F_y, and F_z are the components of a body force per unit volume acting on an element (Timoshenko and Goodier 1970). From these equations, it is obvious that normal stresses and shear stresses are interrelated.

As an example, consider the stresses in different planes of a prismatic bar under tension (fig. 4.10). The normal stress in the direction of the applied force F is given by $\sigma_L = F/A$, where A is the area of a cross section perpendicular to the applied force and L denotes the longitudinal axis. We now introduce a new coordinate system x, y tilted against the longitudinal axis by an angle α. At any point in the cross-sectional plane (x, z), a force F can be decomposed into a component $F_y = F \sin \alpha$ perpendicular to the plane and a force $F_x = F \cos \alpha$ in the plane. The cross-sectional area of the

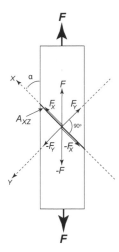

FIGURE 4.10. Transformation of forces in a prismatic bar into a different coordinate system. The force F is decomposed into two components (F_x and F_y). Within the plane A_{xy}, the prismatic bar experiences a normal stress σ_y and a shear stress τ_{yx}.

plane is given by $A_{xz} = A/\sin\alpha$. F_y leads to a normal stress σ_y, which can be expressed using trigonometric relationships as follows:

$$\textbf{(4.23)} \qquad \sigma_Y = \frac{F_Y}{A_{XZ}} = \frac{F}{A}\sin^2\alpha = \sigma_L\sin^2\alpha = \frac{1}{2}\sigma_L(1-\cos2\alpha),$$

while F_x leads to a shear stress τ_{yx}, which is given by the formula

$$\textbf{(4.24)} \qquad \tau_{xy} = \frac{F_x}{A_{XZ}} = \frac{F}{A}\cos\alpha\sin\alpha = \sigma_L\cos\alpha\sin\alpha = \frac{1}{2}\sigma_L\sin2\alpha \cdot$$

Note that for many materials that exhibit elastic behavior across small to modest stresses, there exists a limit beyond which the proportionality between shear stresses and shear strains is not maintained. Beyond this proportional limit many materials yield in shear and undergo plastic deformations. For most materials the shear yield stress is significantly less than the yield stress of the material submitted to tension. Thus, for such materials, the shear yield stress typically defines the safety factor (i.e., a measure of the extent to which the material can experience shearing before it fails). This is true for such diverse materials as structural steel and wood.

4.10 Nonlinear elastic behavior

As noted earlier, materials for which normal stress and normal strain are linearly proportional to each other (those that obey Hooke's law) are called linear elastic materials. Steel and some plant fibers exhibit linear elasticity for a limited range of loadings. Within the linear range, the material is expected to return to its original dimensions once the stresses are removed. Materials such as these are called Hookean materials. Nonlinear elastic materials recover their original dimensions even if strained beyond their proportional limit. Consequently, these elastic materials produce curvilinear stress-strain diagrams within the limits of their elastic behavior, indicating that the ratio of stress to strain changes as a function of the magnitude of stress.

The slope measured anywhere along the stress-strain diagram provides the instantaneous quotient of stress to strain, called the tangent modulus, which is sometimes referred to as the instantaneous elastic modulus (eq. [4.5]). Thus, the tangent modulus changes within the elastic range of behavior of nonlinear elastic materials. Rubber is an excellent example of a material that exhibits nonlinear elasticity. It instantly recovers its dimensions when stresses are removed, but the strains that develop tend to decrease in proportion to the stresses. Thus, the tangent modulus tends to increase as the stress level increases. By contrast, the stress-strain diagrams of some nonlinear elastic materials, such as silk and many primary plant tissues, are convex, indicating that the tangent modulus tends to decrease as the stress increases beyond a certain level. This phenomenon, which is called strain hardening versus strain softening, will be treated in section 4.12.

4.11 Viscoelastic materials

When an ideal linear elastic material is loaded and unloaded, the two portions of its stress-strain diagram will be identical provided the range of loading does not exceed the proportional limit (see fig. 4.3). This behavior is rarely seen for plant materials, even for those that exhibit a linear stress-strain relationship when initially loaded. Rather, the unloading portion of the stress-strain diagram follows a different path (see fig. 4.3). The elastic hysteresis of a material is defined as the amount of energy internally dissipated during a loading-unloading cycle. This energy can be calculated from the area spanned between the curves for loading and unloading in a stress-strain plot, multiplied by the volume of the object; that is,

(4.25) $U_{hysteresis} = V \oint \sigma d\varepsilon,$

where the symbol $\oint$ is required because the integration goes along a closed loop. The second law of thermodynamics implies that for any given stress, the strain is larger in the unloading portion than in the loading portion of the stress-strain curve.

For viscoelastic materials, recovery is time dependent. Thus, the mechanical behavior of viscoelastic materials requires mathematical descriptions involving three (rather than two) parameters: stress, strain, *and time*.

Viscoelastic materials can be classified as either linear or nonlinear in behavior. For both types of materials, strain increases as an applied force increases, and with time, the strain decreases when the applied force is removed. The distinction between a linear and a nonlinear viscoelastic material is that the strains in the former show an additive effect for a given interval. That is, for linear viscoelastic materials, if a stress σ_1 produces a strain ε_1 in time Δt, and if another stress σ_2 produces another strain ε_2 in time Δt, then $\sigma_1 + \sigma_2$ will produce a strain of $\varepsilon_1 + \varepsilon_2$ in time Δt. If these two components are not linearly related, then the material is called a nonlinear viscoelastic material.

Most mathematical descriptions of viscoelastic materials assume that the former condition holds true, much as the theory of elastic solids assumes that a material obeys Hooke's law—that it is an ideal linear elastic material. For nonlinear viscoelastic materials, however, the assumption of linearity is a good approximation only for very small strains ($\leq 1\%$). Since most biological materials are nonlinear viscoelastic materials, and since most function mechanically at high or extremely high strains, the assumption of linearity between the elastic and the viscous responses is not valid. Unfortunately, no mechanistic model exists that can fully treat the behavior of nonlinear viscoelastic materials. Thus, we can draw almost no generalizations about the mechanical behavior of biological viscoelastic materials experiencing large strains.

Viscoelastic materials have much in common with non-Newtonian fluids. Both materials evince time and stress rate dependency. Indeed, aqueous solutions of high-molecular-weight DNA show viscoelastic behavior. To fully characterize the mechanical behavior of viscoelastic materials, we must quantify their responses under varying stress levels and their responses to each stress level over time. Two fundamental types of experiments are used to quantify the mechanical properties of viscoelastic materials: dynamic experiments and transient experiments. Dynamic experiments are those in which either stress or strain is varied cyclically with time and the mechanical response of the material is measured over various frequencies of deformation (Vincent 1990a). They tend to be procedurally and mathematically complex. A material is subjected to a sinusoid time-varying strain with frequency ω. If the material is perfectly elastic, the stresses will always be proportional to the strains, whereas if the material has no elastic response component, the stress-strain plot (called a Lissajous figure) will be a perfect circle. Since a viscoelastic material has both elastic and viscous response components, its Lissajous figure will be elliptical. The viscous modulus, symbolized by E'', can be calculated from the Lissajous figure of a viscoelastic material. It is the ratio of the material's dynamic viscosity μ to the frequency at which strain is varied: $E'' = \mu/\omega$. Accordingly, E'' is a measure of the energy loss within the system resulting from viscous processes. Little energy is lost when the period of oscillation is dissimilar to the characteristic times describing the rates of molecular processes attending mechanical deformations; comparably greater losses of energy occur as ω approaches these characteristic times. Indeed, data from dynamic experiments can shed considerable light on the molecular processes attending the deformations that occur within viscoelastic materials.

In contrast to dynamic experiments, transient experiments are far easier to perform and are far more commonly reported in the literature. These experiments involve deforming a material (either by simple extension or in shear) and following its mechanical response over time. Since there are three variables in a transient experiment (stress, strain, and time), and since one of these variables (time) must always figure as the abscissa, there are two transient experimental formats: either a material is tested under a constant strain level and the decay in stress over time is measured (fig. 4.11A), or the material is tested under a constant stress level and the changes in strain over time are measured (fig. 4.11B). The former type of experiment is called a stress relaxation experiment; the latter is called a creep experiment.

The relaxation time of a viscoelastic material is measured by means of a stress relaxation experiment in which a sample of a material is deformed to a fixed level and the decay in the stress required to maintain this strain level is recorded over time. This kind of behavior can be depicted metaphorically in the form of a Maxwell model in which an elastic spring

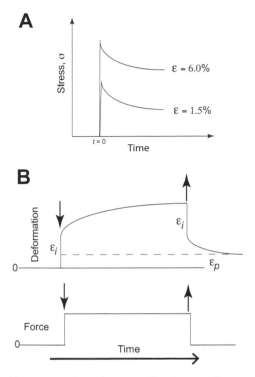

FIGURE 4.11. Graphic representations of a stress relaxation experiment and a creep experiment. (A) In a stress relaxation experiment, a specimen is subjected to an increasing strain up to a certain level (e.g., $\varepsilon = 1.5\%$ or 6.0%), which is then kept constant, and the decay in the stress level is followed over time. (B) In a creep experiment, a specimen is subjected to a constant force over a fixed time interval, after which the force is removed. If the material is viscoelastic, it will show an instantaneous elastic strain ε_i but continue to deform (called "creep") until the force is removed (*see top diagram*). Once the force is removed, the specimen will restore its initial elastic deformation (but may have a permanent, plastic strain ε_p).

element operates in series with a nonelastic (delayed-response) dashpot (fig. 4.12A). In a very real (mathematically precise) way, the relaxation time is the ratio of a material's dynamic viscosity to its modulus of elasticity. For liquids this ratio is very large; for solids it is very low; and for viscoelastic materials it lies somewhere in between.

In stress relaxation experiments, at any time t, the rate of change in strain can be considered equal to zero, and the instantaneous stress (at time t) is given by the formula

(4.26) $$\sigma(t) = \sigma_0 \exp(-t/T_R),$$

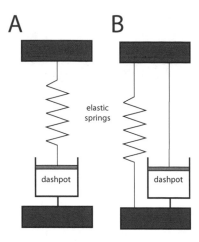

FIGURE 4.12. Diagrammatic representations of a Maxwell model and a Voigt/Kelvin model. (A) In the Maxwell model, an elastic spring operates in series with a dashpot to produce an initial instantaneous elastic response followed by a delayed response. (B) In the Voigt/Kelvin model, the two elements operate in parallel to produce a more complex behavior than the Maxwell model.

where σ_0 is the stress at time zero and T_R the relaxation time. Thus, the relaxation time is the time required for the stress to decrease to $1/e$ the original stress. Unfortunately, most biological materials do not follow this simple relationship. Instead, their behavior is characterized by several relaxation times, or even a continuum of relaxation times. In these cases, the decrease in the applied load is usually plotted as a function of log time. (Stress relaxation for a solid subjected to shear is discussed by Nakajima and Harrell [1986].)

Relaxation is very important to many biological phenomena, not the least of which are cell growth and the expansion of the cell wall. Recall that if the cell wall mechanically operated as an ideal elastic material, given the remarkably high elastic modulus of cellulose, tremendous internal pressures would be required to extend the cell wall, and when those forces were removed, the cell wall would return to its original dimensions and shape, provided the proportional limit of cellulose had not been exceeded. It is not surprising, therefore, that the walls of growing cells operate as either plastic or viscoelastic materials. Indeed, relaxation curves, measured by stretching cell walls to some predetermined strain and measuring the decay in the tensile stress, indicate that both viscoelastic and plastic deformations occur within cell walls.

It is important to note that some of the plastic behavior evinced by growing cell walls results from the addition of new materials to the cell wall rather than from the deformation of preexisting materials. Additionally, all the available information pertaining to plant cell walls indicates that they behave mechanically as nonlinear viscoelastic materials. None theless, the mechanical testing of cell walls by techniques such as stress relaxation provides the opportunity to treat a complex biological phenomenon in terms of a relatively simpler physical phenomenon, permitting considerable insight.

If a viscoelastic material is placed under tension, there is an instantaneous elastic deformation followed by a relatively slow, time-dependent deformation called creep (see fig. 4.11B). The most obvious way to deal with creep is to place a material under a constant uniaxial stress and plot the resulting changes in strain over time. This method is called a transient creep experiment. The rate at which deformations occur is governed by the viscosity of the material being tested. It also depends on the stress and the temperature. Formally, this behavior can be represented by a Voigt/Kelvin model in which an elastic spring and a dashpot operate mechanically in parallel (fig. 4.12B).

The relaxation modulus $E(t)$ as measured in stress relaxation experiments is the ratio of stress to strain at time t, or

$$(4.27) \qquad E(t) = \frac{\sigma(t)}{\varepsilon},$$

where the strain ε is a constant. In creep experiments, the reciprocal quantity is called the compliance, which is defined as

$$(4.28) \qquad D(t) = \frac{\varepsilon(t)}{\sigma},$$

where σ is constant. $E(t)$ and $D(t)$ are usually referred to as the tensile relaxation modulus and the tensile compliance, respectively, since they are measured for a specimen placed in tensile stress. When followed over long times, changes in the tensile relaxation modulus and the tensile compliance can become asymptotic. The ratio of strain to stress immediately after the strain has reached a constant level is called the unrelaxed compliance, symbolized by D_U, while the value approached asymptotically is called the relaxed compliance, symbolized by D_R.

In summary, the single most important parameter influencing our perception of a viscoelastic material's behavior is time. A viscoelastic material will behave as a linear elastic solid under stress when strain is measured instantly; it will behave much like a viscous fluid under stress when strain is measured over a very long period, and it will behave as an intermediate between a solid and a fluid when strain is measured over times on the order of minutes to hours.

4.12 Plastic deformation

Plastic strains are unrecoverable deformations that occur as a result of permanent molecular reorganizations within a simple material or microstructural deformations within a composite material. Some materials exhibit plastic behavior at very low loadings; others behave plastically only after very high loadings. The nonlinear recovery of strains in an elastic material is called anelastic behavior. Two terms, the degree of elasticity and elastic hysteresis (see section 4.11), are useful in describing anelastic behavior. The degree of elasticity is defined as the ratio of the elastic (recovered) deformation to the total deformation when a material is loaded to a given stress level and then unloaded to zero stress. The degree of elasticity of a perfectly elastic material is unity, indicating that the recovered deformation precisely equals the total deformation. By contrast, the degree of elasticity of a perfectly plastic material is zero, indicating that no deformation is recovered when the stress level drops to zero. For elastic-plastic materials, the degree of elasticity can vary widely depending on how far loading has induced plastic deformations.

Beyond their proportional limit, many fabricated and naturally occurring polymeric materials evince one of two trends in the relationship between stress and strain: the tangent modulus may increase or decrease. The former indicates strain hardening; the latter indicates strain softening (fig. 4.13). Strain hardening is typically a physical manifestation of changes in the molecular structure of a material, as when polymeric chains (or constituent crystals within a crystalline, strain-hardening solid) increasingly align parallel to the axis of a uniaxial stress. There is an obvious advantage to strain hardening, since the instantaneous elastic modulus increases as the stress level increases.

Strain hardening can produce some very unusual behavior and can lead to erroneous conclusions about the nature of a material. For example,

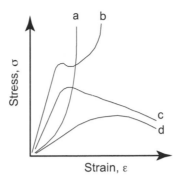

FIGURE 4.13. Representations of strain-hardening (a and b) and strain-softening (c and d) stress-strain behavior. In each case, beyond a certain stress level, the materials either progressively resist deformation (strain-hardening) or deform disproportionately more (strain-softening).

there are materials with a negligible initial elastic range that will nonetheless produce an initially linear stress-strain diagram. The linear portion of the stress-strain diagram is the result of linear strain hardening. Thus, the slope of the initially linear portion of the stress-strain diagram is *not* the elastic modulus. Rather, that slope is the relationship of the change in the generalized equivalent stress σ_h to the change in the generalized plastic strain increment ε_p, which is called the strain hardening coefficient, symbolized by $H' = d\sigma_h/d\varepsilon_p$. Parenchyma isolated from potato tubers displays this type of behavior. Significantly, some plant fibers exhibit tensile strain hardening as a result of the rearrangement of their cellulosic molecular infrastructure. Thus, as the magnitude of externally applied forces increases and exceeds their proportional limits, the instantaneous elastic modulus of these cell walls increases, providing an increased, albeit modest, resistance to further deformations.

In other materials the initiation of plastic deformation is typically seen as a dramatic reduction in the slope of the stress-strain diagram just beyond the proportional limit. Once initiated, deformations continue even under a constant load. In many plant tissues plastic deformations are the product of the rupturing of cell walls and result in permanent strains when the tissue is unloaded. Unlike steel structures, of course, plants have the capacity to add new structural components and repair many forms of mechanical damage.

Compliance under tension or compression is an important biological parameter, particularly for cell walls. Indeed, after some cell walls plasticize under a hydrostatic load, they can be restiffened and resume elastic behavior after the load is removed. Thus, plastic deformations are not always undesirable; many are important to plant growth. Typically, each plant cell is encased in a relatively stiff cell wall, secreted by the protoplast external to its plasma membrane. Further expansion of the protoplast is achieved by the plasticizing of the cell wall, which allows the wall to yield under the internal pressure exerted on it by the expanding protoplast. In young or actively growing tissues, the entire cell wall infrastructure of the tissue may behave as a plastic material, deforming irreversibly as the living protoplast prevents the apoplast from becoming an elastic solid by metabolically decreasing the yield point of the cell wall.

In the case of turgid plant cell walls, creep has been found to have two components: a retarded linear or nonlinear elastic component and a plastic component. When the tensile force is removed, the cell wall undergoes an instant elastic contraction, followed by a relatively slow viscoelastic contraction. The plastic component of deformation, called residual strain, is not recovered. This property of cell walls is vital to growth, since the residual strains reflect an increase in the surface area of the cell wall, which in turn reflects the increase in the volume of the enveloped protoplast. Provided the protoplast can control the material properties of its cell wall and maintain a certain level of plastic-viscoelastic behavior, it can utilize the residual strain in the cell wall to provide room for growth. In this sense, growth provides a biologically unique way to deal with stresses that exceed the proportional limit.

To further understand the importance of the limit of elasticity and strain hardening, we can turn to the work of one of the first biomechanicists to interpret stress-strain diagrams. Simon Schwendener (1874), who was trained as both an engineer and a botanist, discovered that the fibers of monocots lying near the tissue that conducts cell sap (phloem) have limits of elasticity comparable to those of the best iron and steel of his day (table 4.1). For example, the phloem fibers of rye (*Secale cereale*) can withstand a stress of 0.147 to 0.196 GN/m^2 before they begin to show plastic behavior, whereas the iron and steel wires Schwendener examined had limits of elasticity of 0.215 to 0.241 GN/m^2 (modern steels begin to yield at stresses as high as 1.5 GN/m^2; see Timoshenko 1976).

A simple theory of plasticity treats the behavior of materials within their plastic range (Hill 1950), although a variety of more complex models (which cannot be treated here) have been developed to account for elastic-plastic effects. There are five principal assumptions in this simple plastic theory:

TABLE 4.1 **Proportional limits and corresponding strains of plant fibers and some metal wires**

Material	Stress (GN/m²)	Strain (%)
Secale cereale	0.147–0.196	4.4
Lilium auratum	0.186	7.6
Phormium tenax	0.196	13.0
Papyrus antiquorum	0.196	15.2
Molina coerulea	0.216	11.0
Pincenectia recurvata	0.245	14.5
Copper wire	0.119	1.0
Brass wire	0.130	1.4
Iron wire[a]	0.215	1.0
Steel wire[a]	0.241	1.2

Source: Data from Schwendener (1874).
[a]Values given are for metals of Schwendener's day. Strains for modern iron and steel wires are orders of magnitude smaller.

1. The deformations within a structure are small. This assumption permits the use of the geometry of the undeformed structure to compute the equations of equilibrium.

2. Plastic deformations produce no change in the volume of the material.

3. The forces operating on and in the material maintain a constant proportion to one another.

4. The material is ductile; that is, it is capable of plastic deformation without fracture. The stress-strain diagram of the material is assumed to be that of an ideal elastic-plastic material. Accordingly, the consequences of strain hardening and strain history are neglected.

5. Plastic deformations occur when any component within a given cross section develops plastic strains. This assumption neglects the fact that within any cross section there exist structural components that will retain their elastic behavior. Thus, the effects of shear between adjacent elastic and plastic fibers are neglected.

For growing plant tissues, some of these assumptions are relatively reasonable, even the first, which assumes little or no change in shape; for example, the *instantaneous* shape of a tissue or organ can be used to calculate the equations of equilibrium, provided growth is evaluated in a series of small incremental steps. However, assumption 2 may not always be accurate given the cellular nature of plant tissues, particularly porous plant tissues such as wood and cork. Plastic deformations in porous plant tissues are likely to involve changes in tissue volume. Assumption 5, too, may not always apply. The biphasic stress-strain curves and the accompanying plastic changes observed for fresh compression wood and other plant material appear to be the result of shearing between fibers and matrix and of sliding—the so-called Velcro effect (Burgert 2006). Clearly, there are limits to the idealization of elastic, viscoelastic, and plastic behavior, particularly for biological materials, which usually show all of these mechanical traits.

4.13 Strength

The yield stress is the magnitude of the stress beyond which stress and strain are no longer proportionally related. If strained beyond the yield stress, the material is usually weakened. So in the engineering sciences, the yield stress is an important estimate of a safety factor. The proportional limit of a material may not be equivalent when measured in tension and in compression. Unfortunately, the proportional limits of many plant materials in tension and in compression are not reported in the literature, and they cannot be inferred from the elastic modulus of a material. For example, the elastic modulus E of a typical clear-grained specimen of pine wood may be as high as 8.51 GN/m^2, while the maximal stress at which elastic behavior is retained may be as low as 0.045 GN/m^2. In comparison, a mechanical analysis of *Laminaria* (a marine brown alga) reports $E = 17.9$ MN/m^2 and a yield stress of 2.5 MN/m^2 (see Harder et al. 2006). Thus, neither the yield stress nor the maximal stress can be inferred simply from values of elastic moduli.

The maximal stress that a material can tolerate is the strength of the material. This maximal stress should be distinguished from the yield strength and from the breaking strength. Under permanent loads, a material will always fail if strained to exceed its strength. The stress at which the material finally disintegrates is often less than its strength. If the material fails gradually in this way, there is some difficulty in the definition of the maximal strain. It can denote the strain at maximal stress or the strain at which the material ultimately fails. Under a strain-controlled loading-unloading experiment, a material showing strain softening can be strained beyond the point of maximal stress and yet not mechanically fail, provided that the breaking strain is not reached.

Generally, the strength of a material when it is measured under tension or compression will differ from its strength when it is measured in shear. The compressive strength of composite materials in which stiff fibers are embedded in a more compliant matrix will be less than their tensile strength. This difference is easily explained by fiber buckling. The compressive strength of most species of wood is roughly 50% of the tensile strength of the same species, although there is a substantial range among species in the absolute values of both (table 4.2, see also table 8.2).

The strengths of different biological materials tested under tension span a few orders of magnitude: from only a few MN/m^2 for marine algal tissues (e.g., *Durvillaea antarctica*, 0.7 MN/m^2; see Koehl 1979; *Fucus serratus*, 4.2 MN/m^2; see Wheeler and Neushul 1981), to fresh tendon (0.08 GN/m^2), a typical wood measured along the grain (0.1 GN/m^2), and cotton fibers (0.35 to 0.91 GN/m^2). Thus, the difference in the strengths of animal tendons and plant fibers is roughly one order of magnitude, the middle range of which encompasses the tensile strength of nylon (0.45 GN/m^2). By contrast, the tensile strength of nickel-treated steel is substantially greater (1.4–1.6 GN/m^2). However, the tensile strength of plant materials is remarkably high in terms of their density. Venkataswamy et al. (1987) measured the tensile strength of the midrib (the main vascular bundle running the length of each leaf) of the coconut palm. They reported values ranging between 0.17 and 0.30 GN/m^2, significantly greater than the tensile strength of annealed aluminum (0.059 GN/m^2) and very close to that of hard rolled aluminum bronze (0.26 GN/m^2), whose densities are more than twice that of the midrib tissues.

Although the tensile strengths of many plant materials are comparable to those of some metals, most plant materials have the added advantage of considerable elastic extension before they break. Strands of the lichen *Usnea* are capable of elongating elastically from 60% to 100% of their original length. In general, lignification of tissues reduces their capacity for extension. Thus, woody plant stems sustain little deformation under bending stresses near their breaking stress compared with nonwoody stems.

TABLE 4.2 **Some mechanical properties of wood**

Physical property	Linden *Tilia vulgaris*	Oak *Quercus rubra*	Poplar *Populus canescens*
Density at 12% MC (kg/m^3)	561	705	481
MOE (bending) (GN/m^2)	9.2	10.5	7.2
Bending strength (MN/m^2)	54	72	44
Compression strength (MN/m^2)	26.1	28.7	20.1
Shear strength (MN/m^2)	7.3	9.2	5.9

Note: With the exception of density, all properties refer to green wood. MC = moisture content; MOE = modulus of elasticity.

The strength of a *structure*, called the breaking load, is simply the magnitude of the loading that results in breakage. Unlike the strength of a *material*, which has the same units as stress, the breaking load has units of force, and its magnitude can vary significantly even within the same class of structures composed of the same material, such as tree trunks. The breaking load will not be the same for different structures made of the same material, nor will it be the same for the same structures differing in their absolute dimensions. Trees with identical shapes and sizes may have different breaking loads if the material properties of their woods differ. Similarly, trees of the same species that differ in size may have different breaking loads, even if the strength of their wood is the same among all the individuals tested. Even though wood is a cellular structure, it can be relatively uniform in its cellular infrastructure, and within limits we can legitimately speak of its tensile or compressive strength rather than its breaking load. However, the tensile or compressive strength of wood samples possessing internal flaws may differ significantly from that of clear-grain samples. Such structural defects play a significant role in determining how stresses are accommodated by a particular sample and when fractures will be propagated through a specimen placed in tension. The tensile strength of wood species is usually specified for clear-grain samples (i.e., those lacking knots or anomalous growth patterns). Internal flaws have a significant bearing on the mechanics of tree trunks and branches, which can possess many knots or structural defects in the form of fungal or bacterial or animal damage.

An obvious concern arises over the potential for variation in the tensile strength or other material properties of plant tissues. When the tensile strengths of many samples of tissue from a population of plants are measured and plotted as a function of their relative frequencies (the number of individuals, expressed as a percentage, within the sampled population that show each range of strengths), the data typically take the form of a Weibull frequency distribution. This has been shown for the breaking (tensile) strength of garlic flower stalks (*Allium sativum*; Niklas 1990) placed in bending and the tensile strengths of parenchyma from potato tubers (Lin and Pitt 1986) and wood (Woeste et al. 1979). The Weibull frequency distribution is one of a few that have been shown to describe phenomena whose statistical behavior is responsive to environmental variation (Weibull 1939; Ang and Tang 1975). The implication of this observation is that, within a population, the strength of tissues varies in a manner that ensures the survival of some individuals over a broad range of mechanical stress levels. In other words, some individuals are stronger than others and can perpetuate the population or the species after some unpredictable environmental event, such as a storm. The potential for a tissue's mechanical properties to vary among individuals from the same population confers a design factor or margin of safety on the population, ensuring that some will survive to reproduce and continue to occupy a site. Of less theoretical but much practical importance, variation among individuals in the strength of the same tissue highlights the importance of relying on numerous measurements of material properties when considering biomechanical models designed to predict the behavior of a tissue, an organ, or an individual plant.

When considering the strength of elastic structures capable of undergoing large-scale plastic deformations, engineers typically speak of the "collapse" load of the structure, which is another term for the "yield" load. For a simple prismatic bar, the collapse load is given by the product of the uniaxial yield stress and the cross-sectional area of the bar. When dealing with complex structures such as steel frameworks and plant stems, however, some parts of the structure may yield before others because of localized stresses. For complex structures, it is important to know the limits of the elastic load: the highest load the structure can tolerate before unrecoverable deformation occurs at any point within it. In some structures the elastic load may be very small, but for plastic collapse to occur, yielding in the structure must become general and widespread.

4.14 Fracture mechanics

When an object breaks under tension, new surfaces are created in the form of a crack or fracture. To understand the mechanics of fracture, we must return to the concept of strain energy and consider additional concepts such as brittleness, ductility, and toughness. Farquhar and Zhao (2006) provide an overview of the theory of engineering fracture mechanics and discuss its relevance to wood and other plant-based materials.

As already noted, the energy stored within an object in the form of recoverable strains is the energy that can be used to restore an object to its original shape and dimensions after an applied load is removed. When an applied load produces critical stresses, the strain energy stored can be sufficient to fracture the structure. Fracture typically occurs through the propagation of a crack that begins on the surface of a member experiencing

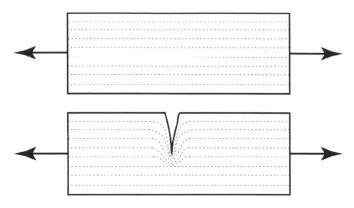

FIGURE 4.14. Fracture propagation in a brittle material. The dotted lines are stress trajectories that converge at the tip of the notch, indicating that stress levels increase in this location.

tensile stresses. Since the creation of new surfaces within a solid requires energy, breakage under tensile stress requires the expenditure of strain energy stored within the volume of an object. Indeed, as the fracture propagates, more and more strain energy must be supplied. An important feature in the expenditure of this energy is that only the strain energy stored in the vicinity of a fracture can be used. Thus, as the fracture propagates, more and more strain energy must be drawn from an increasing volume within the object (see Vincent 1990b).

Fractures initiate when strain energy is released near discontinuities in the force trajectories traversing an object. Force trajectories can be imagined by considering a small notch on a specimen placed in tension. Although the stress in the bulk of the specimen remote from the notch is uniformly distributed, the lines of tensile force operating within the specimen follow trajectories that converge and increase in concentration as they curve around a notch or imperfection (fig. 4.14). These discontinuities in force trajectories are called stress concentrations or stress raisers. Although the applied tensile stress remote from the notch may be substantially less than is needed to cause the material to break, the tensile stress developed near the notch may exceed the tensile strength of the material.

This principle is important to remember when comparing the tensile strengths of various materials, since strength is a property of materials, whereas fracture strength is a measure of the structural homogeneity of a structure. The higher the structural perfection of an object, the greater its fracture strength. Since it is reasonable to predict that the probability of an imperfection, such as a notch or crack, will increase as the size of an organic structure increases, it is reasonable to suppose that fracture strength is a statistical function of the size distribution of a class of structures. Fortunately, this assumption has only limited bearing on the fracture mechanics of tree trunks and branches, which increase in size as they grow. In theory, as these structures get older and grow bigger, the probability of growth defects that will form the nuclei for fracture propagation should increase. Thus, older and larger branches and trees can be assumed to have a higher probability of mechanical failure resulting from "notch stresses." However, as shown by Mattheck (1990), growth and repair mechanisms in trees serve to avoid notch stresses.

The physics underlying fracture mechanics was first dealt with successfully by A. A. Griffith (1893–1963). He initially explored fracture mechanics for brittle linear elastic materials in terms of an energy budget: that is, the difference between the strain energy released by the propagation of a crack and the energy needed to form cracked surfaces. Griffith determined that there exists a critical crack length below which the energy budget is negative, such that the crack will not increase in length. Put differently, even if local stresses are high at the tip of the crack, the structure will not fracture at the site if the crack is below this critical length.

Griffith derived an equation from which the critical crack length can be calculated. The derivation of this equation is based on an estimate of the elastic energy released during cracking. Consider a metal bar with a crack (of length l) oriented transversely to an applied tensile stress (see fig. 4.14). Intuitively, we recognize that near the surfaces of the crack, the local tensile stresses are relaxed to zero, but as we progress away from the crack surfaces, the tensile stress σ approaches the magnitude of the applied tensile stress. Let us assume that a semicircular region with a radius roughly equal to l around the crack will have been relieved of some of its elastic energy. The elastic energy relieved per unit thickness u_{crack} of the metal bar can be estimated using the formula

(4.29) $$u_{crack} = \frac{\pi}{2} l^2 \sigma^2 / E.$$

The surface energy, per unit thickness of the metal bar, required to generate the crack's new surfaces is

(4.30) $$u_{surface} = l\Omega,$$

where Ω is the specific surface energy.

This approach, however, neglects that part of the total energy that has been converted into thermal energy. Furthermore, it is often impossible to determine the surface area exactly. Therefore, it is more appropriate to replace the surface tension Ω with the experimentally determined work of fracture W (see below). Note that $u_{surface}$ is a linear function of crack length, whereas u_{crack} depends on the square of l. Accordingly, energy has to be supplied to produce a small crack, but upon enlargement of the crack, the release of strain energy will eventually dominate the energy balance. Once the crack reaches a critical length, it will self-propagate, potentially leading to catastrophic failure.

The length l of a crack at which the energy required to produce the crack is equal to the strain energy released is called the Griffith length L_G. By precise integration of the true shape of the stress trajectories near a crack, Griffith found the relationship

(4.31)
$$L_G = \frac{2WE}{\pi\sigma^2}$$

(compare eqs. [4.29] and [4.30]). For $l < L_G$, the energy needed for the propagation of a crack is larger than the strain energy released. Any brittle structure having cracks of lengths less than L_G is thus essentially safe from fracturing. However, L_G is not a material constant, but depends on the mechanical loads operating on the structure. Thus, even small cracks are potentially dangerous. This is the reason why the detection of minute cracks in technical constructions is of vital importance for their safety.

Empirical fracture mechanics indicates that, in most circumstances, unreasonably large amounts of surface energy have to be assumed if Griffith's formula is to be used. And it has been proposed that the total energy required to propagate a crack is the sum of the surface energy already discussed and the energy needed to produce plastic deformations. The essential condition for non-brittle, elastic-plastic fracture is that the plastic displacements at the tip of the propagating crack are attended by elastic displacements within regions distant from the crack. This condition is often found in ductile materials.

Expressed differently, Griffith's theory does not apply to highly deformable elastic materials. For ductile materials, such as many types of metals, the energy supplied by an external force is dissipated over a large portion of the volume of the material near the crack's surfaces. Necking is a physical manifestation of this dissipation of energy in ductile materials placed in tension: atoms or molecules slip past one another over a considerable distance from a stress raiser, so the specimen's cross-sectional area is attenuated by localized plastic straining. For example, rubber is a brittle material, but it strains elastically at such low stresses that notches in it can deform into mechanically harmless blunt regions well before the breaking stress is achieved.

This type of behavior is called notch insensitivity, and plants have evolved several ways to achieve it. Some plant materials are ductile and can dissipate stress concentrations in the form of internal molecular slippage. Natural rubber, plant latex, and growing cell walls are only three examples. Another way is to round off naturally occurring indentations or notches. Undulating leaf margins and the rounded surfaces of stems are examples of this design. Another way to be insensitive to abrasions or scratches is to produce many parallel fibers (in a binding matrix) that will shear little. In this arrangement, if a fiber breaks, the stress is distributed evenly among the remaining fibers. This design is seen in leaves possessing parallel vascular bundles running their length, as in grasses.

4.15 Toughness, work of fracture, and fracture toughness

Toughness corresponds to the energy required to fracture a material. The area under a stress-strain diagram provides a measure of a material's toughness in terms of the energy absorbed per unit volume in units of J/m^3. As a general rule, toughness is maximized by an optimal combination of strength and ductility.

A related feature is the work of fracture. This term defines the energy per unit thickness of a specimen divided by the length l of a crack produced. It is most easily measured by initiating a crack with a small cut and observing the energy necessary to enlarge the crack to a certain length (see section 9.2). The work of fracture measured in this way is characteristic for the material investigated and has units of J/m^2.

The work of fracture W allows us to distinguish two very broad categories of brittle materials: weak materials (for which W = the true surface energy) and strong materials (for which $W \gg$ true surface energy). For example, steel and glass are both brittle, and both are comparably hard. But steel is much stronger than glass. The work of fracture for mild steel lies between 10^5 and 10^6 J/m^2, while that for glass is on the order of only 1.0 J/m^2

(Gordon 1976). Strong brittle materials, such as hard steel, retain their strength even if their surfaces are heavily scratched or slightly notched, whereas weak brittle materials, such as glass and many types of stone, retain little of their strength if their surfaces are scored. Stonemasons utilize this characteristic when they score a block of granite and cleave it with a relatively small percussion. By the same token, steel girders can sustain much superficial scratching and still support large, heavy structures. Indeed, Cottrell (1964) made the observation that the advance of humanity from the Stone Age to the Iron Age was based on advancement in the understanding of fracture mechanics.

Engineers and botanists frequently refer to the fracture toughness of a material or plant tissue. Fracture toughness, symbolized by K_c and given in units of J/m$^{3/2}$, is a property that describes the ability of a certain material to resist fracture. For isotropic materials, K_c is related to the elastic modulus E, the work of fracture W, and the Poisson's ratio v, such that

$$(4.32) \qquad\qquad K_c = \frac{(EW)^{1/2}}{(1-v)}.$$

For these materials, K_c decreases as the Poisson's ratio decreases. For non-brittle and anisotropic materials, determining fracture toughness tends to be very complex procedurally. Thus, considerable efforts have been made to determine fracture toughness from more easily measured tensile stress-strain diagrams. A more detailed discussion of fracture toughness is given by Bodig and Jayne (1993).

4.16 Composite materials and structures

Thus far, we have used the term "material" in a very cavalier manner, but it has a very precise meaning in engineering. The term refers to either a pure substance or an alloy that can be approximated as essentially homogeneous in composition. When more than one substance or material are combined, and when this combination has some internal structural heterogeneity, the term *composite material* is used (Bodig and Jayne 1993).

Composite materials can have either a periodic or a nonperiodic ultrastructure. That is, their heterogeneity can be reiterative, with the various component materials distributed in a geometrically predictable manner, or not. In either case, the mechanical properties of composite materials depend on the structural relationships among the various component materials from which they are fabricated as well as the material properties of each component (Mura 1982). Plant cell walls, which have a highly ordered arrangement of polymers (principally cellulose) embedded within a more or less amorphous matrix (see chap. 8), can be viewed as periodic composite materials. Some tissues, such as parenchyma, which consists of a more or less geometrically ordered arrangement of cell walls holding together the fabric of the protoplast, may also be viewed as periodic composite materials. Furthermore, each tissue type within a plant organ is precisely arranged and distributed, but each tissue may differ from the others in its mechanical properties (see Kutschera 1989, for example). Accordingly, the material properties of plant cell walls and some tissues and organs can be approximated by the behavior of composite materials.

The first basic theory for the mechanical behavior of composite materials was developed in the nineteenth century by James Maxwell, who considered two limiting models currently called the equal stress or Reuss model and the equal strain or Voigt model (fig. 4.15). The Voigt model sets the upper limit for the magnitude of the composite elastic modulus. It assumes that all the material elements within the composite material lie parallel to the direction of the externally applied force and that the strains within each element are equivalent in direction and magnitude so that the deformational compatibility between adjacent layers is maintained.

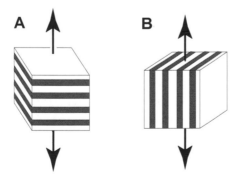

FIGURE 4.15. Schematic representations of (A) the Reuss model and (B) the Voigt model.

According to the Voigt model, the elastic modulus for a two-component composite material is given by the formula

(4.33) $$E_{comp} = E_1 v_1 + E_2 v_2,$$

where E is the elastic modulus and v is the volume fraction of each of the components, with $v_1 + v_2 = 1$.

The Reuss model sets the lower limit for the magnitude of the composite elastic modulus. It assumes that the material elements within the composite are aligned normal to the direction of the externally applied force and that the stresses developed within each element are equivalent in magnitude, whereas the strains may differ depending on the elastic modulus of each element. According to the Reuss model, the elastic modulus for a two-component composite is given by the formula

(4.34) $$\frac{1}{E_{comp}} = \frac{v_1}{E_1} + \frac{v_2}{E_2}.$$

Clearly, however, for real composites, the elastic modulus falls somewhere between these limits.

Biologically, the most interesting composites are those in which stiff fibers are embedded in a more compliant matrix. In plants, we usually find stiff cellulose fibers arranged in microfibrils and a matrix consisting of glycoproteins, hemicelluloses, pectins, and lignins. One of the most important structural parameters in determining the mechanical behavior of a cell wall is the microfibrillar angle: the spiral angle of the microfibrils with respect to the longitudinal axis, as observed in the secondary cell wall (see box 5.5 and chaps. 8–9).

Although the equal strain condition is observed in composites containing continuous fibers, this is not the case if the fibers in the composite are discontinuous. The mechanical properties of such composites are largely determined by the bonding between fibers and matrix. Upon straining, the tensile forces are transmitted between fibers and matrix by means of shear forces that develop at the fiber-matrix interface. At high tensile stresses, the shear forces can become large enough to lead to shear separation. Fibers and matrix can slide past each other in the manner of a plastic flow. In some materials, this behavior is apparent in a biphasic stress-strain curve. A steep slope at small strains up to the yield point is followed by a second linear range with much smaller slope. Upon returning to zero overall stress, a permanent deformation is observed, but the material has regained its original stiffness, indicating that it has not suffered any significant damage. This behavior has been explained by a molecular stick-slip mechanism (Köhler and Spatz 2002; Keckes et al. 2003; Burgert 2006).

4.17 The Cook-Gordon mechanism

Composite materials consisting of stiff fibers embedded in a compliant matrix combine stiffness with toughness. Due to their lower specific weight, their toughness often compares favorably with that of metals. The best-known example is fiberglass, which consists of thin glass fibers glued together by a resin. Glass is very brittle, as is the resin. However, when the two are put together, the result is an extremely tough material. Although fiberglass was invented only at the beginning of the twentieth century, the same design principle is manifested in many biological materials, another observation that indicates the adaptive and innovative power of evolution.

The important advantage of such composites is that cracks will not propagate through them when they are subjected to tension. Two conditions must be met to prevent cracks from propagating: the diameter of the stiff fibers must be smaller than the Griffith length, and the matrix in which the fibers are embedded must be ductile.

Cook and Gordon (Gordon 1978) have analyzed the shapes of cracks formed in brittle and ductile materials under tension and in composite materials with alternating brittle and ductile components. They show that an initial crack with a small tip radius through a brittle section will enlarge its tip radius as it reaches a ductile section. In this way, the stress concentration at the tip is greatly reduced, and the strain energy is distributed over a large area, such that the crack will not propagate further (fig. 4.16). This mechanical "invention" has been reported in the stems of some everyday plants, such as the common dandelion (Niklas and Paolillo 1998).

Interestingly, for the actual stress distribution at the tip of the crack, it does not matter very much how the load is applied, whether by a tensile force, by bending, or by the application of a wedge. Crack propagation and crack stopping are therefore general principles of fracture mechanics. Understanding these principles has led to the stunning success of composite materials in modern industrial design. It has also given us insight into optimization principles in the evolution of biological materials.

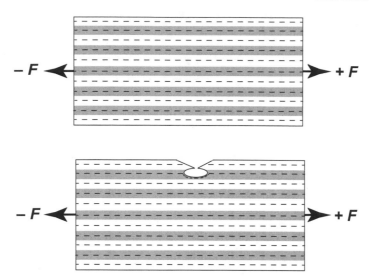

FIGURE 4.16. Crack stopping in a composite material. An initial crack with a small tip radius will enlarge its tip radius as it reaches a ductile section. The shaded areas represent ductile material. The dotted lines are stress trajectories.

Literature Cited

Ang, A. H.-S., and W. H. Tang. 1975. *Probability concepts in engineering planning and design.* New York: John Wiley.

Bodig, J., and B. A. Jayne. 1993. *Mechanics of wood and wood composites.* Malabar, FL: Krieger.

Burgert, I. 2006. Exploring the micromechanical design of plant cell walls. *Amer. J. Bot.* 93:1391–1401.

Chappell, T. W, and D. D. Hamann. 1968. Poisson's ratio and Young's modulus for apple flesh under compressive loading. *Amer. Soc. Agr. Eng. Trans.* 11:608–11.

Cottrell, A. H. 1964. *The mechanical properties of matter.* New York: John Wiley.

Farquhar, T., and Y. Zhao. 2006. Fracture mechanics and its relevance to botanical structures. *Amer. J. Bot.* 93:1449–54.

Gibson, L. J., M. F. Ashby, and B. A. Harley. 2010. *Cellular materials in nature and medicine.* Cambridge, MA: Cambridge University Press.

Gordon, J. E. 1976. *The new science of strong materials or why you don't fall through the floor.* 2nd ed. London: Penguin.

———. 1978. *Structures or why things don't fall down.* London: Penguin.

Harder, D. L., C. L. Hurd, and T. Speck. 2006. Comparison of mechanical properties of four large, wave-exposed seaweeds. *Amer. J. Bot.* 93:1426–32.

Hill, R. 1950. *The mathematical theory of plasticity.* London: Oxford University Press.

Keckes, J., I. Burgert, K. Frühmann, M. Müller, K. Kölln, M. Hamilton, M. Burg-hammer, S. V. Roth, S. E. Stanzl-Tschegg, and P. Fratzl. 2003. Cell-wall recovery after irreversible deformation of wood. *Nature Mater.* 2:810–14.

Koehl, M. A. R. 1979. Stiffness or extensibility of intertidal algae: A comparative study of modes of withstanding wave action. *J. Biomech.* 12:634.

Köhler, L., and H.-C. Spatz. 2002. Micromechanics of plant tissues beyond the linear elastic range. *Planta* 215:33–40

Kutschera, U. 1989. Tissue stresses in growing plant organs. *Physiol. Plant.* 77:157–63.

Lakes, R. 1987. Foam structures with negative Poisson's ratio. *Science* 235:1038–40.

Lin, T.-T., and R. E. Pitt. 1986. Rheology of apple and potato tissue as affected by cell turgor pressure. *J. Text. Stud.* 17:291–313.

Mark, R. E. 1967. *Cell wall mechanics of tracheids.* New Haven, CT: Yale University Press.

Mattheck, C. 1990. Design and growth rules for biological structures and their application to engineering. *Fatigue Fract. Eng. Mater. Struct.* 13:535–50.

Mura, T. 1982. *Micromechanics or defects in solids.* The Hague: Martinus Nijhoff.

Nakajima, N., and E. R. Harrell. 1986. Stress relaxation as a method of analyzing stress growth, stress overshoot and steady-state flow of elastomers. *J. Rheol.* 30:383–408.

Niklas, K. J. 1989. Mechanical behavior of plant tissues as inferred from the theory of pressurized cellular solids. *Amer. J. Bot.* 76:929–37.

———. 1990. Determinate growth of *Allium sativum* peduncles: Evidence for determinate growth as a design factor for biomechanical safety. *Amer. J. Bot.* 7:762–71.

Niklas, K. J., and D. J. Paolillo Jr. 1998. Preferential states of longitudinal tension in the outer tissues of *Taraxacum officinale* (Asteraceae) peduncles. *Amer. J. Bot.* 85:1068–81.

Schwendener, S. 1874. *Das mechanische Prinzip im anatomischen Bau der Monocotylen mit vergleichenden Ausblicken auf die übrigen Pflanzenklassen.* Leipzig: Engelmann.

Skotheim, J. M., and L. Mahadevan. 2005. Physical limits and design principles for plant and fungal movements. *Science* 308:1308–10.

Timoshenko, S. 1976. *Strength of materials.* Part 1. 3rd ed. New York: Krieger.

Timoshenko, S., and J. N. Goodier. 1970. *Theory of elasticity.* 3rd ed. New York: McGraw-Hill

Venkataswamy, M. A., C. K. S. Pillai, V. S. Prasad, and K. G. Satyanarayana. 1987. Effect of weathering on the mechanical properties of midribs of coconut palms. *J. Mater. Sci.* 22:3167–72.

Vincent, J. F. V. 1989. Relationship between density and stiffness of apple flesh. *J. Sci. Food Agr.* 47:443–62.

————. 1990a. *Structural biomaterials*. Princeton, NJ: Princeton University Press.

————. 1990b. Fracture in plants. *Adv. Bot. Res.* 17:235–82.

Vogel, S. 1998. *Cat's paws and catapults*. New York: Norton.

Weibull, W. 1939. *A statistical theory of the strength of materials*. Stockholm: Royal Swedish Institute.

Wheeler, W. N., and M. Neushul. 1981. The aquatic environment. In *Physiological plant ecology*, vol. 1, edited by O. L. Lange, P. S. Nobel, C. B. Osmond, and H. Ziegler, 229–47. Encyclopedia of Plant Physiology, n.s., vol. 12A. New York: Springer Verlag.

Woeste, F. E., S. K. Suddarth, and W. L. Galligan. 1979. Simulation of correlated lumber properties data: A regression approach. *Wood Sci.* 12:73–89.

The Effects of Geometry, Shape, and Size

Es ist dafür gesorgt, dass die Bäume nicht in den Himmel wachsen.
It is so arranged, that the trees do not grow into the heavens.
—Johann Wolfgang von Goethe, *Dichtung und Wahrheit*

The distinctions drawn among elastic, viscous, and viscoelastic materials in chapter 4 provide a starting point from which to explore the mechanical behavior of plants. Nevertheless, plants are not "materials." They are also structures whose geometry, shape, and size contribute in important (and often complex) ways to their mechanical performance. Thus far, we have paid little attention to the effects of geometry, shape, and size on biophysical phenomena, except with regard to the way stresses and strains are calculated. The objective of this chapter is to redress this omission by turning our attention to solid mechanics and the fundamental problem addressed by the theory of elasticity, which helps us to understand the spatial distribution and magnitudes of stresses within an object.

The difficulty that faces us when confronted with this task is formidable because plant structures are rarely geometrically simple, whereas most of engineering theory deals with highly idealized structures, such as columns, beams, and shafts that have uniform (homogeneous) material properties. In practice, engineers specify the material properties, geometry, shape, size, and loading conditions of real objects such that their properties largely conform to those of idealized beams, columns, or shafts. Our ability to treat the more complex situations encompassing the behavior of plants requires an understanding of what happens when an ideal support member is loaded and subsequently deforms in bending, compression, or torsion. With due regard to the extent to which material and geometric properties vary biologically, the mechanical behavior of many plant organs can be *approximated* by analogy with that of the idealized columns, beams, or shafts treated by elementary beam theory, which is the foundation of solid mechanics.

We will begin our treatment of elementary beam theory by considering the mechanical behavior of beams subjected to static and dynamic loads. Static loads are those that have long durations of application and tend not to change in magnitude or direction of application. In contrast, dynamic loads may have very short durations of application and typically change in magnitude and direction of application. Elementary beam theory treating static loading conditions generally focuses on practical and relatively simple situations, such as cantilevers or simply supported beams subjected to point loads or uniformly distributed loads. For plants, it is also useful to consider self-loading because the minimum state of stress for any object denser than its surrounding medium results from the mechanical forces exerted on and within it that are generated directly by the pull of gravity. Plants also experience natural dynamic loading conditions resulting from the mechanical forces generated by the motion of water or air. Dynamic beam theory tends to emphasize the harmonic motion of beams resulting from cyclic exchanges between kinetic and potential energy. Mathematically, this is a most demanding topic because it involves the consideration of differential equations. Fortunately, powerful tools such as the software program Mathematica are available to solve these equations.

5.1 Geometry and shape are not the same things

Before we begin our treatment of the effects of geometry, shape, and size on the mechanical behavior of plants, it is important to remember that geometry and shape are not the same things. Consider a cylinder with a circular cross section. All cylinders have the same geometry, which results when a cross section is translated along a straight line to produce a solid. However, not all objects with this geometry have the same shape. By *shape*, we mean the outward aspect of any object as defined numerically by a dimensionless ratio or quotient constructed using the dimensions of the object itself. Put differently, shape can be quantified without reference to any external measurement or standard. For example, the shape of any circular cylinder with length L and diameter D can be expressed by the quotient L/D. We can call this quotient the slenderness factor for any terete cylinder. Notice that this descriptor is numerically independent of size because it is dimensionless. Thus, two cylinders differing in size (measured, say, by volume or surface area) can have the same shape. Likewise, two cylinders with the same size can differ considerably in shape.

The foregoing may appear elementary, or even trivial, until one realizes that it shows unequivocally that it is possible for any organism, other than one that is perfectly spherical, to simultaneously change its geometry, shape, and size, or to change any one of these parameters independently, over the course of its growth and development. Similarly, in theory, any or all three variables can change over the course of a lineage's evolution, collectively or individually. The significance of this observation will become clearer as we proceed with the various topics treated in this chapter.

5.2 Pure bending

As discussed in chapter 4, the relationships among tensile, compressive, and shear forces and the stresses they produce are governed solely by the cross-sectional area of the object through which these forces operate. That is not the case in bending or torsion, in which the specific attributes of geometry, shape, and absolute size come into play. For example, a hollow tube with a ringlike cross section has a higher flexural stiffness than a solid bar with a circular cross section of the same area. As we will see in section 5.3, the geometric quantity that is responsible for this phenomenon is called the second moment of area.

We will start with the consideration of a cylindrical beam of length l with a circular cross section of radius r bent into a section of a circle by a moment M (fig. 5.1A,B). This condition is referred to as *pure bending*, provided that the beam does not experience simultaneous axial, shear, or torsional forces. In principle, we can bend the beam into a full circle with a radius R (fig. 5.1C). The convex surface of the beam will be under tension, while the concave surface of the beam will be in compression. However, if we examine any cross section through the beam, we can locate a line through the cross section where neither tension nor compression is experienced. Three-dimensionally, all of these "neutral lines" define a surface called the neutral plane, usually but not quite correctly referred to as the "neutral axis."

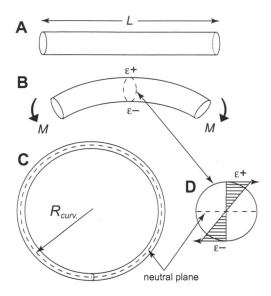

FIGURE 5.1. Tensile and compressive bending strains generated when a cylindrical beam composed of a homogeneous material (A) is subjected to a bending moment M (B). Tensile strains ($\varepsilon+$) occur along the convex surface; compressive stresses ($\varepsilon-$) occur along the concave surface. (C) If the beam with length L is bent into a perfect circle with a radius of curvature R_{curv}, the centroid axis of the beam precisely aligns with the neutral plane. The neutral plane is the surface running through the beam along which tensile and compressive strains are zero (the magnitudes of the tensile and compressive strains are indicated by the lengths of the arrows shown in D).

The strains (fig. 5.1D) that will develop anywhere within any cross section through the bent beam are determined by specifying the distance a from the neutral plane for the point at which the strain ε is measured. Since $L = 2\pi R$ and $-r \le a \le r$, it follows that the strain ε is given by the formula

(5.1)
$$\varepsilon = \frac{2\pi(R+a)-2\pi R}{2\pi R} = \frac{a}{R},$$

where R is the radius of the centroid axis, which is defined as the longitudinal axis created by connecting the center of mass of sequential cross sections through the beam.

If the beam consists of a homogeneous material that behaves as a linear elastic solid, the stress σ can be expressed as

(5.2)
$$\sigma = E_L \varepsilon = E_L \frac{a}{R},$$

where E_L is the modulus of elasticity in the longitudinal direction. In this notation, tensional stresses are positive ($0 < a \le r$) and compressive stresses are negative ($-r \le a < 0$).

Bending a beam into a full circle is not a necessary condition for demonstrating these principles because we can bend the beam into some arc of a circle and still define mathematically the resulting stresses and strains by replacing the radius of the circle with the radius of curvature, which is the inverse of curvature. The arc resulting from bending the beam, which is called the bending line, may be represented in two dimensions as $y = f(x)$, where the x-axis points in the longitudinal direction and the y-axis points in the direction of the applied bending force. In this coordinate system, which will be used throughout this chapter, the curvature is defined as

$$(5.3) \qquad curvature = \frac{1}{R_{curvature}} = \frac{\left| \dfrac{d^2 f(x)}{dx^2} \right|}{\left[1 + \left(\dfrac{df(x)}{dx} \right)^2 \right]^{3/2}},$$

where $R_{curvature}$ is the radius of curvature. For beams extending in the x-direction and experiencing small displacements as a result of an applied load (i.e., small deflections), we realize that $df(x)/dx \ll 1$, such that the curvature can be approximated by the second derivative.

5.3 The second moment of area

In order to relate stress and strain to the applied bending moment, we have to consider the force and the moment acting on an infinitesimally small element dA in the cross section at any distance a from the neutral plane (fig. 5.2). Applying equation (5.2), this force is given by the formula

$$(5.4) \qquad force = \sigma \, dA = E_L \frac{a}{R} dA,$$

and the moment of this force about the x-axis is given by

$$(5.5) \qquad moment = force \cdot a = E_L \frac{a^2}{R} dA.$$

In turn, the total moment M over the entire cross section is given by integrating over the entire area of the cross section:

$$(5.6) \qquad M = \frac{E_L}{R} I = \frac{E_L}{R} \int_{area} a^2 dA.$$

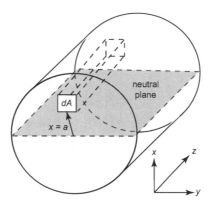

FIGURE 5.2. Section of a cylindrical beam showing the neutral plane and an infinitesimally small unit area (dA) located at a distance $x = a$ from the neutral plane. The contribution made by this unit area to the second moment of area (I) equals a^2dA.

This integral is called the second moment of area, denoted conventionally by I. By definition, I is given by the formula

(5.7)
$$I = \int_{area} a^2 dA,$$

which can be generalized for any cross-sectional area. Notice that I invariably has units of length raised to the fourth power (fig. 5.3).

The product of the elastic modulus E and the second moment of area I is flexural stiffness EI, which is sometimes called flexural rigidity. EI has units of force times area. For beams that consist of different materials with different moduli of elasticity, such as plant stems, overall stiffness can be evaluated as the sum of the flexural stiffnesses of the individual structural elements, tissues, or cell types:

(5.8)
$$E_{av} \sum_i I_i = \sum_i E_i I_i.$$

It may not always be possible, however, to determine the moduli for all of the individual components i. The measurement of a segment of a plant stem may therefore yield only an average modulus of elasticity E_{av}, which has been called the structural modulus of elasticity (see also the terminology of Niklas 1989a). Under some very fortuitous circumstances, the mechanical properties of individual components can be determined by subtraction.

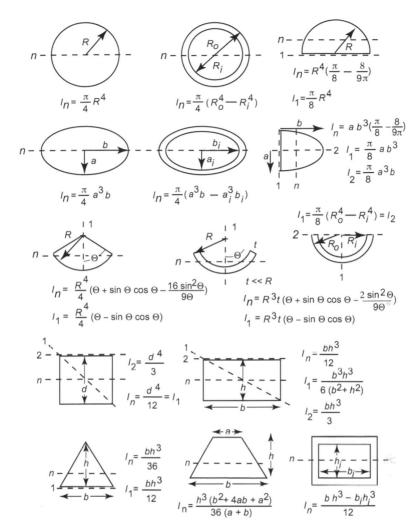

FIGURE 5.3. Formulas for computing the second moment of area (I) for an assortment of simple geometries. Note that the formula for I depends on the plane of bending (indicated by dashed lines, where n refers to the neutral plane, which is defined as the plane for which I is at its minimum. For symmetrical cross sections, the neutral plane is also the plane of symmetry). Additional formulas for I are provided by Niklas (1992).

5.4 Simple bending

Combining equations (5.1) and (5.6) leads to a relationship between the bending moment and its resulting strain:

$$(5.9) \qquad\qquad M = E_L I \frac{\varepsilon}{a}.$$

Equation (5.9) is of fundamental importance when dealing with bending. In the linear elastic range, it can be expressed in terms of stresses:

$$(5.10) \qquad\qquad M = I \frac{\sigma}{a}.$$

Pure bending is observed only if the radius of curvature for the segment of a beam under investigation is constant, which, provided that E_L and a are constant, implies that σ is also constant (see eq. [5.2]). Pure bending, however, can be realized approximately when slender beams or bars with a large span-to-depth quotient (i.e., those with large slenderness factors) are bent. In three-point bending (see figure 9.4A), span is measured as the distance between two supports, and depth is taken as the dimension measured in the direction of the applied force. For a cylindrical beam, depth is simply diameter D. Some typical cases of bending of slender beams are treated in boxes 5.1 and 5.2.

BOX 5.1 **Bending of slender cantilevers**

A cantilever is a beam fixed in a solid support at one end and free to deflect at the other. The position of the support is taken as the origin. The cantilever may be loaded with a point force F at its free end. If l is the length between the support and the point of loading, the bending moment M at any point along the cantilever at distance x from the support is given by

$$(5.1.1) \qquad\qquad M = F(l - x).$$

Thus, the moment increases linearly from the point of loading to the fixed end. If the flexural stiffness of the beam is uniform, the curvature will also increase linearly. Assuming, for example, that the unloaded beam was horizontal, the curvature of the beam is approximated by the second derivative of the bending line (see eq. [5.3]) such that

BOX 5.1 **(Continued)**

(5.1.2)
$$M = \frac{EI}{R} = EI\frac{d^2 f(x)}{dx^2} = F(l-x).$$

The bending line $f(x)$ is found by integration:

(5.1.3)
$$f(x) = \frac{F}{EI}\left(\frac{lx^2}{2} - \frac{x^3}{6}\right).$$

The maximum deflection y_{max} is found at the free end, where $x = l$:

(5.1.4)
$$y_{max} = f(l) = \frac{F}{EI}\frac{l^3}{3}.$$

The same cantilever loaded with a uniform weight per unit length w can be treated by the same integration method. Here, we also find the maximum deflection at the free end, as given by the formula

(5.1.5)
$$y_{max} = f(l) = \frac{w}{EI}\frac{l^4}{8}.$$

An example of this case, which we intentionally oversimplify (by assuming that EI is uniform), is a horizontal branch that is self-loaded by its own weight. Numerical integration may be necessary to compute the deflection of a real branch.

BOX 5.2 **Three-point bending of slender beams**

One method of exploring the mechanical properties of plant specimens, or segments thereof, is three-point bending (see chap. 9). In engineering, this type of arrangement is known as "a simply supported beam" with a central concentrated load. That is, the beam rests on two supports with span length l and a force F applied centrally between the two supports. Taking $l/2$ as the origin and noting that each of the supports takes a load of $F/2$, the bending moment at every point between the supports is given by the formula

BOX 5.2 **(Continued)**

(5.2.1)
$$M = EI \frac{d^2 f(x)}{dx^2} = \frac{F}{2}\left(\frac{l}{2} - x\right).$$

Double integration of equation (5.2.1) yields the formula

(5.2.2)
$$f(x) = -\frac{F}{EI}\left(\frac{lx^2}{8} - \frac{x^3}{12}\right).$$

If the beam is placed horizontally before the load is applied, the midpoint deflection is the vertical displacement from the original location of the beam at the point at which the force is applied:

(5.2.3)
$$y_{max} = f(x = l/2) - f(x = 0) = \frac{Fl^3}{48EI}.$$

When the same beam is uniformly loaded, we find that the displacement is given by the formula

(5.2.4)
$$y_{max} = \frac{5wl^4}{384} \frac{F}{EI}.$$

Additional examples are described by Stephens (1970).

Thus far, we have considered the tensile and compressive stresses that result in bending using the abstraction called pure bending to understand bending when shearing and torsion are ignored (i.e., simple bending). However, when the slenderness factor is small, as in the case of short or thick beams, shear components have to be considered. This topic will be treated in the next section, where it will become apparent what large or small slenderness factors actually mean.

5.5 Bending and shearing

The preceding section drew attention to the fact that shear stress components can be largely or totally neglected when beams with large slenderness factors are bent, but that is not true in the case of beams with small slenderness factors, in which shear stresses can become highly significant. This is a consequence of the conditions of equilibrium (discussed in section 4.9). In the case of simple bending, the equilibrium condition (eq. [4.22]) can be simplified to

(5.11)
$$\frac{\partial \sigma_x}{\partial x} + \frac{\partial \tau_{xy}}{\partial y} = 0.$$

Accordingly, if the bending stresses are not constant along the length of the beam, shear stresses will also appear.

To appreciate this phenomenon, consider a beam with a rectangular cross section with breadth b and width w, and an element $dx \cdot da \cdot b$ within the beam at a distance a from the neutral plane (fig. 5.4). When a bending moment M acts on one end of this element and a bending moment $M + dM$ acts on the other end, the corresponding normal stresses are given according to equation (5.10) as

(5.12a)
$$\sigma_x = \frac{M}{I} a$$

and

(5.12b)
$$\sigma_x + d\sigma_x = \frac{M + dM}{I} a.$$

Equation (5.12) leads to an expression for the gradient of the normal stress,

(5.13)
$$\frac{d\sigma_x}{dx} = \frac{dM}{dx} \frac{1}{I} a,$$

whereas using equation (5.11) leads to an expression for the gradient of the shear stress,

(5.14)
$$\frac{d\tau_{xa}}{da} = -\frac{d\sigma_x}{dx} = -\frac{dM}{dx} \frac{1}{I} a.$$

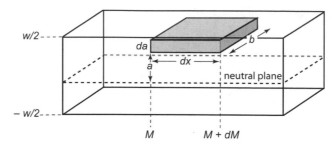

FIGURE 5.4. An infinitesimally small element with length dx, width da, and breadth b, placed at a distance a from the neutral plane, is subjected to a bending moment M at position x and to a bending moment $M + dM$ at position $x + dx$ in the longitudinal direction.

The shear stress in the x–z plane at height h is obtained by integration:

$$(5.15) \qquad \tau_{xa} = -\frac{dM}{dx}\frac{1}{I}\int_{-w/2}^{h} a\, da = \frac{dM}{dx}\frac{1}{2I}[(w/2)^2 - h^2],$$

where h is a fixed position on the a-axis with $-w/2 \le h \le w/2$ and $h = 0$ at the middle of the beam.

Introducing the appropriate expression for the second moment of area of a rectangular cross section (see fig. 5.3), we see that

$$(5.16a) \qquad \tau_{xa} = \frac{dM}{dx}\frac{6}{bw^3}[(w/2)^2 - h^2].$$

For a beam with a circular cross section with radius r, the shear stress in the x–z plane at height h is given by the formula

$$(5.16b) \qquad \tau_{xa} = \frac{dM}{dx}\frac{4}{3\pi}\frac{1}{r^4}(r^2 - h^2)$$

(Stephens 1970). Equations (5.15) and (5.16) show that the shear stresses in a beam have a parabolic distribution over its width, with a maximum at the neutral plane in the middle of the beam. Two examples illustrating the foregoing principles are described in boxes 5.3 and 5.4.

The situation becomes a little more complex when we consider the stresses that occur in a fiber that is helically wound within a beam that is bent. This situation is important not only in the material sciences dealing with fibrous composites, but also in plant biology, because the walls of

BOX 5.3 **Bending and shearing of a cantilever**

Consider the shear deflection y_s of a cantilever with a rectangular cross section having length l with breadth b and width w in the direction of a force F acting on its free end. The bending moment M is simply $M = Fx$, where x is the distance from the free end in the longitudinal direction. Therefore, the gradient of the bending moment is given by

(5.3.1)
$$\frac{dM}{dx} = F.$$

Inserting this relation into equation (5.16a), we see that the shear stress in an element with volume $dV = lbdh$ at distance h from the neutral plane is given by the formula

(5.3.2)
$$\tau = F \frac{6}{bw^3}\left[(w/2)^2 - h^2\right],$$

and that the shear energy in the same element is given by the formula

(5.3.3)
$$energy = \tau y\, dV = \tau y l b\, dh.$$

Within the linear elastic range, this energy can be expressed as

(5.3.4)
$$energy = \frac{\tau^2}{2G} l b\, dh,$$

where G is the shear modulus (see eq. [4.16]). Combining equations (5.3.2) and (5.3.4) leads to the formula

(5.3.5)
$$energy = \left\{F \frac{6}{bw^3}\left[(w/2)^2 - h^2\right]\right\}^2 \frac{l b}{2G}\, dh.$$

Finally, the total shear energy in the cantilever is obtained by the integration of equation (5.3.5):

BOX 5.3 **(Continued)**

(5.3.6) $total\ energy = 2F^2 \dfrac{l}{2G}\dfrac{36}{bw^6} \displaystyle\int_0^{w/2}\left[(w/2)^2 - h^2\right]^2 dh$

or

(5.3.7) $total\ energy = \dfrac{3}{5}\dfrac{F^2}{G}\dfrac{l}{bw}.$

This energy is equal to the work done by the load F moving through the shear deflection y_s:

(5.3.8) $\dfrac{1}{2}F\,y_s = \dfrac{3}{5}\dfrac{F^2}{G}\dfrac{l}{bw}.$

Thus, the deflection of the cantilever due to shear is given by the formula

(5.3.9) $y_s = \dfrac{6}{5}\dfrac{F}{G}\dfrac{l}{bw}.$

Because shear deflections and bending deflections are additive, the total deflection at the free end is found by combining equations (5.1.4) and (5.3.9). We find that the deflection is given by the formula

(5.3.10) $y_{sum} = y_{bend} + y_{shear} = 4\dfrac{F}{E}\dfrac{l^3}{bw^3} + \dfrac{6}{5}\dfrac{F}{G}\dfrac{l}{bw},$

where the second moment of area I is expressed according to equation (5.7).

In passing, equation (5.3.10) takes the following form if the cantilever has a circular cross section with a uniform radius r:

(5.3.11) $y_{sum} = y_{bend} + y_{shear} = \dfrac{4}{3\pi}\dfrac{F}{E}\dfrac{l^3}{r^4} + \dfrac{5}{9\pi}\dfrac{F}{G}\dfrac{l}{r^2}.$

BOX 5.4 **Bending and shearing of a simply supported beam**

In three-point bending, the shear deflection of a simply supported beam with the same rectangular cross section as for the cantilever in box 5.3 subjected to a central concentrated load F is given by

(5.4.1)
$$y_s = \frac{3}{10} \frac{F}{G} \frac{l}{bw},$$

where the second moment of area I is expressed according to equation (5.7).

As indicated in box 5.3, shear and bending deflections are additive. Therefore, the deflection in three-point bending is obtained by combining equations (5.2.3) and (5.4.1). We find that the total deflection is given by the formula

(5.4.2)
$$y_{sum} = y_{bend} + y_{shear} = \frac{1}{4} \frac{F}{E} \frac{l^3}{bw^3} + \frac{3}{10} \frac{F}{G} \frac{l}{bw}.$$

For large span-to-depth quotients, the first term in this formula will dominate; the second term will contribute significantly only for small span-to-depth quotients. The following formula shows the relationship between l/w and E/G for which shear deflection contributes only 5% of the total deflection:

(5.4.3)
$$l/w = 4.8\sqrt{E/G}.$$

For composites, in which the ratio E/G can be on the order of 20, it is advisable to use span-to-depth quotients larger than 20 to avoid significant contributions from shear deflections.

On the other hand, a three-point bending experiment can be used to determine both E and G by noting that equation (5.4.2) can be rearranged into the form

(5.4.4)
$$\frac{1}{E_{app}} = \frac{1}{E} + \frac{6}{5} \frac{1}{G} \frac{w^2}{l^2} \text{ with } E_{app} = \frac{F}{y_{sum}} \frac{l^3}{4bw^3}.$$

Thus, by changing the distance between the supports holding a beam and plotting $1/E_{app}$ according to equation (5.4.4) as a function of $(w/l)^2$, E can be determined by extrapolating $(w/l)^2$ to 0, and G can be calculated from the slope of this linear plot (Spatz et al. 1996).

BOX 5.4 **(Continued)**

For a beam with a circular cross section and uniform radius r, equations (5.4.2)–(5.4.4) take the following forms:

$$(5.4.5) \qquad y_{sum} = y_{bend} + y_{shear} = \frac{1}{3\pi} \frac{F}{E} \frac{l^3}{r^4} + \frac{4}{27\pi} \frac{F}{G} \frac{l}{r^2},$$

$$(5.4.6) \qquad l/w = 2.95\sqrt{E/G},$$

$$(5.4.7) \qquad \frac{1}{E_{app}} = \frac{1}{E} + \frac{4}{9} \frac{r^2}{Gl^2} \quad \text{with } E_{app} = \frac{F}{y_{sum}} \frac{l^3}{3\pi r^4}.$$

many cell types contain helically wound strands of polymers such as cellulose. Consider the simple biological case of a sclerenchyma cell's wall, in which cellulose fibers are embedded and wound around the cell in the form of a helical spiral. An important structural parameter determining the mechanical performance of this kind of cell wall is the microfibrillar angle α; that is, the angle between the fiber and the longitudinal axis of the cell. For small microfibrillar angles ($\alpha \ll 1$), the stiffness of a sclerenchyma cell depends, among other things, on the ratio $1/\alpha^4$ (see box 5.5). Consequently, a very small difference in the microfibrillar angle can change the stiffness of this kind of cell significantly.

BOX 5.5 **The influence of the microfibrillar angle on the stiffness of a cell**

Consider a cylindrical cell with length L, radius R, and helically spiraling inextensible (cellulose) fibers of length B within its wall such that the number of helical turns is given by Ω. The fibers may be the main strengthening component of the cell wall. If these fibers are constrained so that they cannot twist, the number of turns along each fiber cannot change upon straining of the cell.

We will first consider a single helix. The geometric relationships

$$(5.5.1) \qquad \frac{B^2}{\Omega^2} = \frac{L^2}{\Omega^2} + (2\pi R)^2 \quad ; \quad 2\pi R \frac{\Omega}{B} = \sin \alpha$$

BOX 5.5 **(Continued)**

are those described in box 4.2 (see fig. 4.2.1). The curvature of each fiber *Curv*, which is the reciprocal of the radius of curvature, is given by

$$Curv = \frac{1}{R_{curv}} = \frac{R}{R^2 + \left(\dfrac{1}{2\pi}\dfrac{L}{\Omega}\right)^2} = 4\pi^2 \frac{\Omega^2}{B^2} R.$$

(5.5.2)

Upon extension of the cell, the spiral will increase in length from L to $(L + \Delta L)$, and the radius will decrease from R to $(R - \Delta R)$. Likewise, the curvature will decrease from *Curv* to $(Curv - \Delta Curv)$, with

(5.5.3)
$$\Delta Curv = 4\pi^2 \frac{\Omega^2}{B^2} \Delta R.$$

Noting that, in its undeformed state, the spiral has length L and radius R, we see that the force F necessary to extend the spiral by ΔL is given by

(5.5.4)
$$F = -EI\frac{1}{R}\Delta Curv = -4\pi^2 EI\frac{\Omega^2}{B^2}\frac{\Delta R}{R},$$

where EI is the stiffness of the fiber. According to equation (4.2.7), the change in radius is related to the change in length by

(5.5.5)
$$\frac{\Delta R}{R} = -\frac{\Delta L}{L}\frac{1}{\tan^2\alpha}.$$

By combining equations (5.5.4) and (5.5.5), we can express the force as a function of the strain $\varepsilon = \Delta L/L$ such that

(5.5.6)
$$F = 4\pi^2 EI\frac{\Omega^2}{B^2}\frac{1}{\tan^2\alpha}\frac{\Delta L}{L} = 4\pi^2 EI\frac{\Omega^2}{B^2}\frac{1}{\tan^2\alpha}\varepsilon.$$

In the case of a cell with N fibers, equation (5.5.6) can be extended to take the form

(5.5.7)
$$F = 4\pi^2 NEI\frac{\Omega^2}{B^2}\frac{1}{\tan^2\alpha}\varepsilon.$$

BOX 5.5 **(Continued)**

Dividing the force by the cross-sectional area of the cell yields the tensile stress $\sigma_{tensile}$ in the cell wall, which has to be distinguished from the bending stress in the fiber:

$$(5.5.8) \qquad \sigma_{tensile} = \frac{F}{\pi R^2} = 4\pi NEI \frac{\Omega^2}{B^2} \frac{1}{R^2} \frac{1}{\tan^2 \alpha} \varepsilon.$$

Using the geometric relationship given by equation (5.5.1), the radius R can be expressed as a function of B and the microfibrillar angle α:

$$(5.5.9) \qquad \sigma_{tensile} = 16\pi^3 NEI \frac{\Omega^4}{B^4} \frac{1}{\sin^2 \alpha} \frac{1}{\tan^2 \alpha} \varepsilon.$$

The modulus of elasticity of the cell is given by

$$(5.5.10) \qquad E_{cell} = \frac{\sigma_{tensile}}{\varepsilon} = 16\pi^3 NEI \frac{\Omega^4}{B^4} \frac{1}{\sin^2 \alpha} \frac{1}{\tan^2 \alpha}.$$

For small microfibrillar angles, $\sin \alpha$ and $\tan \alpha$ can be approximated by α. With an error of less than 5% for $\alpha < 20°$, the modulus of elasticity for uniaxial extension of the cell can therefore be expressed as

$$(5.5.11) \qquad E_{cell} = 16\pi^3 NEI \frac{\Omega^4}{B^4} \alpha^{-4}.$$

Formula 5.5.11 is derived here for extension of the cell; however, for reasons of symmetry, it also applies to uniaxial compression.

5.6 Fracture in bending

When bending extends beyond the linear range of elastic behavior, plant materials will usually undergo plastic deformation. In green wood, it is easy to observe that this deformation is accompanied by the buckling of fibers on the compression side. The final fracture initiated on the tension side, however, is a complex process, which can be understood only if the

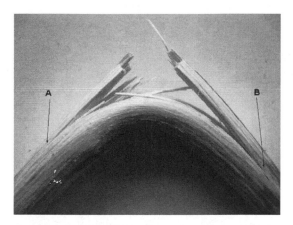

FIGURE 5.5. Fracture in bending of a 4-year-old axis of *Corylus avelana*. Cracks in longitudinal direction extend to points A and B. Between A and B, the second moment of area is smaller than in the intact sections of the axis.

anisotropy and the heterogeneity of wood (particularly the mechanical differences between early and late wood) are taken into consideration.

A crack on the tension side is usually initiated at the site of a discontinuity. It will propagate a short distance in the radial direction, but it will change direction and run some distance in the longitudinal direction, whereby the material splits open to an extent depending on the magnitude of the bending moment and the material properties of the axis (fig. 5.5). Upon further bending, the crack may run further in a radial direction, again followed by an extension in the longitudinal direction.

Shearing is not the primary cause of this mechanical behavior because the aforementioned sequence of events is observed even under conditions in which shear stresses are negligible. As expected from the Cook-Gordon mechanism (see section 4.17), crack propagation in the radial direction is likely to be stopped at the interface between late wood and early wood. Crack propagation in the direction of the fibers requires much less energy than crack propagation orthogonal to the fibers and acquires energy from the release of bending energy from those parts of the axis that split open.

The process of crack propagation inevitably leads to a much smaller second moment of area I in the section of the axis where parts have already split compared with that in the intact axis. This reduction in I, in turn, causes a change in the bending line. The fractured section has a larger curvature, and the adjoining parts of the axis are less curved, than immediately before the initiation of fracture. This change has two consequences. First, the bending energy in the adjoining parts of the fracture is reduced and will eventually be insufficient to propagate the crack in the longitudinal direction. Therefore, the crack will stop at some distance from where the fracture was initiated. Second, due to the larger curvature at the site of fracture, tension will be sufficient to initiate additional radial cracks such that the process will continue as long as the bending moment is applied.

Young axes of *Salix fragilis* are one exception to this generality. Upon bending, they display brittle fracture with a smooth fracture surface in a tightly circumscribed plane near their point of origin on larger stems (Beismann et al. 2000). To our knowledge, the structural basis of this anomalous mechanical behavior has not been elucidated.

5.7 Torsion

A different type of loading is experienced if an object is subjected to a torque. The stem of a tree with an asymmetric crown may be twisted in the wind, or a branch may be twisted when self-loading operates eccentrically. If a beam is subjected to a pure torque, not accompanied by bending, tension, or compression, every cross section is in a state of pure shear. This condition can be seen by inspecting figure 5.6. The angle of twist θ is

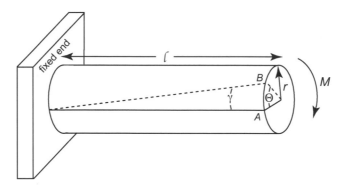

FIGURE 5.6. A bar with a circular cross section and uniform radius r, fixed at one end and twisted at the other through an angle θ by a torsional moment (M). The distortion can be expressed as a shear strain γ in a tangential-longitudinal coordinate system.

equivalent to a shear strain γ, by which the longitudinal-tangential surface is distorted. The circular arc between points A and B has a length given either by $l\gamma$ or $r\theta$; that is,

(5.17) $$l\gamma = r\theta.$$

To relate shear strains and the angle of twist to the applied torque, we have to make four assumptions:

1. The material is homogeneous and elastic.
2. Strains and stresses do not exceed the limit of proportionality.
3. Radial lines remain radial after twisting.
4. Plane cross sections remain planar after twisting.

Assumption 4 is justified only for shafts with the geometry of a cylindrical column. This case is treated first.

In the linear elastic range (assumption 2), shear stresses and shear strains are related through the shear modulus G (eq. [4.16]). Together with equation (5.17), this leads to the following relationship between shear stress and angle of twist:

(5.18) $$\frac{\tau_{TL}}{\theta} = G\frac{r}{l},$$

where τ_{TL} is the stress immediately beneath the surface of the column.

From assumption 3, it follows directly that the strain is proportional to the distance from the neutral axis (i.e., the center of rotation), which in our case is the line connecting the midpoints of the cross sections. If γ_r is the strain at radius r, then the strain in any element dA at a distance $a < r$ from the neutral axis is given by the formula

(5.19) $$\gamma_a = \frac{a}{r}\gamma_r,$$

and in the linear elastic range, the shear stress τ_a in the same element is given by the formula

(5.20) $$\tau_a = \frac{a}{r}\tau_{TL}.$$

Noting that the element dA is subjected to a torque dT,

$$(5.21) \qquad dT = \tau_a \, a \, dA = \tau_{TL} \frac{a}{r} a \, dA,$$

we see that the total torque T is obtained by integrating over the entire area of the cross section:

$$(5.22) \qquad T = \frac{\tau_{TL}}{r} \int_{area} a^2 \, dA = \frac{\tau_{TL}}{r} J.$$

The integral denoted by J is called the polar moment of inertia. It is an analogue to the second moment of area I (see eq. [5.7] and fig. 5.3). The difference between J and I revolves around the meaning of a. In equation (5.22), a denotes the distance from the center of rotation, whereas in equation (5.7), a is the distance from the neutral plane. Combining equation (5.17) with the definition of J (eq. [5.22]) gives a general equation for torsion:

$$(5.23) \qquad \frac{T}{J} = \frac{\tau}{r} = G\frac{\theta}{l}.$$

By analogy to flexural stiffness, the product of the shear modulus G and the polar moment of inertia J is called the torsional stiffness (symbolized by C). Like the flexural stiffness EI, the torsional stiffness $C = GJ$ has units of force times area. Within the linear elastic range, the energy of torsion can be calculated from the equation

$$(5.24) \qquad Energy = \frac{1}{2}T\theta = \frac{\tau^2}{2G}J\frac{l}{r^2}.$$

For beams with no circular symmetry, planar cross sections are warped upon torsion (Timoshenko and Goodier 1970, 292). Therefore, J in equations (5.23) and (5.24) must be replaced by an effective polar moment of inertia J_{eff} (Young 1989). With the exception of cylindrical beams, J_{eff} is invariably smaller than J. For beams with rectangular cross sections with small and large sides of length a and b, the numerical value of J_{eff} can be computed from the formula

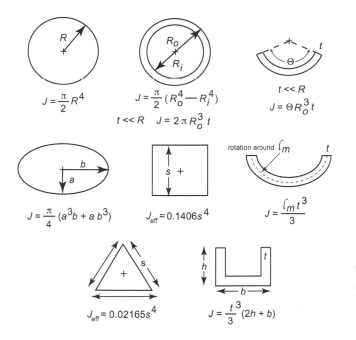

FIGURE 5.7. Formulas for computing the polar moment of inertia J for some simple geometries. Formulas for some additional geometries are provided by Niklas (1992).

(5.25a) $$J_{eff} = k_1 a^3 b.$$

The function k_1 can be approximated as

(5.25b) $$k_1 = -0.1296[\log(b/a)]^2 + 0.3168 \log(b/a) + 0.1406$$

(see Timoshenko and Goodier 1970, 313). For a quadratic cross section with $a = b$, we obtain $k_1 = 0.1406$. Additional values for J_{eff} are provided in figure 5.7 (see also Young 1989).

5.8 Static loads

For terrestrial plants, mechanical loads are in most cases gravitational forces and drag forces. For aquatic plants, gravitational forces can be largely neglected; drag forces in moving water are the main loads experienced. While it is obvious that gravitational forces are static loads, wind loads generally have static and dynamic components with a whole spectrum of frequencies. Static loads are defined here as loads having a much lower frequency than the eigenfrequency of the plant. The responses of plants to dynamic loads and the meaning of eigenfrequency will be treated in section 5.12.

A comparison of the forces on upright or slightly tilted stems of terrestrial plants resulting from self-loading and those resulting from a wind-induced bending moment shows that under most circumstances, wind forces are far more important in terms of mechanical stability and survival. This can be easily demonstrated using a simplified model of an average tree (box 5.6). More detailed accounts of such considerations lead to the same conclusions (Spatz and Brüchert 2000). An account of aerodynamic drag on trees and their resistance to stem breakage and uprooting in relation to forest management is given by Peltola (2006). Some details of root anchorage were treated in section 2.9.

BOX 5.6 **Comparison of forces on a tree trunk resulting from self-loading with those experienced in bending**

Consider a tree tilted at an angle α with a trunk of constant radius r up to the midpoint of the crown at height H. The crown has a radius R and weight W_{crown}. For simplicity, the force represented by the crown weight W_{crown} is approximated as a point load acting on the crown's midpoint. The stem has a weight per unit length w of

(5.6.1)
$$w = \rho g \pi r^2,$$

where ρ is bulk trunk density. The compressive force acting on the trunk at any point below the midpoint of the crown is given by the formula

(5.6.2)
$$F_{weight} = [W_{crown} + w(H-h)]\cos\alpha,$$

where h is the height at this point above ground. For a tree of 20 m stem height with a stem radius $r = 0.25$ m, we obtain $w \approx 2{,}000$ N/m, assuming a green wood density of $\rho \approx 1{,}000$ kg/m³ and a total trunk weight $W_{stem} \approx 40{,}000$ N. If we let the crown weight $W_{crown} = W_{stem}/2$, we find that at the trunk's base, the compressive force equals $60{,}000 \cos\alpha$ (in units of N) with a corresponding stress of $\sigma_{weight} = F_{weight}/\pi r^2 \approx 3 \times 10^6$ $\cos\alpha$ (in units of N/m²).

BOX 5.6 **(Continued)**

The bending moment below the midpoint of the crown at height h exerted by W_{crown} is

(5.6.3) $$M_{crown} = W_{crown}(H-h)\sin\alpha,$$

and the bending moment exerted at height h by a trunk segment of width dy at height y is

(5.6.4) $$moment = w\,dy\,(y-h)\sin\alpha, \text{ with } H > y > h.$$

Because the weight-force of the crown is taken as a point load, the total bending moment $M_{bending}$ at height h is found by integration:

(5.6.5) $$M_{bending} = \left(M_{crown} + w\frac{(H-h)^2}{2}\right)\sin\alpha.$$

At ground level, where $h = 0$, equation (5.6.5) simplifies to

(5.6.6) $$M_{bending} = \left(W_{crown}H + w\frac{H^2}{2}\right)\sin\alpha.$$

According to equation (5.12), this bending moment must be balanced by a bending moment within the stem,

(5.6.7) $$M_{stem} = \sigma\frac{I}{r} = \sigma\frac{\pi}{4}r^3,$$

where σ is the maximum stress at the outside of the stem. At ground level, $M_{bending} = 0.8 \times 10^6 \sin\alpha$ (in units of Nm) and $\sigma = 65 \times 10^6 \sin\alpha$ (in units of N/m²). Comparing the stresses, we find a quotient $\sigma_{bending}/\sigma_{weight} \approx 20\sin\alpha/\cos\alpha$. Thus, in this simplified example, we see that the maximum bending stresses in a leaning tree are larger than the compressive stresses due to the tree's weight when $\alpha \geq 3°$. For tapered stems, the forces due to the weight of the stem can be found by integration and the moments by double integration, following the same line of reasoning (Spatz and Brüchert 2000).

Finally, we compare the stresses resulting from self-loading with those resulting from wind loads acting on the crown of the same tree. In this simplified model, we neglect wind speed changes with height (see section 6.9) and assume a steady wind speed U that results in a drag force of $F_{wind} = 0.5\,\rho\,A_{sail}U^2C_D$ (compare

BOX 5.6 **(Continued)**

eq. [2.15]) acting at the midpoint of the crown. The density of air at atmospheric pressure and 20°C is ρ = 1.205 kg/m³. The drag coefficient can conservatively be estimated as C_D = 0.5, and the sail area can be calculated as $A_{sail} = \pi R^2$, where R = 10 m is the radius of the crown, taken here as half the stem height. Numerically, this leads to F_{wind} = 95U^2. This force will exert a bending moment at the base of the stem of

(5.6.8) $M_{wind} = F_{wind} H \cos \alpha = 1.9 \times 10^3 U^2 \cos \alpha.$

For small angles α and moderate wind speeds of 10 m/s, we obtain a wind-induced bending moment of $M_{wind} \approx 0.19 \times 10^6$ in units of Nm and a corresponding stress at the surface of the stem base of $\sigma \approx 15 \times 10^6$ N/m², comparable to the bending moment if the same tree were to lean with an angle $\alpha = 15°$.

For tapered stems and vertical gradients of wind speed, the calculations would have to be modified. Nevertheless, the general conclusion is that under most circumstances, wind forces on a tree are the most important mechanical loads.

A particularly interesting mechanical adaptation to wind forces is seen in the sedge *Carex acutiformis* (Ennos 1993). The lowland form of this species has a stem with a triangular cross section. Lignified material around the vascular bundles provides stiffness against bending. The slender, tapered stem carries the seed head, but the tip sags appreciably under this weight. Because the strands of strengthening tissues are separated from one another, the stem has very low torsional rigidity. The ratio of the flexural stiffness (EI) to the torsional stiffness (GJ) of stems ranges between 20 and 100. This allows stems to twist in the wind in a way that reduces drag forces, which might otherwise lead to failure by breakage or local buckling. Another case of extensive and reversible reconfiguration in high wind is reported for the lightweight petioles of bananas, which possess enough rigidity to hold huge leaves, yet possess low torsional stiffness, which allows twisting of the leaves. In addition, the two halves of the leaves can fold together downwind once the petiole has twisted away from the wind (Ennos et al. 2000). Algae have nonlignified tissues and, correspondingly, a modulus of elasticity in the range of only a few MN/m^2. Yet their breaking strains can be as high as 50% (Harder et al. 2006). This flexibility confers a considerable advantage (Vogel 1998) in that the plants can reconfigure, and thus significantly reduce, the drag forces and bending moments in the flow.

In summary, even though the polar moment of inertia for most geometries is significantly larger than the second moment of area (i.e., $J > I$), the torsional stiffness of most structures is much less than their flexural stiffness (i.e., $JG \ll EI$) because shear moduli are very much smaller than elastic moduli for almost any kind of material (i.e., $G \ll E$) (Niklas 1992).

5.9 The constant stress hypothesis

Very early on in the history of plant biomechanics, Metzger (1893), using measurements of wind profiles and tree taper, concluded that the magnitudes of wind-induced stresses are relatively constant along the lengths of tree branches and trunks. (It is implicitly understood that these stresses are averaged over days, month, or years, since wind speed and direction change frequently.) This constant stress hypothesis is still entertained in the literature in the context of both wind loads and self-loads (see McMahon 1973; King and Loucks 1978; Morgan and Cannel 1994). Mattheck (1992) has used the constant stress hypothesis to establish boundary conditions for simulations of tree growth using finite element analyses (see section 10.5). However, it is difficult to examine this approach critically without knowing how the stresses (particularly wind-induced stresses) experienced by trees are time-averaged. By comparing different trees in the same location, and presumably under the same wind regimes, several investigators have shown that the observed shapes of trees are not invariably compatible with the hypothesis of constant stress (e.g., Bertram 1989; Niklas and Spatz 2000). Nevertheless, it seems fair to say that at present, the hypothesis can be neither proved nor disproved.

From a theoretical perspective, it is interesting that the assertion of constant stress leads to predictions of optimal tree design, provided that the distribution of mechanical loads, particularly wind loads, is known. This is demonstrated by calculating the shapes of stems for two different kinds of trees under the premise of a constant stress distribution (box 5.7). These calculations show that the local gradient of wind loads is the determining factor for the optimal tapering of stems. This dependence on the local wind profile shows again that the application of the constant stress hypothesis requires knowledge of the distribution of loads averaged over an as yet unknown span of time. It can be speculated that the adequate averaging time will increase with the age of the tree and will be different for different species.

BOX 5.7 **Predictions for the geometry of a tree trunk obeying the constant stress hypothesis**

We will first consider the model tree already introduced in box 5.6, but modified such that its trunk radius tapers according to the power function

(5.7.1) $$r = r_B \left(1 - \frac{h}{H} \right)^\alpha = r_B z^\alpha,$$

where r_B is the radius at the trunk base, h is distance above ground, H is tree height at the crown midpoint, and z is a dimensionless variable ($z = 0$ at the top and $z = 1$ at the base of the trunk) such that the exponent α describes trunk geometry (for a cylinder $\alpha = 0$, for a cone $\alpha = 1$). Assuming that the wind speed is constant over the height of the tree, the wind load can be modeled as a point load acting on the midpoint of the crown at a height H. The wind-induced bending moment at any level with distance $l = (H - h)$ from the top is given therefore by the formula

(5.7.2) $$M_{wind} = F_{wind}\, l = F_{wind} H z.$$

The counteracting moment induced in the trunk is given by equation (5.6.7). Combining equations (5.6.7) and (5.7.1) gives the stress at height Hz at the trunk surface:

(5.7.3) $$\sigma = \frac{4}{\pi} F_{wind} \frac{l}{r^3} = \frac{4}{\pi} F_{wind} \frac{H}{r_B^3} \frac{z}{z^{3\alpha}}.$$

Note that a constant stress σ requires that l/r^3 is constant, which is valid only for a tapering mode of $\alpha = \frac{1}{3}$ along the portion of the trunk below the crown.

BOX 5.7 **(Continued)**

When dealing with forces and moments acting on the portion of the trunk within the crown, we can no longer consider point loadings, but must rather consider distributed loads. Consider a spruce tree with branches all the way down to ground level. Its silhouette may have the form of an isosceles triangle with a base D and a height H, identical to the tree's height. Again we use the simplification of a constant wind profile. The distributed wind load can be expressed in the form of a line load as a function of $z = (1 - h/H)$:

(5.7.4) $$q_{wind}(z) = 0.5\,\rho_{air}A_{sail}(z)\,U^2C_D = 0.5\,\rho_{air}DzU^2C_D,$$

where A_{sail} is the sail area per unit length, which, for a triangular tree silhouette, can be replaced by Dz. The equivalent force at point p_1 along the z axis can be obtained by integration from the top downward:

(5.7.5) $$F_{wind}(p_1) = 0.5\rho_{air}U^2C_DDH\int_0^{p_1} z\,dz = 0.5\rho_{air}U^2C_DDHp_1^2/2.$$

The bending moment at position p_2 is obtained by a second integration:

(5.7.6) $$M_{wind}(p_2) = 0.5\rho_{air}U^2C_DDH^2\frac{1}{2}\int_0^{p_2} p_1^2\,dp_1 = 0.5\rho_{air}U^2C_DDH^2p_2^3/6,$$

with $p_2 \geq p_1 \geq z$.

The wind-induced moment must be compensated for by a bending moment within the trunk, which is related to the stress in the trunk surface at position p_2 by the formula

(5.7.7) $$M_{stem}(p_2) = \sigma\frac{I}{r} = \sigma\frac{\pi}{4}r^3 = \sigma\frac{\pi}{4}r_B^3p_2^{3\alpha}.$$

Because $M_{stem}(z) = M_{wind}(z)$, we see that the stress at any point along the z axis is given by the formula

(5.7.8) $$\sigma = \frac{1}{3\pi}\rho_{air}U^2C_D\frac{DH^2}{r_B^3}\frac{z^3}{z^{3\alpha}}.$$

BOX 5.7 **(Continued)**

Noting that a constant wind profile and a triangular tree silhouette have been assumed, the stress σ can be constant along the length of the trunk only if $\alpha = 1.0$.

For different silhouettes, or when wind speed $U = U(z)$ varies with height, different exponents are required for the constant stress hypothesis to be true. Their numerical values can be computed by incorporating the appropriate function for the particular silhouette and the function describing the wind speed profile explicitly in the integral in equation (5.7.4) and implicitly in equation (5.7.5). The effects of silhouette streamlining on the calculation of drag forces should also be taken into account (see section 6.8).

5.10 Euler buckling

Provided that flexural stiffness is known, engineering theory can be used to calculate the extent to which an idealized stem can grow before it will deflect from the vertical under an applied axial compressive load. The smallest axial compressive load that produces this deflection is called the critical load, symbolized here by W_{crit}. If the compressive load on a stem is less that the critical load, it will remain perfectly vertical and undergo only axial compression. This loading condition is said to be stable; that is, if a small deflection should occur as a result of a transient applied force, the stem will return to its vertical position once the transient force is removed. However, if the compressive load is increased, a point is reached at which the stem becomes unstable and deflects from the vertical (fig. 5.8).

The elementary theory of elastic buckling of columns is credited to Leonhard Euler (1707–1783), who applied his "calculus of variation" to describe this buckling phenomenon mathematically. As an example of this type of analysis, we will consider a tall and slender vertical column of length L, anchored at its base, free to move at its top, and loaded through its centroid axis with a top weight-force F. For simplification, the weight of the stem is considered negligible compared with the top load. If the column is deflected from the vertical by a small increment Δs measured at the position of the top load F, a bending moment $M_{bend} = F\Delta s$ will act on the stem. The restoring moment is given by $M_{restor} = EI/R_{curvature}$. If the

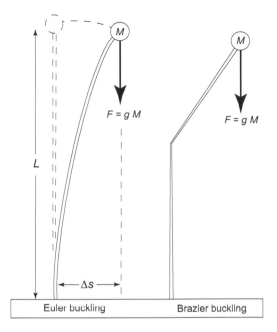

FIGURE 5.8. Illustrations of Euler and Brazier buckling, either of which can result when a slender column is subjected to a force F (here illustrated by a ball with mass M located at the free end). Euler (or long-wave) buckling results in a curved column; Brazier (or short-wave) buckling results in a localized crimp.

flexural stiffness EI is constant along the length of the column, the radius of curvature at the top is given by

$$(5.26) \qquad \frac{1}{R_{curvature}} = \frac{\pi^2}{4} \frac{\Delta s}{L^2}.$$

The column is stable under the condition $M_{bend} < M_{restor}$, but unstable under the condition $M_{bend} > M_{restor}$. A critical top load F_{crit} is reached when $M_{bend} = M_{restor}$. That condition is fulfilled when

$$(5.27) \qquad F \Delta s = \frac{\pi^2}{4} EI \frac{\Delta s}{L^2}, \quad \text{or} \quad F_{crit} = \frac{\pi^2}{4} \frac{EI}{L^2}.$$

The essence of the "calculus of variation" is that F_{crit} proves to be independent of Δs as long as the deflection is small. For a given top load F, the maximum length at which stability is maintained is given by

(5.28)
$$L_{max} = \left(\frac{\pi^2}{4} \frac{EI}{F} \right)^{1/2}.$$

A different case was treated by Greenhill (1881), who considered a slender column buckling under its own weight. He reduced the problem to a differential equation that could be solved by Bessel functions and found a formula for the maximum length of very slender columns:

(5.29)
$$L_{max} = C \left(\frac{E}{\rho g} \right)^{1/3} r^{2/3}.$$

For a column with a constant radius r, the parameter C equals 1.26, whereas for a cone with a basal radius r, $C = 1.96$. When dealing with columns composed of elements differing in density and flexural stiffness, Greenhill's differential equation can be solved using computer software packages such as Mathematica (Spatz and Speck 2002).

It is worth noting that the maximum height of a column does not depend on the strength of the material used to fabricate it, but only on the radius and the density-specific elastic modulus of the material (i.e., $E/\rho g$). The immunity of elastic buckling to the strength of materials has a number of implications for the elastic stability of trees whose woods differ in strength. More important is the role played by shape in dictating the way a vertical column fails under a compressive load. When the critical load F_{crit} is reached, an ideal column either undergoes crushing failure or elastically deforms by bending from the vertical. Whether crushing or buckling failure occurs depends on the ratio of the column's radius to its length (a variant of the inverse of the slenderness factor, L/D). By comparing equation (5.27) with the definition of the compressive strength σ_{crit}, we see that the value of this ratio characterizing the transition from one mode of failure to the other is given by the formula

(5.30)
$$\left(\frac{r}{L} \right)_{transition} = \frac{4}{\pi} \left(\frac{\sigma_{crit}}{E} \right)^{1/2}.$$

Accordingly, tall and slender columns will deflect from the vertical rather than crush under their critical axial load, whereas short and wide columns will tend to undergo crushing failure when their loading conditions exceed a critical level. For most plant materials, the quotient of σ_{crit} and E is on the order of a few percent, whereas the quotient r/L that marks the transition from Euler deflection to crushing failure is on the order of 0.1.

Euler buckling (in the configuration presented here, with one end fixed and the other end free) is a self-enhancing process. Once a column deflects from the vertical, the bending moment continues to increase and thereby results in greater and greater buckling (and often to straining of the material beyond its elastic limit). Therefore, in many cases, Euler buckling can lead to catastrophic failure.

Greenhill (1881) used his formula to assess the maximum height to which a tree can grow. MacMahon (1973) advanced the concept that Euler buckling provides the limit to plant growth, at least in an environment of low wind speeds. Young trees and some tropical trees have indeed been found to grow to the mechanical limit predicted by Greenhill's formula (Jaquen et al. 2007). The majority of adult trees, however, stay well below this limit. This difference can be interpreted as a safety factor in mechanical design. Safety factors can be calculated as the quotient of the load-bearing capacity and the maximal load a structure can be expected to experience. Whether plants "calculate" in a similar fashion is highly problematic. From the perspective of mechanoperception (see chap. 2), it is difficult to envision how a plant could sense the probability for Euler buckling upon additional growth. A broader perspective on the limits of plant height is provided by juxtaposing the hydraulic and mechanical requirements of plant life (Koch et al. 2004; Niklas and Spatz 2004; Niklas 2007).

5.11 Hollow stems and Brazier buckling

Galileo Galilei (1638) was the first to note that hollow tubes have the advantage of resisting bending and torsion at a relatively low weight per unit length compared with a solid cylinder composed of the same material. This advantage can be expressed as the quotient of the second moment of area I and the cross-sectional area A, which for a solid cylinder with radius r equals $I/A = r^2/4$. For a hollow tube with an outer radius r_a and an inner radius $r_i = r_a - t$, the quotient equals $I/A = (r_a^2 + r_i^2)/4 = (2 r_a^2 - 2 r_a t + t^2)/4$. Therefore, in theory, the smaller the wall thickness t, the better the relation between bending stability and weight. We say "in theory" because the cross section of a thin-walled tube can ovalize upon bending, which can lead to a local collapse of the tube. This form of failure is called local buckling or Brazier buckling (see fig. 5.8).

Brazier and Euler buckling are interrelated phenomena because Euler buckling can initiate Brazier buckling. If a bending moment is continuously applied, Brazier buckling can become a self-perpetuating process. Ovalization leads to a reduction in the second moment of area and therefore to a local increase in curvature, which in turn leads to higher lateral stresses, further ovalization, and finally to catastrophic local failure. This phenomenon was first treated by von Karman (1911) and later by Brazier (192 7), both of whom restricted their considerations to isotropic materials. The critical bending moment for a long tube with an outer radius r_a and wall thickness t is given by the formula

(5.31) $$M_{crit} = 0.99 \frac{E}{(1-v^2)^{1/2}} r_a t^2 \text{ for } r_a > 10\,t,$$

where v is the Poisson's ratio (Young 1989). For hollow tubes of anisotropic material, such as many plant stems, a numerical treatment that can give approximations for heterogeneous materials must be applied (Spatz et al. 1997).

Many plants, such as *Equisetum*, possess hollow stems that have various nodal diaphragms located along their lengths. These structures reduce the effective length of tubular stems by amplifying end-wall effects; that is, they restrict the deformations at the ends of each tubular segment by regionally increasing stiffness. Although these diaphragms contribute as little as 2% to the weight of a stem, they can increase the stability of the entire stem by as much as 16%–20% (Niklas 1989b). Different species of *Equisetum* have different additional anatomical configurations that increase the mechanical stability of stems. A particularly impressive example in which stability against Brazier buckling is realized with a minimum of biological expenditure is seen in *E. hyemale* (fig. 5.9). The stems of this species have an outer ring of non-sclerenchymatous strengthening tissue (called the hypodermal sterome) and an inner double ring of endodermis connected by bridges of strengthening tissue and vascular bundles. These bridges resemble I-beams much like those proposed by Schwendener (1874).

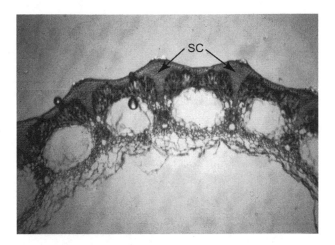

FIGURE 5.9. A transverse section through a stem internode of *Equisetum hyemale*, showing the large central (pith) canal, the vallecular canals (in the cortex), and rods of sclerenchymatous tissue (SC) running longitudinally just beneath the external ridges of the fluted internode.

Additional anatomical configurations that establish and maintain mechanical stability are seen throughout the plant kingdom. Cereals, for example, have a thin outer ring of sclerenchyma enclosing an inner ring of parenchyma. If fully turgid, this parenchyma provides an inner lining that calculations show contributes significantly to the stability of hollow stems. *Arundo donax* has a very thin outer ring of sclerenchymatous tissue enclosing a partially sclerenchymatous parenchyma with a gradient of stiffness decreasing from the outside to the inside. The extreme stability of the hollow stems of bamboo species results from a fully sclerenchymatous tissue in the tubular stem wall, again with a gradient of stiffness in the radial direction (Liese 1998). In contrast, some plants, such as *Juncus effusus*, have an outer ring of strengthening tissue enclosing aerenchyma, which contributes little to stability against Brazier buckling. The stems of trees grow in girth as well as length, thereby increasing their second moment of area and polar moments of inertia as they grow in height. With very few exceptions (e.g., *Cecropia*), tree stems are hollow only after extensive fungal infestation. However, even if the ratio of the effective wall thickness t to the outer radius t/r_a is as small as 0.3, bending stability is reduced by only 24%.

5.12 Dynamics, oscillation, and oscillation bending

The time evolution of physical processes such as the swaying of a tree due to wind is called dynamics. The dynamics of tree sway are important because trees and other large plants are more endangered by dynamic than by steady winds. The reason for this is that a tree can get into resonance with frequency components of the wind that approach the eigenfrequency of the tree. The eigenfrequency of an upright beam or plant fixed at its base and free at the other end is defined here as the frequency with which it undergoes free bending oscillations. Even moderate dynamic winds can excite a tree to oscillate with very large amplitudes, which may ultimately lead to failure in a so-called resonance catastrophe. As shown by Kerzenmacher and Gardiner (1998), energy from the wind is most effectively transferred to a tree in its lowest (fundamental) frequency of oscillation and less effectively in higher frequencies (overtones). It is necessary to understand the physics of free bending oscillations to comprehend the complex interaction of wind and trees (Mayer 1987; Gardiner 1995; Guitard and Castera 1995). The relevant physics is outlined in box 5.8.

BOX 5.8 Derivation of eigenfrequencies

For an upright slender stem with negligible mass and an apical load Mg, oscillations in an x–y plane can be described as an equilibrium of bending moments,

(5.8.1)
$$EI\frac{d^2y}{dx^2} + Mg(y - y_T) - (\omega^2 + \delta^2)My_T(x - x_T) = 0,$$

where M is the apical load mass, g is gravitational force ($g = 9.80665$ m/s²), ω the circular frequency, δ the decay constant, x is a coordinate along the beam length L (such that $x = 0$ at the top and $x = L$ at the base), x_T is the position of the apical load, y is the deflection from the resting position, and y_T is the deflection at $x = x_T$. The first term in equation (5.8.1) gives the moment on the stem (see eq. [5.14]), the second term is the bending moment induced by the apical load, and the third term is the moment due to the acceleration of the mass M. This last term takes into account the fact that during oscillations, the apical load will change its height above ground, being highest at $y = 0$ and lower at $y = y_T$. This

BOX 5.8 **(Continued)**

---

equation is valid only for small amplitudes of oscillation. In addition, for simplicity, we will treat only slightly damped oscillations, such that $(\omega^2 + \delta^2) \approx \omega^2$.

For plant stems, it is reasonable to assume that the bending stiffness varies smoothly and monotonously over length and can therefore be approximately represented by the power function $EI = E_B I_B z^{4\alpha + \beta}$, where E_B is the modulus of elasticity at the base; I_B the second moment of area at the base; $z = x/L$ is a dimensionless variable along the stem length such that $z = 0$ at the top and $z = 1$ at the base; α is the tapering mode, with $r = r_B z^\alpha$ and, correspondingly, $I = I_B z^{4\alpha}$; and β is the mode of dependence of the modulus of elasticity, with $E = E_B z^\beta$.

When we introduce the dimensionless variable $\psi = y/y_T$, the differential equation (5.8.1) can be written in the form

(5.8.2a)
$$\frac{d}{dz}\left(z^{4\alpha + \beta}\frac{d^2\psi}{dz^2}\right) + \Theta\frac{d\psi}{dz} - \Omega = 0,$$

with

(5.8.2b)
$$\Theta = L^2 \frac{Mg}{E_B I_B} \quad \text{and} \quad \Omega = L^3 \frac{M}{E_B I_B}\omega^2.$$

The term Ω stands for the acceleration of the apical load mass, and the term Θ accounts for the influence of gravity in the vertical orientation of the stem. With appropriate boundary conditions,

(5.8.2c)
$$\psi[1] - 0, \quad \frac{d\psi}{dz}[1] - 0, \quad \frac{d^2\psi}{dz^2}[z_T] - 0,$$

the differential equation can be solved by Mathematica 4.0 to yield numerical values for Θ and Ω and for α and β. Details of the computation are given by Spatz and Speck (2002).

The differential equation for free vibrations of an upright slender rod or stem with finite mass, but without an apical load, can be formulated as the equilibrium of line loads,

(5.8.3)
$$\frac{d^2}{dx^2}\left(EI\frac{d^2y}{dx^2}\right) + \frac{d}{dx}\left\{\left[g\int_0^x \rho(\xi)A(\xi)d\xi\right]\frac{dy}{dx}\right\} - \omega^2\rho A y = 0,$$

BOX 5.8 (Continued)

where A is the cross-sectional area, ρ is the density of the stem's material, and ξ is an integration variable, with $0 \le \xi \le x$. The cross-sectional area A can be approximated by the power function $A = A_B z^{2\alpha}$, where A_B represents the cross-sectional area at the base. A smooth and monotonous gradient of density can be approximated by $\rho = \rho_B z^\gamma$, where ρ_B is the density at the base. Introducing the power functions into equation (5.8.4), we obtain the differential equation

(5.8.4a)
$$\frac{d^2}{dz^2}\left(z^{4\alpha+\beta}\frac{d^2 y}{dz^2}\right) + G\frac{d}{dz}\left(\frac{z^{2\alpha+\gamma+1}}{2\alpha+\gamma+1}\frac{dy}{dz}\right) - Hz^{2\alpha+\gamma}y = 0$$

(5.8.4b)
$$\text{with } G = L^3\frac{\rho_B A_B g}{E_B I_B} \text{ and } H = L^4\frac{\rho_B A_B}{E_B I_B}\omega^2.$$

The term H stands for the acceleration of the mass of the stem and the term G for the influence of gravity in its vertical orientation. This equation can be solved with appropriate boundary conditions, which are

(5.8.4c)
$$y[1] = 0, \quad \frac{dy}{dz}[1] = 0 \quad \frac{d^2 y}{dz^2}[0] = 0, \quad \frac{d^3 y}{dz^3}[0] = 0$$

when $G = 0$ or $H = 0$. Good approximations can be found for finite values of G and H.

As an example, for $G = 0$, the calculations yield a solution for the eigenfrequency of an upright cone ($\alpha = 1, \beta = 0, \gamma = 0$) with a numerical value of $H_0 = 76.025$ and a value of $G_0 = 30.530$ for $H = 0$. This leads to an approximation for the eigenfrequency corrected for the effect of gravity ($G \ne 0$) of

(5.8.5)
$$\omega = \left(76.025\frac{E I_B}{\rho g A_B}\frac{1}{L^4} - 2.45\frac{g}{L}\right)^{1/2}.$$

Different numerical values are found for different α, β, and γ (Spatz and Speck 2002).

The corresponding equations have been tested by comparing the calculated eigenfrequencies with those observed for 25 Sitka spruce trees (Brüchert et al. 2003). Of course, the properties of real trees never follow power laws exactly. In particular, the mass of the side branches has to be included in the effective density,

BOX 5.8 **(Continued)**

---

whose gradient is only poorly described by a power law with an exponent γ. In addition, the dependence of the modulus of elasticity on the height above ground deviates somewhat from a power law with an exponent β. However, these variables are far less important than the parameter α (which describes the actual taper of these trees reasonably well), such that the calculated values for the eigenfrequencies differed, on average, by only 7% from those observed.

The interaction of gusty and turbulent winds with a swaying tree poses a particularly difficult problem (Mayer 1987). The exact differential equation for one-dimensional wind-induced tree sways (Finnigan and Mulhearn 1978; see equation 13 in Mayer 1987) includes terms for oscillation damping and for the aerodynamic drag acting over the length of the tree, taking the movement of the tree into account. This equation can only be solved numerically. An alternative is the spectral method (Holbo et al. 1980), in which the power spectrum of the tree's response results from a frequency analysis of the wind speed with respect to the wind load:

(5.32)
$$P_y(f) = |H_m(f)|^2 P_K(f),$$

where f is the frequency of the wind speed, $P_y(f)$ is the power spectrum of the tree's response, and $P_K(f)$ is the power spectrum of the wind load (fig. 5.10). $H_m(f)$ is a mechanical transfer function. Nieser (1979) has shown that for conifers treated as linear systems, $H_m(f)$ depends primarily on the eigenfrequency in the fundamental mode of oscillation (and to a lesser degree, on the first harmonic and the damping characteristics of the tree). The power spectrum of the wind load can be derived from the power spectrum of the wind velocity using an aerodynamic transfer function. The relationship for the horizontal wind component U_h can be given in the form

(5.33)
$$P_k(f) = \rho_{air}^2 C_D^2 A_{sail}^2 \bar{U}_h^2 [H_a(f)]^2 P_{U_h}(f)$$

(Boos and Schueller 1982), where A_{sail} is the projected area of the tree perpendicular to the wind flow (called the sail area), $\bar{U}_h$ the mean wind speed in the horizontal direction, $H_a(f)$ the aerodynamic transfer function,

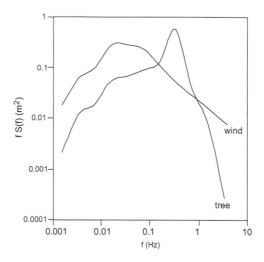

FIGURE 5.10. Power spectra of wind velocity in a forest near Hartheim, Germany, and a Scots pine's simultaneous response to wind recorded over a period of 1 hour. The power spectra are derived from Fourier analyses of the velocities of the horizontal component of the wind and the respective movement of the tree. In the case of the wind velocity spectrum, the product of frequency and the spectral density of velocity is plotted against the mean frequency of the interval. In the case of the tree's response, the product of frequency *f* and the spectral density of trunk displacement *S(t)* is plotted against the mean frequency of the interval. (Courtesy of Dr. D. Schindler, Freiburg, Germany.)

and P_{U_h} (*f*)the power spectrum of the horizontal wind component. Mayer (1987) has shown that for spruce trees in the fundamental mode of oscillation, $[H_a(f)]^2 \approx 1$. For large trees, for which the aerodynamic drag cannot be studied in wind tunnels, an uncertainty remains because, due to stream-lining of the branches at high wind speeds, the sail area A_{sail} and the drag coefficient C_D are not known.

An important aspect of the transfer of energy from the wind to a tree, or to any large plant, is the damping of oscillations. Damping causes a decrease in the amplitudes of free oscillations and thus reduces the danger of a resonance catastrophe in dynamic winds. Oscillation and oscillation damping in trees have been widely studied (see Mayhead 1973; Milne 1991; Peltola et al. 1993; Moore and Maguire 2004; Jonsson et al. 2007). If friction among different trees, or among different branches, and dissipative mechanisms in the root-soil system are set aside, there are two principal sources of damping: fluid damping and viscous damping within the material (i.e., wood in the case of trees). Fluid damping (dissipation of energy to the surrounding medium) depends on the square of the velocity of the object's movement relative to the surrounding medium (see eq. [2.15]). With estimates of the effective sail area and the drag coefficient, fluid damping can be calculated by iteration of the loss of energy during each cycle of the oscillation. Viscous damping in the material (conversion of mechanical energy into heat) is linearly related to the velocity of relative movements between adjoining branches (or systems of branches that move in consort with one another). It can be determined in loading-unloading experiments by measuring the loss of energy in a hysteresis loop (see fig. 4.3). In order to relate these measurements to a real tree, they should be performed using green wood.

In complex structures such as trees, certain processes can enhance damping. In gusty winds, branches do not necessarily sway in line or in phase with their subtending stems. Rather, they can perform independent movements relative to one another. In this way, energy is distributed among branches and twigs and is dissipated more effectively than in a structure too stiff to allow relative movements between its elements. This phenomenon, which is referred to as structural damping (Niklas 1992), is not a different mode of damping. It merely emphasizes the enhancement of overall damping by the relative movements of structural elements (branches and twigs), which affect both fluid and viscous damping—processes that are most effective in the periphery of the tree canopy. Structural damping can be caused by the loose coupling mass (termed mass damping; James 2003) or by the distribution of mechanical energy through resonance phenomena within the tree (termed multiple resonance damping by James et al. 2006; Spatz et al. 2007), a phenomenon well known in the engineering sciences (Holmes 2001). As shown for a Douglas fir, a tree can react to dynamic wind loads like a system of coupled damped oscillators. This concept has been confirmed theoretically by Rodriguez et al. 2008. The interaction between the different elements leads to a higher damping ratio, and additionally to less strain on the stem as compared with a structure with much stiffer side branches. Multiple resonance damping is therefore essential for the survival of trees and other large plants growing in windy environments.

Literature Cited

Beismann, H., H. Wilhelmi, H. Baillères, H.-C. Spatz, A. Bogenrieder, and T. Speck. 2000. Brittleness of twig bases in the genus *Salix*—fracture mechanics and ecological relevance. *J. Exp. Bot.* 51:617–33.

Bertram, J. E. A. 1989. Size-dependent differential scaling in branches: The mechanical design of trees revisited. *Trees Struct. Funct.* 4:241–53.

Boos, G., and G. I. Schueller. 1982. Die Beschreibung der Windlast im Rahmen eines probabilistischen Sicherheitskonzepts. *Beitr. Anwend. Aerolast. Bauwes.*, Heft 17:63–112.

Brazier, L. G. 1927. On the flexure of thin cylindrical shells and other thin sections. *Proc. Roy. Soc. London*, ser. A, 116:104–14.

Brüchert, F., H.-C. Spatz, and O. Speck. 2003. Oscillations of plant stems and their damping: Theory and experiments. *Phil. Trans. Roy. Soc. London*, ser. B, 358:1487–92.

Ennos, A. R. 1993. The mechanics of the flower stem of the sedge *Carex acutiformis*. *Ann. Bot.* 72:123–27.

Ennos, A. R., H.-C. Spatz, and T. Speck. 2000. The functional morphology of the petioles of the banana *Musa textilis*. *J. Exp. Bot.* 51:2085–93.

Finnigan, J. J., and P. J. Mulhearn. 1978. Modelling waving crops in a wind tunnel. *Boundary-Layer Meteorol.* 14:253–77.

Galilei, G. 1638. *Discorsi e dimonstrazioni matematiche, intorno a due nuove scienze.* Leida, Italy: Appresso gli Elsevirii.

Gardiner, B. A. 1995. The interaction of wind and tree movement in forest canopies. In *Wind and trees*, edited by M. P. Coutts and J. Grace, 41–59. Cambridge: Cambridge University Press.

Greenhill, G. 1881. Determination of the greatest height consistent with stability that a vertical pole or mast can be made, and the greatest height to which a tree of given proportions can grow. *Proc. Camb. Phil. Soc.* 4:65–73.

Guitard, D. G. E., and P. Castera. 1995. Experimental analysis and mechanical modelling of wind-induced tree sways. In *Wind and trees*, edited by M. P. Coutts and J. Grace, 182–94. Cambridge: Cambridge University Press.

Harder, D. L., C. L. Hurd, and T. Speck. 2006. Comparison of mechanical properties of four large, wave-exposed seaweeds. *Amer. J. Bot.* 93:1426–32.

Holbo, H. R., T. C. Corbett, and P. J. Horton. 1980. Aeromechanical behavior of selected Douglas-fir. *Agr. Meteorol.* 21:81–91.

Holmes, J. D. 2001. *Wind loading of structures.* London: Spon Press.

James, K. R. 2003. Dynamic loading of trees. *J. Arboricult.* 29:165–71.

James, K. R., N. Haritos, and P. Ades. 2006. Mechanical stability of trees under dynamic loads. *Amer. J. Bot.* 93:1361–69.

Jaquen, G., T. Almeras, C. Coutand, and M. Fourier. 2007. How to determine sapling buckling risk with only few measurements. *Amer. J. Bot.* 94:1583–93.

Jonsson, M. J., A. Froetzi, M. Kalberer, T. Lundström, W. Ammann, and V. Stöckli. 2007. Natural frequencies and damping ratios of Norway spruce (*Picea abies* [L.] Karst) growing on subalpine forested slopes. *Trees* 21:541–48.

Kerzenmacher, T., and B. Gardiner. 1998. A mathematical model to describe the dynamic response of a spruce tree to wind. *Trees* 12:385–94.

King, D., and O. L. Loucks. 1978. The theory of tree bole and branch form. *Radiat. Env. Biophys.* 13:141–65.

Koch, G. W., S. C. Sillet, G. M. Jennings, and S. D. Davis. 2004. The limits to tree height. *Nature* 428:851–54.

Liese, W. 1998. *The anatomy of bamboo culms.* Leiden: E. J. Brill.

Mattheck, C. 1992. *Design in der Natur. Der Baum als Lehrmeister.* Freiburg: Rombach Verlag.

Mayer, H. 1987. Wind-induced tree sways. *Trees* 1:195–206.

Mayhead, G. J. 1973. Sway periods of forest trees. *Scott. For.* 27:19–23.

McMahon, T. A. 1973. The mechanical design of trees. *Science* 233:92–102.

Metzger, K. 1893. Der Wind als maßgebender Faktor für das Wachstum der Bäume. *Mündener forstl. Hefte* 3:35–86.

Milne, R. 1991. Dynamics of swaying of *Picea sitchensis. Tree Physiol.* 9:383–99.

Moore, J. R., and D. A. Maguire. 2004. Natural sway frequencies and damping ratios of trees: Concepts, review and synthesis of previous studies. *Trees* 18:195–203.

Morgan, J., and M. G. R. Cannel. 1994. Shape of tree stems—a re-examination of the uniform stress hypothesis. *Tree Physiol.* 14:49–62.

Nieser, H. 1979. Spektralmethode—Erläuterung des Davenport'schen Konzepts. *Beitr. Anwend. Aerolast. Bauwes.,* Heft 13:103–11.

Niklas, K. J. 1989a. Mechanical behavior of plant tissues as inferred from the theory of pressurized cellular solids. *Amer. J. Bot.* 76:929–37.

———. 1989b. Nodal septa and the rigidity of aerial shoots of *Equisetum hyemale. Amer. J. Bot.* 76:521–31.

———. 1992. *Plant biomechanics: An engineering approach to plant form and function.* Chicago: University of Chicago Press.

———. 2007. Maximum plant height and the biophysical factors that limit it. *Tree Physiol.* 27:433–40.

Niklas, K. J., and H.-C. Spatz. 2000. Wind-induced stresses in cherry trees: Evidence against the hypothesis of constant stress levels. *Trees Struct. Funct.* 14:230–37.

———. 2004. Growth and hydraulic (not mechanical) constraints govern the scaling of tree height and mass. *Proc. Nat. Acad. Sci. USA* 101:15661–63.

Peltola, H. M. 2006. Mechanical stability of trees under static loads. *Amer. J. Bot.* 93:1501–11.

Peltola, H., S. Kellomäki, A. Hassinen, M. Lemittinnen, and J. Aho. 1993. Swaying of trees as caused by wind: Analysis of field measurements. *Silva Fenn.* 27:113–26.

Rodriguez, M., E. de Langre, and B. Moulia. 2008. A scaling law for the effects of architecture and allometry on tree vibration modes suggests a biological tuning to modal compartmentalization. *Amer. J. Bot.* 95:1523–37.

Schwendener, S. 1874. *Das mechanische Prinzip im anatomischen Bau der Monocotylen mit vergleichenden Ausblicken auf die übrigen Pflanzenklassen.* Leipzig: Engelmann.

Spatz, H.-C., H. Beismann, F. Brüchert, A. Emanns, and T. Speck. 1997. Biomechanics of *Arundo donax. Phil. Trans. Roy. Soc. London,* ser. B, 352:1–10.

Spatz, H.-C., and F. Brüchert. 2000. Basic biomechanics of self-supporting plants: Wind loads and gravitational loads on a Norway spruce tree. *For. Ecol. Mgmt.* 135:33–44.

Spatz, H.-C., F. Brüchert, and J. Pfisterer. 2007. Multiple resonance damping or how do trees escape dangerously large oscillations? *Amer. J. Bot.* 94:1603–11.

Spatz, H.-C., and O. Speck. 2002. Oscillation frequencies of tapered plant stems. *Amer. J. Bot.* 89:1–11.

Spatz, H.-C., E. J. O'Leary, and J. F. V. Vincent. 1996. Young's moduli and shear moduli in cortical bone. *Proc. Roy. Soc. Lond. B* 263: 287–94.

Stephens, R. C. 1970. *Strength of materials, theory and examples.* London: Edward Arnold.

Timoshenko S., and J. N. Goodier. 1970. *Theory of elasticity.* 3rd ed. New York: McGraw-Hill.

Vogel, S. 1998. *Cat's paws and catapults.* New York: Norton.

Von Karman, T. 1911. Über die Formänderung dünnwandiger Rohre, insbesondere federnder Ausgleichsrohre. *Z. Ver. deutsch. Ing.* 55:1889–95.

Young, W. C. 1989. *Roark's formulas for stress and strain.* 6th ed. Singapore: McGraw-Hill.

Fluid Mechanics

You have to know a little about fluid mechanics to realize what your organisms must contend with and to recognize the opportunities open for them to be the cleverly adapted rascal in whom we take delight.—Steven Vogel, *Life in Moving Fluids*

Every organism operates physiologically and mechanically in a fluid that is either gaseous or liquid. Therefore, understanding the physical properties and behavior of fluids is essential to virtually every aspect of biological enquiry. Some of these properties were considered in earlier chapters, but only in very cursory ways. Furthermore, the kinetic properties of fluids—the physics of fluid motion—were not examined in sufficient detail to be meaningful in understanding fluid mechanics as it relates to plant life. The goal of this chapter is to provide a deeper foundation with which to explore the behavior of plants in moving fluids.

6.1 What are fluids?

The defining property of fluids is the ease with which and the extent to which they can be deformed. Fluids lack definite shape and are easily distortable, whereas solids have definite shape and respond to mechanical forces by resisting them. Recall from chapter 4 that the shear modulus G of a solid is given by the slope of its shear stress versus deformation curve, $G = \tau/\gamma$ (eq. [4.16]). The higher the shear modulus, the greater the solid's ability to resist an externally applied force. The relationship between stresses and strains for a Newtonian fluid, however, is given by the formula that defines dynamic viscosity,

(6.1)
$$\mu = \frac{\tau}{\left(\dfrac{dU}{dx}\right)},$$

where τ is the shear stress and dU/dx is the shear rate (the local velocity gradient). This formula indicates that fluids resist the *rate* of shear. By *Newtonian*, we mean a fluid that exhibits a linear proportional relationship between applied shear stresses and the resulting rates of shear. Notice that because of this condition, we can ascribe a single numerical value to a fluid's dynamic viscosity (for a given temperature or pressure). A Newtonian fluid is therefore analogous to the ideal elastic solid, known as a Hookean solid. One way to visualize the dynamic viscosity of a Newtonian fluid is to think of μ as a measure of the "stickiness" between adjoining moving layers of a gas or liquid. Typically, gases have low μ and liquids have higher μ; for example, for air (at 20°C), $\mu = 18.08 \times 10^{-6}$ kg m^{-1} s^{-1}, and for pure water (at 20°C), $\mu = 1.003 \times 10^{-3}$ kg m^{-1} s^{-1} (see table 2.1 for values given in Pa s).

Naturally, if Newtonian fluids exist, so must non-Newtonian fluids. These fluids have flow properties that are not described by a single constant dynamic viscosity; that is, the relationship between shear stresses and strain rates is nonlinear, in a manner analogous to that in which stress and strain are not linearly related to one another in the case of nonlinear elastic materials. Examples of non-Newtonian fluids are paint, blood, ketchup, and phloem sap.

The distinction between solids and fluids, however, is not as sharp as it may appear. One of the definitions of a solid is that it possesses a crystalline structure. In this sense, glass is a liquid of extremely high viscosity. Although it has the properties of a solid when examined over a short time, glass flows when subjected to gravity over long periods. There are many other materials that behave as solids under some conditions and as fluids under other conditions. For example, jellies (one of many kinds of thixotropic materials) behave as elastic solids if they are allowed to stand and congeal, but they behave like fluids when they are mechanically shaken or if they are exposed to a sustained mechanical force.

In terms of the physics of motion, the distinction between the gaseous and the liquid state is even less sharp than that between a fluid and a solid. For reasons that relate to the nature of intermolecular forces, most substances can exist in either of two stable states that exhibit "fluidity" (easy deformability). Although the density of the gaseous state is normally less than that of the liquid state, this feature is not a sufficient basis for distinguishing between these two states of matter because it merely relates to the magnitudes of the mechanical forces required to produce given magnitudes of acceleration, rather than to differences in the type of motion produced. Rather, the feature that truly distinguishes a gas from a liquid is a property called bulk elasticity; that is, compressibility. Gases can be compressed or expanded to a much greater degree than liquids or solids. As a consequence, significant variations in pressure exerted on a gas result in far more significant variations in its specific volume (and thus density) compared with a liquid. Another, perhaps more intuitive, way of saying this is that liquids have a well-defined volume, whereas gases fill any space provided to them.

Yet another way of visualizing what distinguishes a solid, liquid, or gas is to consider the distances over which intermolecular forces operate among neighboring molecules in these three states of matter. At small distances (on the order of 10^{-8} cm) from the centers of nonionized molecules, mutual intermolecular forces result. These attractive or repulsive forces work at small distances on the order of 3×10^{-8} cm for most simple molecules. At larger distances, on the order of 10^{-7} to 10^{-6} cm, mutual intermolecular forces are weakly attractive or repulsive. However, these forces begin to fall off at 10^{-7} cm and become negligible at 10^{-6} cm. Calculations indicate that the average spacing of molecules in gases (under normal earth temperatures and pressures) is on the order of 3×10^{-7} cm, while the average spacing in liquids and solids is on the order of 3×10^{-8} cm. Thus, in gases, molecules are so far apart that only very weak cohesive forces act among neighboring molecules. In contrast, the molecules in a liquid or in a solid experience comparatively strong force fields generated by their neighbors.

The science of fluid mechanics is normally concerned with the behavior of gases and liquids at macroscopic scales that exceed atomic dimensions by many orders of magnitude. That is, fluid mechanics supposes that fluids are perfectly continuous in their structure and that physical features, such as density and viscosity, are steadily and evenly distributed throughout the volume of a fluid. This supposition is called the *continuum hypothesis*. Its general validity is evident when we observe the behavior of air or water or, indeed, almost any other normally occurring fluid. Another assumption common to most treatments of fluid mechanics (albeit not universal) is that fluids behave as Newtonian fluids. The foundations of hydrodynamic theory, which rest on these two simplifying assumptions, were already well developed in the nineteenth century by Claude-Louis Navier (1785–1836) and George Gabriel Stokes (1819–1903). The Navier-Stokes equations describe the equilibrium of the forces on an infinitesimal fluid element (see box 6.1).

BOX 6.1 **The Navier-Stokes equations**

Fluid flow in space and time can be seen as a field of the three-dimensional velocity vector $\vec{U}$. Starting from Newton's mechanics of motion, the equilibrium of the forces on an infinitesimal fluid element is described by the Navier-Stokes equations. Here we consider only incompressible Newtonian fluids with constant density ρ and constant viscosity μ (which implies a condition of constant temperature). The simplification of incompressibility is valid for liquid water and for air from subsonic velocities up to roughly Mach 0.3.

For such fluids, the Navier-Stokes equation can be written in vector form as

$$(6.1.1) \qquad \rho \left(\frac{\partial \vec{U}}{\partial t} + \vec{U} \cdot \vec{\nabla} \vec{U} \right) = -grad\,P + \mu\,\nabla^2 \vec{U} + \rho\,\vec{g},$$

where $\vec{\nabla}$ is the nabla operator. The acceleration is expressed as the sum of the two terms $\frac{d\vec{U}}{dt} = \frac{\partial \vec{U}}{\partial t} + \vec{U} \cdot \nabla \vec{U}$, where the term $\frac{\partial \vec{U}}{\partial t}$ stands for unsteady acceleration and the term $\vec{U} \cdot \nabla^2 \vec{U}$ for convective acceleration due to inertia. The term $grad\,P$ is the gradient of the pressure force per unit volume, $\mu \nabla^2 \vec{U}$ accounts for the viscous force per unit volume, and $\rho \vec{g}$ is the gravity force per unit volume. All symbols with an arrow superscript denote a vector or a vector operator.

In a three-dimensional Cartesian coordinate system, the velocity vector $\vec{U}$ can be written as (u_x, u_y, u_z); likewise, the gravitational vector can be written as (g_x, g_y, g_z). In this connotation, equation (6.1.1) can be broken down into three differential equations:

$$\rho \left(\frac{\partial u_x}{\partial t} + u_x \frac{\partial u_x}{\partial x} + u_y \frac{\partial u_x}{\partial y} + u_z \frac{\partial u_x}{\partial z} \right) = -\frac{\partial P}{\partial x} + \mu \left(\frac{\partial^2 u_x}{\partial x^2} + \frac{\partial^2 u_x}{\partial y^2} + \frac{\partial^2 u_x}{\partial z^2} \right) + \rho g_x,$$

$$\rho \left(\frac{\partial u_y}{\partial t} + u_x \frac{\partial u_y}{\partial x} + u_y \frac{\partial u_y}{\partial y} + u_z \frac{\partial u_y}{\partial z} \right) = -\frac{\partial P}{\partial y} + \mu \left(\frac{\partial^2 u_y}{\partial x^2} + \frac{\partial^2 u_y}{\partial y^2} + \frac{\partial^2 u_y}{\partial z^2} \right) + \rho g_y,$$

$$\rho \left(\frac{\partial u_z}{\partial t} + u_x \frac{\partial u_z}{\partial x} + u_y \frac{\partial u_z}{\partial y} + u_z \frac{\partial u_z}{\partial z} \right) = -\frac{\partial P}{\partial z} + \mu \left(\frac{\partial^2 u_z}{\partial x^2} + \frac{\partial^2 u_z}{\partial y^2} + \frac{\partial^2 u_z}{\partial z^2} \right) + \rho g_z.$$

(6.1.2)

Together with the continuity equation for incompressible fluids,

$$(6.1.3) \qquad \frac{\partial u_x}{\partial x} + \frac{\partial u_y}{\partial y} + \frac{\partial u_z}{\partial z} = 0,$$

BOX 6.1 **(Continued)**

---

the four components of equation (6.1.2) form four nonlinear partial differential equations for the four unknowns: P, u_x, u_y, and u_z.

It has not yet been proved mathematically that global solutions of this set of differential equations exist. The Navier-Stokes equations are, therefore, the domain of numerical computations. A number of special cases are of particular importance:

1. Stationary flow: $\dfrac{\partial \bar{U}}{\partial t} = 0$
2. Laminar flow: Viscous forces dominate inertial forces
3. Turbulent flow: Inertial forces dominate viscous forces
4. Inviscid flow: $\mu = 0$, characterizing ideal fluids
5. Inviscid and irrotational flow

6.2 The Reynolds number

We introduced the concept of the Reynolds number previously, but it is time to explore this concept in greater detail. In this chapter we are concerned with the differences between viscid and inviscid flow regimes, and the Reynolds number tells us which of these two regimes governs fluid flow in any particular case.

Specifically, theory and observation both show that the viscous properties of a fluid dominate when flow speeds are slow or when objects obstructing the flow are small. Under these circumstances, we can speak of viscous flow. Likewise, when flow speeds are high or obstructions are large, inertial (inviscid) properties dominate fluid flow. These rules are extremely useful. Under normal conditions, however, large volumes of fluids are in constant motion because they are incessantly subjected to externally applied forces. This raises the question of when we should adopt the "large and inertial" versus the "small and viscous" perspective on fluid behavior.

This question can be resolved by calculating the magnitude of the Reynolds number. This dimensionless quotient was first introduced in chapter 2, in which we considered boundary layers in a cursory way (see section 2.2). However, the Reynolds number is so important to the study of fluid mechanics that it is well worth deriving it conceptually using two basic physical principles: Newton's second law, which describes inertial forces F_I, and the definition of dynamic viscosity μ, which describes the nature of viscous forces F_V:

(6.2a) $$F_I = 0.5\rho A_\perp U^2,$$

(6.2b) $$F_V = \mu\left(\frac{A_\| U}{\ell}\right),$$

where ρ is the density of the fluid, $A_\perp$ denotes the surface area facing (perpendicular to) the fluid flow, $A_\|$ designates the area parallel to the direction of flow, U is the ambient fluid's speed, and ℓ is the distance between moving fluid layers. Taking the quotient of $A_\perp$ and $A_\|$ to be 1.0 and neglecting the numerical factor of 0.5, the quotient of F_I and F_V gives the Reynolds number (compare with eq. [2.13]):

(6.3) $$\mathrm{Re} = \frac{\rho A U^2}{\mu A U/\ell} = \frac{\rho \ell U}{\mu} = \frac{\ell U}{v},$$

where ℓ now denotes a reference (characteristic) dimension stipulating the size of the object obstructing flow and $v = \mu/\rho$ is the kinematic viscosity.

Inspection of equation (6.3) shows that the same numerical values of Re can be achieved by manipulating the numerical value of any of three variables. Increasing ℓ by a factor often has the same affect as increasing U or decreasing v by the same factor. As a consequence, Re values are "impartial" to whether a fluid's kinematic viscosity varies or whether the size of the object submerged in it increases or decreases. For this reason, we can substitute water for air, provided we manipulate ℓ or U to achieve the same Re (and provided that we also account for differences in drag, which is a topic treated later in this chapter). By the same token, Re can change dramatically as an organism grows in size, even if its ambient fluid environment remains relatively constant. For example, consider the fertilized egg (zygote) of an alga measuring 20 μm in diameter attached to a rock. Even if the mainstream flow over this rock reaches $U = 2.0$ m/s, this hypothetical zygote lives in the boundary layer, where the flow velocity is close to zero regardless of ambient flow speeds (the "no-slip" condition); in other words, the zygote lives in a viscid world where Reynolds numbers are extremely small. Even if we use $U = 2.0$ m/s to estimate an upper limit for Re, we see that the Reynolds number is still very small; that is, Re =

$(20 \times 10^{-6} \text{ m}) (2.0 \text{ m/s})/(1.003 \times 10^{-3} \text{ kg m}^{-1} \text{ s}^{-1}) \approx 0.04$. Yet when this cell grows into a large and long plant, such as a kelp, with a characteristic dimension of 1.0 m, the Reynolds number becomes $(1.0 \text{ m}) (2.0 \text{ m/s})/(1.003 \times 10^{-3} \text{ kg m}^{-1} \text{ s}^{-1}) \approx 2,000$. Clearly, the same individual can begin its life in the "small and viscous" world of low Re only to enter later the "large and inertial" Re world as it grows in size.

Our example of how Re changes as an organism's size increases raises a question: How do we select the characteristic dimension referred to in equation (6.3)? When dealing with simple geometries and shapes, such as spherical zygotes and cylindrical algal morphologies, standard practice defines the dimension that should be used. For flow around spheres (or cylinders aligned orthogonal to the direction of ambient flow), ℓ is diameter; for disks (or flat leaves) aligned parallel to fluid flow, ℓ is length (or average length); and for flow through tubes, ℓ equals the inner tube diameter. However, for many biological structures with more complicated shapes and geometries (e.g., flowers, tapered and square stems, or ruffled leaves), the choice of the linear dimension used as ℓ can be highly problematic and, to some extent, even arbitrary. This presents a potentially serious problem. Unless flow regimes are measured directly and empirically, the extent to which any reference dimension correctly reveals the actual flow patterns remains uncertain. In addition, the fluid flow measured at different points around the surfaces of even geometrically simple objects can vary, often significantly, especially at high ambient flow speeds. Indeed, this "heterogeneity" becomes particularly relevant when we examine boundary layers (see section 6.5). For these and other reasons, it is often useful (and convenient) to calculate Reynolds numbers using different ℓ based on local biological landmarks. These "local" Reynolds numbers are very informative when ambient flow regimes result in different flow regimes over different parts of the same organ or organism.

Regardless of how ℓ is selected, it is important to remember that the numerical values of Re provide only a rough indication of when one flow regime changes into another. They do not provide precise values for the quotients of inertial and viscous forces. For example, experiments have repeatedly shown that the transition from laminar to turbulent flow through pipes always occurs at Re $\approx 2,300$. However, this does not mean that inertial forces are 2,300 times as large as viscous forces because, in these experiments, the dimensions of pipes were used to characterize velocity gradients (which were not measured directly). In addition, some flow regimes remain unchanged over substantial numerical ranges of Re. For example, at $40 < \text{Re} < 100,000$, the downstream surfaces of a vertical cylinder generate alternating eddies or vortices that are shed farther downstream in a predictable and periodic manner (often referred to as a von Karman street; see section 6.5). Thus, the same flow regime can persist over three orders of magnitude of Re.

As noted, Reynolds numbers (rather than fluid density, viscosity, or flow rate individually) characterize the flow regime generated by and formed around any object obstructing fluid flow. Understandably, therefore, one of the earliest efforts in fluid mechanics was the description of how flow regimes change around objects over a large range of Re for "ordinary" (simple) geometries moving through air or water. Another early task was the determination of the drag forces generated by different flow regimes. Here, the objective was to relate the magnitudes of drag forces to changes in flow regimes using Re as an index. In a few limited cases, closed-form solutions were formulated for both of these purposes. However, these solutions pertain to very simple geometries (e.g., flat plates, spheres, and untapered circular cylinders), which limits their biological applicability. Fortunately, it is not difficult to visualize changes in flow fields directly by photographing the motion of particulates suspended in fluids forced to move around different geometries placed in a wind tunnel or a flume. Nor is it difficult to measure directly the drag forces these geometries experience as flow regimes change. Therefore, the description of how flow fields change as a function of Re, and how these changes affect the magnitude of drag forces, can (and often must) rely on observation.

6.3 Flow and drag at small Reynolds numbers

Living in a world of large Reynolds numbers dominated by inertial forces, it is difficult for us to envision life existing in flow regimes dominated by viscous forces. (A lucid and entertaining account of the problems encountered by small organisms such as bacteria is given by Purcell [1977].) Nevertheless, in this section, we will explore the behavior of fluids moving either very slowly or around small objects.

For small stationary objects (represented here by spheres or circular cylinders) and for $\text{Re} \ll 1$, the steady state flow regime is laminar and symmetrical fore and aft. A drag results from the velocity gradient of the fluid layers around the object, while the layer adhering to the object has zero velocity (which is known as the no-slip condition). The total force resulting from the tangential stresses is typically opposite to the direction of ambient flow. This force is called the friction drag because it is entirely and directly the result of viscosity (friction within the fluid). For the special case of a small sphere, an elaborate calculation leads to Stokes's formula for the drag force F,

(6.4) $$F = -6\pi \mu r U,$$

where μ is the dynamic viscosity, r is the radius of the sphere, and U is the velocity of the unperturbed flow some distance away from the sphere. Note that the same force is experienced whether the fluid moves with a velocity U around a stationary sphere or the sphere is dragged with a velocity U through a resting fluid. In addition, notice that the drag force F is linearly related to U. An example of estimating the terminal settling velocities of small biological objects is presented in section 6.10.

Another case of laminar flow, which was discussed very briefly in chapter 3, is flow through a narrow capillary tube with a circular inner cross section as described by the Hagen-Poiseuille equation,

(6.5) $$\frac{\Delta V}{\Delta t} = -\frac{\pi}{8\mu}\left(\frac{\partial P}{\partial x}\right)r^4$$

(compare eq. [6.5] with eq. [3.12] or [3.13]). Box 6.2 gives the mathematical derivation of this formula by considering the equilibrium between a pressure gradient and the frictional resistance to the flow.

BOX 6.2 **Derivation of the Hagen-Poiseuille equation**

Here, we present a detailed derivation of the Hagen-Poiseuille equation, which is important for understanding the laminar flow of a fluid with viscosity μ in a narrow tube of length ℓ with a circular cross section of inner radius r experiencing a pressure difference $\Delta p = p_1 - p_2$ (fig. 6.2.1). The forces acting on any part of the fluid in the form of a hollow cylinder with an outer radius $a + da$ and an inner radius a are given by the formula

(6.2.1) $$F_{pressure} = 2\pi a\, da\, (p_1 - p_2).$$

BOX 6.2 **(Continued)**

These forces are counterbalanced by frictional forces resulting from shearing on the outside and on the inside of the hollow cylinder. These frictional forces are given by the formula

$$(6.2.2) \qquad F_{friction} = \mu \left[\left(A \frac{dU}{da} \right)_{outside} - \left(A \frac{dU}{da} \right)_{inside} \right].$$

where $A = 2\pi a \ell$ is the area of the cylindrical wall, μ is the dynamic viscosity, and U is the flow velocity. Neglecting higher-order terms, the equilibrium of forces can be written as

$$(6.2.3) \qquad 2\pi \, a \, da \, (p_1 - p_2) = 2\pi \, \ell \mu \left[\left(a \frac{dU}{da} \right)_{a+da} - \left(a \frac{dU}{da} \right)_a \right],$$

which can be simplified to

$$(6.2.4) \qquad \frac{1}{\mu} \frac{(p_1 - p_2)}{\ell} = \frac{1}{a} \frac{d}{da} \left(a \frac{dU}{da} \right).$$

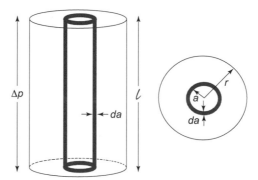

FIGURE 6.2.1. A segment of a tube with length ℓ and radius r over which a pressure difference Δp causes laminar fluid flow. The hollow cylinder (shaded) indicates an element of the fluid with inner radius a and thickness da, within which the flow velocity is taken as uniform.

In a narrow tube with constant diameter and no obstacles to fluid flow, the pressure gradient $(p_1 - p_2)/\ell$ is constant. If we also assume that μ is independent of the flow velocity (i.e., that we have a Newtonian fluid) such

BOX 6.2 **(Continued)**

---

that μ is independent of a, it follows that the differential equation (6.2.4) can be fulfilled only if

(6.2.5)
$$\frac{dU}{da} = \frac{1}{2\mu}\frac{(p_1 - p_2)}{\ell}a.$$

An essential assumption in the derivation of these relationships is that the flow velocity at the inner wall of the tube is zero. This condition, which is called the no-slip condition, holds true not only if the fluid adheres to the tube's material (e.g., water in a glass capillary), but even if the fluid does not wet the inner lining of the tube (e.g., mercury in a glass capillary). Given the boundary condition $U(r) = 0$, equation (6.2.5) can be solved to give the distribution of the flow velocity in the tube:

(6.2.6)
$$U(a) = \frac{1}{2\mu}\frac{(p_1 - p_2)}{\ell}\int_r^a a'\,da' = -\frac{1}{4\mu}\frac{(p_1 - p_2)}{\ell}\left(r^2 - a^2\right).$$

Equation (6.2.6) describes a parabolic distribution of flow velocities with a maximum velocity achieved in the middle of the tube ($a = 0$).

The flow through the tube, expressed as volume per time, is obtained by integration over the entire cross section:

(6.2.7)
$$\frac{\Delta V}{\Delta t} = \int_0^r U(a)\,2\pi a\,da.$$

Inserting equation (6.2.6) into the integral gives

(6.2.8)
$$\frac{\Delta V}{\Delta t} = -\frac{1}{4\mu}\frac{(p_1 - p_2)}{\ell}2\pi\int_0^r \left(r^2 - a^2\right)a\,da,$$

which yields

(6.2.9)
$$\frac{\Delta V}{\Delta t} = -\frac{\pi}{8\mu}\frac{(p_1 - p_2)}{\ell}r^4.$$

Notice that in its differential form, equation (6.2.9) is identical to equation (6.5) in the main text.

6.4 Flow of ideal fluids

A fluid that has no resistance to shear stress is called an ideal fluid. Such a hypothetical substance is modeled as manifesting what is called inviscid flow. That is to say, its flow regime is treated as if it was governed exclusively by inertial forces, and viscous effects are disregarded completely. The concept of inviscid flow leads to some useful relationships, such as the Bernoulli equation (see below), but it fails totally to describe drag forces. This failure is highlighted by what is known as d'Alembert's paradox, which states that an object such as a sphere or a cylinder moves without friction through an ideal fluid. Figure 6.1A illustrates the paradox that emerges from this assumption; namely, that when drag forces are neglected, the decelerating inertial forces on the foreside of an object are compensated for by accelerating inertial forces on the aft side in a symmetrical inviscid flow regime. This conclusion clearly flouts physical reality, because any *real* rigid body experiences a resistance to motion through any *real* fluid that is moving at any detectable speed.

Nevertheless, there is much to be learned by considering an ideal fluid. Consider, for example, a volume V of an incompressible fluid with mass M. At a height h above some reference height, this volume has a potential energy of PV, a gravitational energy of Mgh, and a kinetic energy of $0.5\rho V U^2$, where P denotes pressure. In the absence of frictional losses, the conservation of energy requires that the sum of these energies equals some constant. Therefore, if we divide both sides of this relationship by the volume, we obtain the Bernoulli equation,

$$(6.6) \qquad\qquad P + \rho g h + 0.5\rho U^2 = \text{constant}.$$

An example of the application of the Bernoulli equation is the calculation of the pressure around a cylinder of radius r in an inviscid fluid. If the fluid moves relative to the cylinder and in an orthogonal direction to the cylinder's length, the tangential velocity u_Θ of the flow at the surface of the cylinder is given by the formula

$$(6.7) \qquad\qquad u_\Theta = 2U_\infty \left| \sin\Theta \right|,$$

where Θ is the angle against the direction of flow and U_∞ represents the ambient (unobstructed) velocity of the ideal fluid at some great distance

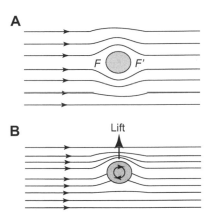

FIGURE 6.1. Graphic illustrations of streamlines in the flow of an ideal fluid around a cross section of a sphere or cylinder (A) and the Magnus effect (B). In an ideal fluid, the pressure force on the foreside is exactly compensated by the pressure force on the aft side ($F = -F'$). With regard to the Magnus effect, a rotating cylinder in an inviscid fluid experiences a drag force F at right angle to the fluid flow that generates lift.

from the cylinder's surface. In this example, the velocities in the radial and longitudinal directions with respect to the cylinder's orientation are zero. Together with the Bernoulli equation, in which the gravitational term is neglected, the foregoing leads to

(6.8) $$P_\infty + 0.5\rho U_\infty^2 = P + 0.5\rho \, (2U_\infty \sin\Theta)^2,$$

where P_∞ is the pressure in the region of undisturbed flow. Solving for the distribution of the pressure on the surface of the cylinder, we obtain the relationship

(6.9) $$P - P_\infty = 0.5\rho U_\infty^2 (1 - 4\sin^2\Theta).$$

Using this formula, we see that, at $\Theta = 0°$ (head on) and at $\Theta = 180°$ (on the aft side), $P - P_\infty = 0.5\rho U_\infty^2$, and that at a right angle to the direction of flow, the negative pressure difference is $P - P_\infty = -1.5\rho U_\infty^2$.

Even though drag is neglected when dealing with inviscid flow, if the cylinder rotates with an angular velocity ω, equation (6.7) has to be expanded by adding another term:

(6.10) $$u_\Theta = 2U_\infty \left|\sin\Theta\right| + \omega \, r\sin\Theta.$$

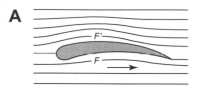

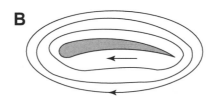

FIGURE 6.2. Streamlines around an airfoil seen (A) in a system of coordinates stationary with respect to the airfoil in which the pressure above the airfoil is less than the pressure below it ($F > F'$) and (B) in a system that moves with the fluid. The fluid passing over the upper surface of the airfoil has a higher flow speed than the fluid passing under the airfoil. This leads to a lower pressure on the upper surface, which generates lift (see eq. [6.6]).

Correspondingly, equation (6.9) will take the form

(6.11) $P - P_\infty = 0.5 \rho U_\infty^2 \left(1 - 4 \sin^2 \Theta - 4 |\sin \Theta| \dfrac{\omega r}{U_\infty} \sin \Theta - \dfrac{\omega^2 r^2 \sin^2 \Theta}{U_\infty^2} \right)$

(for a detailed treatment, see White 1979). Accordingly, equation (6.11) reduces to

$$P - P_\infty = 0.5 \rho U_\infty^2 \qquad \text{at } \Theta = 0°, 180°,$$

$$P - P_\infty = 0.5 \rho U_\infty^2 \left(-3 - 4 \dfrac{\omega r}{U_\infty} - \dfrac{\omega^2 r^2}{U_\infty^2} \right) \qquad \text{at } \Theta = 90°,$$

$$P - P_\infty = 0.5 \rho U_\infty^2 \left(-3 + 4 \dfrac{\omega r}{U_\infty} - \dfrac{\omega^2 r^2}{U_\infty^2} \right) \qquad \text{at } \Theta = 270°.$$

These equations show that there is a pressure difference between the left and right sides (as seen in the direction of flow) of the rotating cylinder (fig. 6.1B). This pressure difference results in "lift." It is equivalent to a drag force operating at a right angle to the direction of ambient flow. This phenomenon, which is called the Magnus effect (after H. G. Magnus, 1802–1870), has been utilized by outfitting sailing ships with a pair of rotating cylinders instead of normal sails. The Magnus effect is the basis for the curved path of a tennis ball that has been given a spin. It also motivates the spiral trajectory of the autogyroscopic descent of maple fruits (see Vogel 1988).

The consequences of the Bernoulli equation can also be illustrated with the example of an airfoil in a moving incompressible fluid (fig. 6.2). Due to the special form of the airfoil, the fluid has a higher flow speed above the convex upper side than below the concave lower side. According to equation (6.6), this leads to a lower pressure on the upper side than on the lower side, and thus to the generation of lift. The flow field can be seen from a different perspective as a superposition of the undisturbed inviscid flow and a vortex around the airfoil. The concept of vorticity will be discussed in the next section.

6.5 Boundary layers and flow of real fluids

The concept of the boundary layer, which was first advanced by Ludwig Prandtl (1875–1953), was mentioned in chapter 2, where we discussed a few basic transport laws. However, no formal definition was provided, and many subtleties of the concept were ignored.

As noted earlier, the boundary layer is a comparatively thin layer of fluid adhering to a solid object's surface in the fluid flow. By definition, it is the fluid layer in which the flow speed relative to the object changes from zero directly at the surface of the object to 99% of the speed in the unperturbed flow (fig. 6.3). In this boundary layer, even at high Reynolds numbers, shear, and consequently friction due to the viscosity of the fluid, play important roles. In the majority of studies, the quest is to know the boundary layer thickness δ. This parameter is particularly important for understanding heat conduction and convection and the diffusion of substances into and out of plant organs.

As noted in chapter 2 (section 2.2), the thickness of the boundary layer varies over a solid's surface. Intuitively, we expect δ to increase in the direction away from the solid's leading edge because we anticipate that frictional forces acting over the solid's surface will decrease flow speeds progressively in that direction. We also expect the boundary layer thickness

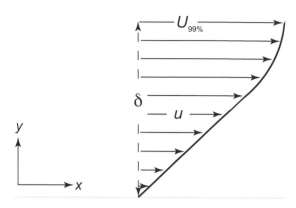

FIGURE 6.3. Representation of a boundary layer created at the surface of a stationary solid plate over which a fluid is moving with an ambient velocity $U_{100\%}$. Within this fluid layer, the local fluid velocity u rises from zero (no-slip condition) to the ambient velocity. The distance on the y-axis at which U reaches 99% of the ambient velocity is defined as the thickness of the boundary layer (δ).

to decrease everywhere over a solid's surface as Re increases. These expectations are consistent with the equations of laminar fluid motion, which show that $\delta \sim x/\mathrm{Re}^{0.5}$, where x is the distance from a solid's leading edge and Re relates to the ambient flow speed. Unfortunately, this relationship holds true only for large Re, which is not in contradiction to the concept of laminar flow, since flow in the boundary layer is laminar up to Reynolds numbers of roughly 10^5. Yet it confirms our intuition by showing that no single numerical value for δ holds true everywhere over a solid's surface. For this reason, most equations for determining the thickness of boundary layers give average values for δ.

Nevertheless, there are closed-form equations that provide estimates of the maximum thickness of boundary layers around very simple geometries. For example, the velocity profile in the boundary layers surrounding very thin flat plates oriented parallel to the flow (taken as the x-direction) is predicted to have a parabolic shape (fig 6.4), with local fluid speed u increasing in the direction away from the plate's surface (taken as the y-direction). If we take the distance y, where the fluid speed reaches 99% of the undisturbed flow speed U, as a reasonable definition of the thickness of the boundary, we find that

(6.12)
$$\delta = 4.75\left(\frac{vx}{U}\right)^{0.5},$$

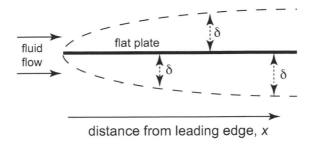

FIGURE 6.4. The thickness of the boundary layer δ as a function of the distance from the up-stream (leading) edge. The dashed parabolic line represents the 99% criterion (known as the Prandtl criterion) for the boundary layer thickness.

where v is the kinematic viscosity. Thus, setting $U = 1.0$ m/s and $x = 0.1$ m, this formula gives $\delta \approx 0.006$ m for air and $\delta \approx 0.0015$ m for water (both at 20°C).

In passing, it is worth noting that theory also shows that the frictional force per unit area σ_f of the plate at distance x from the leading edge is given by the formula

(6.13a)
$$\sigma_f = 0.33\rho U^2 \left(\frac{xU}{v}\right)^{-0.5},$$

and that the drag force D_f exerted on the two sides of a plate of length L and breadth B is

(6.13b)
$$D_f = 1.33\rho U^2 LB \left(\frac{LU}{v}\right)^{-0.5}.$$

Equations for the boundary layer thicknesses around "bluff bodies" (such as spheres or cylinders) are somewhat more complex than equations for those around flat plates because δ depends on where backflow occurs over downstream surfaces. These equations become even more complex if we consider what happens when the flow around a bluff geometry changes. Space precludes deriving the foregoing equations analytically (but see eq. [2.12] in section 2.2). Nor is there sufficient space to explore the many detailed studies devoted to testing them empirically. Suffice it to say that (1) many experiments have examined the local fluid velocity profiles

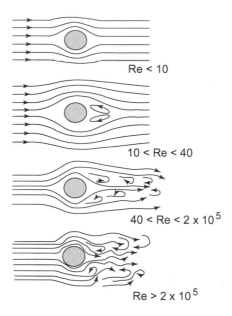

Re < 10

10 < Re < 40

$40 < Re < 2 \times 10^5$

$Re > 2 \times 10^5$

FIGURE 6.5. Flow regimes around a sphere or circular cylinder at increasing Reynolds numbers.

over the surfaces of flat plates, spheres, cylinders, and similarly simple geometries; (2) these experiments confirm that theoretical expectations and observations agree reasonably well; and (3) all estimates of boundary layer thickness based on closed-form solutions should be regarded as estimates of orders of magnitudes (much as Re provides an index for the relative importance of inertial versus viscous forces).

As depicted in figure 6.5, the steady state flow regime for Re ≪ 1 is a symmetrical fore-and-aft laminar flow around spheres and cylinders, although flow asymmetries may occur at distances much greater than the characteristic dimension away from either geometry (Tritton 1977). At slightly larger Re, the fore-and-aft symmetry begins to break down, and shear in the boundary layer leads to the generation of vortices. In the case of a circular cylinder, these phenomena become more and more noticeable, and by Re ≈ 9, a region of slowly circulating fluid appears immediately downstream of the cylinder. At Re ≈ 13, the circulating fluid takes on a definite pattern consisting of two eddies rotating in opposite directions. As Re increases further and approaches 40, the two attached counter-rotating eddies increase in length, their rotational axes increasingly align

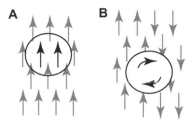

FIGURE 6.6. Graphic illustrations of (A) homogeneous flow and (B) vorticity in an inhomogeneous flow. A small circular element of a real fluid is shown in both systems. In the inhomogeneous system, drag forces will exert a torque on the circular element and thus lead to a rotation of the fluid element.

with the direction of ambient flow, and C_D decreases. At $40 < \text{Re} < 10^5$, counter-rotating eddies detach from the aft surface in an alternating pattern, producing a wake of counter-rotating vortices that move downstream (called a von Karman street). As Re approaches 10^5, the periodicity in the shedding of these eddies increasingly breaks down as turbulence develops in the boundary layer downstream of the aft surface.

As shown by Hermann von Helmholtz (1821–1894), vortices can originate or disappear only within a real fluid. Within an inviscid fluid, they are generated only at the boundary of objects obstructing flow and thereby causing inhomogeneity of the flow field and thus vorticity.

The concept of vorticity is important to many aspects of fluid mechanics, but it requires fairly sophisticated mathematics to comprehend fully (box 6.3). However, it is useful to conceptualize vorticity without the aid of mathematics. One way to visualize vorticity is to imagine a flowing fluid, concentrate on a small part of the fluid, and analyze its movement in a coordinate system that moves along with the flow of the rest of the fluid (fig. 6.6). If the small, "isolated" part of the fluid is rotating, rather than simply moving in the direction of fluid flow, it is said to have vorticity.

BOX 6.3 **Vorticity**

Vorticity is an important kinematic property of fluid flow. It is often referred to as the curl of the fluid velocity. Like the velocity, it can vary according to its location in three-dimensional space. Vorticity is defined as a three-dimensional vector,

(6.3.1) $$\vec{\Omega} = rot\vec{U} = \vec{\nabla} \times \vec{U}.$$

BOX 6.3 **(Continued)**

This expression relates to the term in the Navier-Stokes equation describing convective acceleration, which can be shown using a mathematical identity,

$$(6.3.2) \qquad (\vec{U}\cdot\vec{\nabla})\vec{U} = (\vec{\nabla}\times\vec{U})\times\vec{U} + \frac{1}{2}\,grad(U^2).$$

In a Cartesian coordinate system (x, y, z), vorticity can be described as the vector

$$(6.3.3) \qquad \vec{\Omega} = \left(\omega_x, \omega_y, \omega_z\right) = \left(\frac{\partial u_z}{\partial y} - \frac{\partial u_y}{\partial z}, \frac{\partial u_x}{\partial z} - \frac{\partial u_z}{\partial x}, \frac{\partial u_y}{\partial x} - \frac{\partial u_x}{\partial y}\right).$$

By its definition (and somewhat counterintuitively), vorticity is twice the angular velocity of a fluid element.

We can describe the "evolution" of vorticity in any fluid element by means of the vorticity equation expressed in vector form:

$$(6.3.4) \qquad \frac{d\vec{\Omega}}{dt} = \frac{\partial\vec{\Omega}}{\partial t} + \vec{U}\,(\vec{\nabla}\vec{\Omega}),$$

where $d\vec{\Omega}/dt$ describes the total change in the angular velocity of the fluid element, $\partial\vec{\Omega}/\partial t$ is the change in angular velocity as seen from the perspective of the fluid element, and $\vec{U}(\vec{\nabla}\vec{\Omega})$ is the change in the angular velocity with respect to the fluid system in which the element is moving.

One peculiarity of flow is the "potential vortex," a flow with circular streamlines, yet with a vorticity of zero. It is approximately realized by the flow near a drain hole in a container (and often observed as a swirl in the bathtub). The conditions are ($u_r = 0$, $u_\theta = C/r$, and $u_z = 0$), as formulated in cylindrical coordinates. $C = u_0/r_0$ is a constant, and u_0 and r_0 are reference quantities. As can easily be shown by expressing the nabla operator, also in cylindrical coordinates, $rot\vec{U}$ is zero everywhere, except for the very center, which marks a singularity.

In the absence of vortices, unless they are "potential vortices" (see box 6.3), we speak of an irrotational velocity field. Mathematically speaking, in an irrotational velocity field, the integration of the velocity over a closed path is always zero. This implies that the velocity can be seen as a gradient of a scalar function, called the velocity potential, which simplifies the mathematics considerably. Although the concept of inviscid irrotational potential flow is quite unrealistic, it has been used successfully to describe the flow near the foreside of an obstructing body but outside the boundary layer. If there is boundary layer separation at the aft side, the flow in the wake of the object will modify the streamlines of inviscid flow. The mathematical problem is modeling all of the interactions and patching together all of the different solutions. This can be done only by numerical methods.

6.6 Turbulent flow

Sir Horace Lamb (1849–1934) reportedly said, "When I die and go to heaven there are two matters on which I hope for enlightenment. One is quantum electrodynamics, and the other is the turbulent motion of fluids. And about the former I am rather optimistic." While laminar flow is smooth and steady, turbulent flow is characterized by random fluctuations of velocity and pressure at any given point outside the boundary layer and, at Reynolds numbers in excess of 10^5, even within the boundary layer. The transition between laminar and turbulent flow depends on many factors, such as the shape of the object obstructing flow and its surface roughness. However, the main determinant (or more precisely, the main indicator) is the Reynolds number. As of now, there is no analysis, nor even a numerical solution, that can simulate the fine-scale random fluctuations of turbulent flow. Therefore, the theory of turbulent flow is semiempirical and is concerned with the mean properties of flow and the mean fluctuations in flow, rather than their rapid variations (White 1979).

As shown in figure 6.7, there are three regions in the turbulent boundary layer above a flat surface extending in the x-, z-direction:

1. A wall layer, where viscous shear dominates
2. An overlap layer, where viscous and turbulent shear are important
3. An outer layer, where turbulent shear dominates

In the immediate vicinity of the surface, the averaged velocity $\bar{u}_x(y)$ in the direction of the overall flow (arbitrarily defined as the x-direction) is linearly related to the height y above the surface:

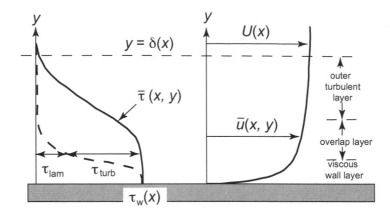

FIGURE 6.7. Average velocity and shear stress distribution in turbulent flow near a flat surface. The symbols are explained in the main text and in box 6.4.

(6.14)
$$\frac{\bar{u}_x(y)}{u^*} = \frac{yu^*}{v},$$

where u^* is the friction velocity as defined in box 6.4. This so-called "wall law" is valid up to a ratio of flow velocity to friction velocity of about 5. Thereafter, the function curves over to meet another function that obeys a logarithmic law,

(6.15)
$$\frac{\bar{u}_x(y)}{u^*} = \frac{1}{\kappa}\ln\left(\frac{yu^*}{v}\right) + B,$$

at a ratio of about 30. With a von Karman constant κ set at 0.41 and with $B = 5.0$, this *logarithmic-overlap layer equation* can be used as an excellent approximation for many turbulent flow problems, such as flow in a pipe or the vertical velocity profiles treated in section 6.9.

In the outer layer, the difference of the averaged velocity $\bar{u}_x(y)$ from the overall stream velocity U_x can be expressed by the relation

(6.16)
$$\frac{U_x - \bar{u}_x(y)}{u^*} = G\left(\frac{y}{\delta}\right),$$

where G is a function that goes asymptotically to zero as y approaches δ.

BOX 6.4 **Turbulent stresses and friction velocities**

Here, we present a semiempirical theory that is based on the suggestion of Osborne Reynolds involving time-averaged turbulent variables such as

$$(6.4.1) \qquad \overline{u}_x = \frac{1}{T} \int_0^T u_x \, dt,$$

where u_x is the velocity component in the x-direction, t is time, and T is an averaged time interval. For turbulent air or water flow, an averaging time $T \approx 5$ s is usually adequate. In a three-dimensional Cartesian coordinate system, the fluctuation in local flow speed u'_x is defined as the deviation of u_x from the averaged value $u'_x = u_x - \overline{u}_x$, and likewise, $u'_y = u_y - \overline{u}_y$, $u'_z = u_z - \overline{u}_z$, and $P' = P - \overline{P}$. The continuity equation emerging from this definition takes the form

$$(6.4.2) \qquad \frac{\partial \overline{u}_x}{\partial x} + \frac{\partial \overline{u}_y}{\partial y} + \frac{\partial \overline{u}_z}{\partial z} = 0,$$

whereas the Navier-Stokes equation for the velocity in the x-direction (eq. [6.1.2]) can be expressed as

$$(6.4.3) \qquad \begin{aligned} \rho \frac{d\overline{u}_x}{dt} &= -\frac{\partial \overline{P}}{\partial x} + \rho g_x + \frac{\partial}{\partial x}\left(\mu \frac{\partial \overline{u}_x}{\partial x} - \rho \overline{u'_x u'_x} \right) \\ &+ \frac{\partial}{\partial y}\left(\mu \frac{\partial \overline{u}_x}{\partial y} - \rho \overline{u'_x u'_y} \right) + \frac{\partial}{\partial z}\left(\mu \frac{\partial u_x}{\partial z} - \rho \overline{u'_x u'_z} \right), \end{aligned}$$

where μ is the dynamic viscosity, which, like density ρ, is considered a constant for most applications.

The three correlation terms $-\rho \overline{u'_x u'_x}$, $-\rho \overline{u'_x u'_y}$, and $-\rho \overline{u'_x u'_z}$ can be thought of as turbulent stresses because they have the same units as stress. However, these three terms actually describe convective accelerations, which must be determined experimentally by flow visualization methods. Visualizations of flow within tubes and close to flat surfaces have shown that the stress $-\rho \overline{u'_x u'_y}$ is dominant (White 1979), so equation (6.4.3) can be simplified to

$$(6.4.4) \qquad \rho \frac{d\overline{u}_x}{dt} = -\frac{\partial \overline{P}}{\partial x} + \rho g_x + \frac{\partial \tau}{\partial y},$$

BOX 6.4 **(Continued)**

where the shear stress τ can be expressed as the sum of two terms:

(6.4.5)
$$\tau = \mu \frac{\partial \bar{u}_x}{\partial y} - \rho \overline{u'_x u'_y} = \tau_{lam} + \tau_{turb},$$

with $\tau_{turb} = -\rho \overline{u'_x u'_y}$. Figure 6.7 shows that τ_{lam} goes to zero outside the overlap layer and that τ_{turb} goes to zero outside the boundary layer. The shear stress τ extrapolated to $y = 0$ is called the wall shear stress τ_w.

A convenient way to express the relationship between flow velocity and shear stresses in a turbulent flow regime is the introduction of the friction velocity u^*. This parameter, which has units of m/s (even though it is not strictly a velocity), is defined as

(6.4.6)
$$u^* = \left(\frac{\tau_w}{\rho} \right)^{1/2}.$$

When flow occurs over a flat surface, the velocity in the direction of the overall flow $\bar{u}_x(y)$ (which is arbitrarily defined to be in the x-direction) is a function of the height y above the surface (fig. 6.4.1),

(6.4.7)
$$\frac{\bar{u}_x(y)}{u^*} = \frac{y u^*}{v} \text{ for } \frac{y u^*}{v} \leq 5.$$

Unfortunately, there exists no analytical expression for the velocity $\bar{u}_x(y)$ in the range $5 < \frac{y u^*}{v} \leq 30$. However, in the region $\frac{y u^*}{v} > 10^3$, defined as the outer turbulent layer, the difference from the overall stream velocity U_x is a function G of the quotient of y and the boundary layer thickness δ,

(6.4.8)
$$\frac{U_x - \bar{u}_x(y)}{u^*} = G\left(\frac{y}{\delta} \right) \text{ for } \frac{y u^*}{v} > 10^3.$$

The function G can be different for different flow profiles (see fig. 6.4.1), but it will asymptotically go to zero as y approaches δ. The two relationships between $\bar{u}_x$ and y (see eqs. [6.4.7] and [6.4.8]) must overlap smoothly, which will be true if a logarithmic law governs the relationship between $\bar{u}_x$ and y in the overlap region:

BOX 6.4 **(Continued)**

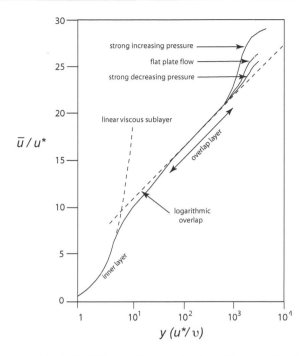

FIGURE 6.4.1. The averaged velocity $\bar{u}(y)$ normalized as a function of the normalized height y above a flat surface in a turbulent boundary layer. (Adapted from White 1979.)

(6.4.9)
$$\frac{\bar{u}_x(y)}{u^*} = \frac{1}{\kappa}\ln\left(\frac{yu^*}{\upsilon}\right) + B \quad \text{for } 30 < \frac{yu^*}{\upsilon} \leq 10^3,$$

where κ is the von Karman constant.

Surface roughness has very little effect on laminar flow, but turbulent flow regimes are strongly affected by this parameter. Experimentally, it can be shown that the constant B in equation (6.15) is numerically reduced by roughness elements, such that the flow velocities in the overlap layer are reduced. Fully rough flow occurs when $\varsigma u^*/u > 70$, where ς denotes the height of the roughness elements. The change in the constant B with respect to changes in roughness is given by the empirical formula

(6.4.10)
$$\Delta B \approx \frac{1}{\kappa}\ln\left(\frac{\varsigma u^*}{\upsilon}\right) - 3.5.$$

BOX 6.4 **(Continued)**

Inserting this relationship into equation (6.4.9) leads to

(6.4.11)
$$\frac{\bar{u}_x(y)}{u^*} = \frac{1}{\kappa} \ln\left(\frac{yu^*}{v}\right) + B - \Delta B \approx \frac{1}{\kappa} \ln\left(\frac{y}{\zeta}\right) + 8.5.$$

It is interesting to see that the kinematic viscosity v is eliminated in the right-hand side of this equation, implying that sufficiently large roughness elements counter-act the establishment of a viscous wall layer.

As explained in more detail in box 6.4, equations (6.14)–(6.16) provide a framework for all theoretical descriptions of turbulent-shear flows (White 1979). (As a historical aside, it should be noted that none of these equations was derived by solving the Navier-Stokes equations. Instead, Prandtl, von Karman, and Millikan deduced these relations from dimensional analysis and experimental observations.)

6.7 Drag in real fluids

Ludwig Prandtl resolved d'Alembert's paradox by introducing the concept of the boundary layer, which is essential for understanding drag forces. As explained in the previous section, close to the surface of a body moving relative to a fluid (except at very high Reynolds numbers), there is a layer of the fluid in which the flow is laminar. In this layer, viscous (shear) forces are prevalent, which invariably lead to drag.

Drag forces were introduced in section 2.3 (see eq. [2.15]), but were neither sufficiently defined nor explored in detail. It is useful to do so now before considering how flow regimes change as a function of Re.

To understand drag, one must appreciate that a fluid produces shear stresses and normal stresses over the surfaces of any rigid body obstructing its flow. The total force resulting from normal stresses has three components:

1. Lift, as described in the previous section, is the force component normal to the direction of fluid motion that owes its existence to the generation of pressure differences around rigid surfaces resulting from vorticity.

2. Form drag is the resultant pressure force parallel and opposed to the direction of fluid motion.
3. Induced drag results when a rigid body supplies kinetic energy to a surrounding fluid as vortices and eddies are produced downstream.

Closed-form solutions for these force components have been developed and tested empirically for a variety of simple geometries.

For bluff bodies, a general formula exists for the total drag force, which shows that drag is proportional to the area A that an object projects toward the oncoming fluid and, in contrast to Stokes's formula (eq. [6.4]), to the square of ambient fluid velocity U:

(6.17) $$F_{Drag} = 0.5\rho\, A U^2 C_D.$$

We have seen the term $0.5\rho U^2$ in the Bernoulli equation (see eq. [6.6]). It is called "impact pressure" or "dynamic pressure." Notice that U appears in the formula for Re (see eq. [6.3]), and that one of the linear dimensions for A appears there as well. In this sense, Re "contains" some of the salient variables defining drag. However, the formula for drag also contains a dimensionless coefficient, the drag coefficient C_D, which is numerically correlated with Re, but not directly linked to it as are U and A. The drag coefficient depends very much on the shape of an object. A flat plate positioned at a right angle to the direction of flow experiences more than twice the drag of a sphere, while streamlined bodies have substantially lower drag coefficients.

It is particularly instructive to explore the relationship between C_D and Re in the context of flow around well-studied geometries, such as circular cylinders and spheres. Figure 6.8 shows this relationship for a cylinder over a very large range of Reynolds numbers. For Re < 1, we can express Stokes's formula (eq. [6.4]) in terms of the drag equation (eq. [6.17]). For a sphere with $A = \pi r^2$, this yields

(6.18) $$6\pi r\mu U = 0.5\rho\pi r^2 U^2 C_D.$$

Taking r as the characteristic length in the Reynolds number Re $= rU/v$, and realizing that $\mu = v\rho$, equation (6.18) leads to the simple relationship

(6.19) $$C_D = 12/\text{Re} \quad \text{for Re} < 1.$$

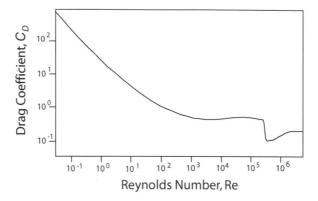

FIGURE 6.8. Drag coefficient C_D for a circular cylinder plotted as a function of the Reynolds number Re. (Adapted from Feynman et al. 1967.)

The drag coefficient continues to decline up to Re ≈ 1,000 (for a cylinder) and Re ≈ 4,000 (for a sphere), thereafter increasing slightly up to Re ≈ 100,000. At even higher Re, the situation becomes more complex. Prandtl was the first to suggest that the reason for the precipitous drop in C_D at $10^5 <$ Re $< 10^6$ lies in the behavior of the boundary layer, where the flow changes from laminar to turbulent. This transition is associated with a shift in the rate of exchange of momentum between different fluid layers, which is much greater in turbulent boundary layers than in laminar boundary layers.

As a fluid moves fore to aft around a bluff body, the pressure in the fluid goes from high to low. However, along this pressure gradient, a region of increased pressure invariably originates aft of the point of lowest pressure. The pressure distribution around a cylinder in an inviscid fluid was described in equation (6.9). The distribution is slightly more complex for laminar or turbulent flow (White 1979). For the fluid to pass beyond the point of lowest pressure, fluid momentum must be lost locally, and at high Reynolds numbers, the loss of this momentum causes the fluid to flow in a pattern that separates from the geometry of the bluff body's aft surface (see fig. 6.5). The point at which separation occurs is aptly called the *separation point of flow*. It also defines the point at which the wall shear stress is zero (see box 6.4) and where the boundary layer around the bluff body begins to thicken dramatically.

6.8 Drag and flexibility

Unlike rigid engineered objects studied in wind tunnels or flumes, many organic structures are sufficiently flexible to change their shape or orientation with respect to the direction of fluid flow. These reconfigurations can dramatically reduce drag by reducing the area an object projects toward fluid flow. They can also have a direct effect on the numerical value of C_D, as seen when trees or other plants in heavy winds deflect and become more streamlined (Vogel 1981, 1988).

Drag coefficients for bluff bodies are almost independent of flow speeds at Reynolds numbers between 1×10^3 and 300×10^3. Trees, on the other hand, can change their shape by bending stems and reshaping leaf laminae. Mayhead (1973) reported data from wind tunnel studies on trees with a height between 5.8 and 8.5 m. As an example, figure 6.9 shows a plot of the drag coefficients of a Scots pine (*Pinus sylvestris*) as a function of flow speed. In a limited range of speeds, the data can be represented by the power function $C_D = \beta U^\alpha$. Noting that it is often difficult to define the projected area of plants, Vogel (1984) analyzed data from various species by using the power function $F_{Drag}/U^2 = \beta U^\alpha$, where β is the regression proportionality constant and the scaling exponent α is a measure of the reduction of drag resulting from the flexibility of the plant body. De Langre (2008) summarized the effects of wind on plants, particularly the complex interactions between wind and tree movement (see section 5.12). Even if the drag coefficient and the scaling exponent α are known, a calculation of the drag and the wind-induced bending moment on a tree in its natural environment is not straightforward. A computation requires, inter alia, knowledge of the wind profile—that is, the wind speed as a function of the height above ground (see section 6.9)—which is rarely known with sufficient accuracy. Yet the practical implications for safety assessment and forest management have motivated several estimates of wind-induced bending moments for real trees in real environments (Peltola and Kellomäki 1993; Milne 1995; Spatz and Brüchert 2000; Sinn 2003; Peltola 2006; see also Schindler et al. 2010).

Although bending moments are often the most critical mechanical loads for plants with self-supporting stems, many benthic plants are critically loaded in tension due to their pronounced flexibility and the large forces exerted on them by moving water. The central position of the load-bearing inner cortical tissue of the stipe of the giant kelp *Nereocystis*

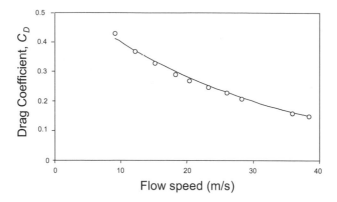

FIGURE 6.9. Plot of the drag coefficients of a Scots pine as a function of the flow speed in a wind tunnel. The drag coefficients relate to the projected area in still air. The data are approximated by a power function. Note that the vertical wind speed profile in a wind tunnel is uniform (U is constant regardless of elevation), such that the relationship between C_D and flow speed does not reflect an ecologically real situation. (Adapted from Mayhead 1973.)

luetkeana permits bending together with high resistance to tension (Koehl and Wainwright 1977). Due to its high extensibility, the work of fracture of the stipe is comparable to that of wood. In passing, it is worth noting that the microfibrillar angle of the cellulose layers in the inner and outer cortical cells of this alga is 60°, which is very close to the microfibrillar angle at which the volume change of a cylinder subjected to small extensions or compressions is zero (see box 4.2). This arrangement is advantageous for the extensibility of a closed cylinder filled with an incompressible fluid, as the change of internal pressure is small upon moderate straining.

By means of elegant towing experiments, Koehl (1999) measured the drag forces on the marine red alga *Chondracanthus exasperatus* as a function of relative flow speed. A low drag coefficient of 0.08 was observed, which was additionally reduced due to considerable reconfigurations of the algal thallus at increased speeds. A list of characteristic values for several other macroalgae is provided by Koehl. In an entirely different range of Reynolds numbers, the submerged water moss *Fontinalis antipyretica* is sufficiently flexible to stretch longitudinally in the direction of flow even at flow speeds as low as 0.1 m/s. Correspondingly, the drag measured in a flume is proportional to the square of the velocity; that is, at flow speeds between 0.1 and 0.5 m/s, the scaling exponent α for the relationship $F_{Drag}/U^2 = \beta U^\alpha$ is zero. Except for a small additive term, the drag is linearly related to the length of this moss (Biehle et al. 1998).

Flexibility is of great importance for organisms in wave-swept environments. The dynamic forces experienced by plants will depend on the ratio of their length to wavelength (i.e., the distance between the maxima of two waves). Short flexible plants will be stretched out fully before the peak water velocity is reached and will thus experience high tensional stresses. A long flexible plant may not be stretched out fully before the direction of the water current is reversed. Therefore, individual body parts actually do not move relative to the flow of water around them. Koehl (1999), measuring flow velocities and forces on two specimens of the kelp *Alaria marginata* measuring 1 and 2.8 m in length, showed that the forces experienced by these two plants under otherwise identical conditions were nearly the same. As mentioned in chapter 2, one way to avoid large stresses is to grow to a size that permits the movement of body parts to accord exactly with the oscillatory movements of local water currents.

6.9 Vertical velocity profiles

As discussed in the previous sections, many biomechanical phenomena operate in the world of large Reynolds numbers, where inertial forces dictate fluid regimes. One very important aspect of this world is the way in which a fluid's velocity profile changes with respect to the surface to which objects are attached; that is, its vertical velocity profile. This aspect of fluid flow can be dealt with directly by measuring ambient flow at different locations (with respect to the substrate) within and above plant communities. In many cases, however, it is convenient to consider velocity profiles theoretically in much the same way in which the boundary layers around small objects are treated. In this section, we briefly review some of the physics relevant to understanding macroscopic velocity profiles for aquatic and terrestrial plant communities. This review is necessarily incomplete because the topic is broad and requires complex mathematics to be fully understood.

Consider first a wave-swept habitat for which the direction of flow is oscillatory. This type of flow environment was first explored by Stokes, who considered the problem in terms of a solid flat plate oscillating back and forth in an infinite body of stationary water. Stokes concluded that the plate has a horizontal velocity u with respect to the moving fluid that equals $u = u_0 \cos(\omega t)$, where ω is the frequency of oscillation and t is time. Given the no-slip condition, which states that $u = 0$ at the surface of the plate ($y = 0$), Stokes showed that the velocity at time t and distance y from the plate in the vertical direction is given by the formula

(6.20a) $\quad u(y,t) = u_0 \exp\left[-y\left(\dfrac{\pi}{vT}\right)^{0.5}\right]\cos\left[\omega t - y\left(\dfrac{\pi}{vT}\right)^{0.5}\right],$

where v is the kinematic viscosity and T is the oscillation period.

This formula requires modification to be useful because the physical reality of a wave-swept substrate involves water moving back and forth over a boundary layer, rather than the system treated by Stokes. This modification requires a shift in the frame of reference so that the plate is made to look stationary. With this shift, the apparent velocity of the body of moving water is $u = -u_0 \cos(\omega t)$, such that equation (6.20a) becomes

(6.20b) $\quad u(y,t) = u_0 \exp\left[-y\left(\dfrac{\pi}{vT}\right)^{0.5}\right]\cos\left[\omega t - y\left(\dfrac{\pi}{vT}\right)^{0.5}\right] - u_0 \cos(\omega t).$

Notice that when $y = 0$ (at the surface of the substrate), equation (6.20b) reduces to $u(0, t) = 0$, a condition that is a requirement of the no-slip condition. In addition, when y is large, equation (6.20b) reduces to $u(y, t) = -u_0 \cos(\omega t)$, which indicates that the fluid is oscillating with respect to the plate's surface. The term $y(\pi/vT)^{0.5}$ in the cosine function indicates a phase shift in the velocity at some distance y from the plate's surface relative to the oscillating flow far from the surface. Finally, equation (6.20b) shows that the reduction in velocity caused by the submerged plate decreases exponentially with increasing y. This is indicated by the term $\exp[-y(\pi/vT)^{0.5}]$. Unfortunately, the phase-shift term $y(\pi/vT)^{0.5}$ makes it difficult to actually calculate the thickness of the boundary layer. However, inspection of equation (6.20b) shows that sessile organisms measuring only a few centimeters in height will experience local velocities comparable in magnitude to those in the mainstream flow in wave-swept environments.

Naturally, Stokes's solution assumes that the surface of the plate is smooth and that it is absolutely parallel to the horizontal back-and-forth oscillation of the fluid. Both of these conditions are rarely, if ever, fulfilled in nature. Efforts to extend Stokes's solution to deal with more realistic flow conditions indicate that fluid flow over plates is not strictly oscillatory because net transport can occur in the direction of wave motion (Longuet-Higgins 1953). These efforts indicate that the speed of transport at distance y from the surface is given by the formula

(6.21) $u_x(y) = \dfrac{2\pi H^2}{16T}\left(\dfrac{k}{\sin h^2(kd)}\right)\left[5 - 8\exp\left(\dfrac{-y}{C}\right)\cos\left(\dfrac{y}{C}\right)\right] + 3\exp\left(\dfrac{2y}{C}\right),$

where H is wave height, k is the wave number (which equals 2π divided by the wavelength λ), d is depth measured from the fluid's surface, and $C = (vT/\pi)^{0.5}$. This formula appears to successfully approximate the behavior of fluid flow over gently sloping beaches.

Similar attempts to model vertical velocity profiles have yielded formulas to deal with different flow conditions or objectives. One frequently encountered formula used to estimate time-averaged changes in fluid speed under turbulent conditions is

(6.22) $$\bar{u}_x = \frac{u^*}{\kappa}\ln\left(\frac{y-d}{z_0}\right) \quad \text{with } y - d \geq z_0,$$

where $\bar{u}_x$ is the time-averaged fluid speed at distance y from the substrate, u^* is the friction velocity, κ is the von Karman constant, d is the zero plane displacement, and z_0 is the roughness height (see box 6.4). This formula has been used by oceanographers to describe turbulent flow and by terrestrial ecologists interested in wind flow patterns within canopies.

Because of its general use, a brief description of the parameters in equation (6.22) is warranted. As defined in box 6.4, the friction velocity u^* is a measure of the magnitude and the correlation of turbulent fluctuations in velocity near the substrate. The zero plane displacement is a correction factor that "lifts" the location directly above the substrate ($y = 0$) to the location where the drag forces per unit height experienced by the plant community are maximal. In most cases, d is set equal to 70% of the plant community height. But the percentage value used should reflect the height at which the greatest number of plant organs causing community drag reside (consider where leaves are concentrated in a cornfield versus a grove of coconut palms). Finally, the roughness height characterizes the friction to horizontal air movement caused by the lower part of the canopy. The numerical value for this parameter can be determined by plotting $\bar{u}_x$ as a function of $\ln(y - d)$ and noting that $(y - d) = z_0$ when $\bar{u}_x = 0$; that is, the regression curve for $\bar{u}_x$ versus $\ln(y - d)$ is extrapolated to determine z_0. In general practice, the roughness height z_0 is typically set at 10% of community height for dense vegetation and less than 10% for sparsely populated communities.

Formulas have also been developed for vertical velocity profiles *within* plant communities. Three are provided here:

(6.23a)   Logarithmic profile: $\bar{u}_x = \dfrac{u_{top}}{\ln\left(\dfrac{h}{z_0}\right)} \ln\left(\dfrac{y}{z_0}\right)$

(6.23b)   Square root profile: $\bar{u}_x = u_{top}\left(\dfrac{y}{h}\right)^{0.5}$

(6.23c)   Square profile: $\bar{u}_x = u_{top}\left(\dfrac{y}{h}\right)^{2}$

In all of these equations, u_{top} is the ambient wind speed above the canopy with the top at height h, and y denotes the height above the substrate. These formulas, two of which neglect the roughness height, predict vertical velocity profiles that differ significantly in their shape, which cautions against using any such formula a priori.

This caveat is warranted in light of empirical measurements of local wind speeds within tree canopies that are best described statistically by third-order polynomial regression curves (Niklas and Spatz 2000). This phenomenology is consistent with the supposition that the flow regimes within plant communities generally consist of three horizontal layers (see Bussinger 1975):

1. A top layer in which the flow regime is dominated by the drag on foliage
2. A middle layer whose behavior is governed by branching architecture, foliage, and ground effects
3. A bottom (understory) layer that can experience gusts at wind speeds approaching those above the canopy, depending on plant density

Although formulas for predicting the flow regime profiles in layers 2 and 3 have not been developed, Bussinger (1975) gives a formula for wind speeds in the top layer:

(6.24) $\bar{u}_x = u_{top} \exp\left[a\left(\dfrac{y}{h} - 1\right)\right],$

where a is an attenuation factor (ranging between 0 for sparse canopies and 4 for dense canopies) and y represents distance above ground level.

6.10 Terminal settling velocity

The behavior of very small objects suspended in air or water is of great interest to a broad spectrum of scientists, ranging from those studying wind-borne pollutants to biologists studying phytoplankton or the dispersal of spores and pollen by wind. It is therefore useful to discuss the physics underlying the behavior of very small objects suspended in fluids, particularly their terminal settling velocities.

When an object is suspended in a fluid that is either more or less dense than itself, the object accelerates either upward or downward. After some time, the object's net body force (weight minus buoyancy) and the drag force exerted on it by its motion relative to the fluid reach equilibrium, and the speed of ascent or descent becomes constant. This constant velocity is called the terminal settling velocity—a phrase that is misleading because it implies that objects always descend, which is not the case when they are less dense then the fluid surrounding them.

To illustrate the concept of terminal velocity, consider a small sphere with density ρ_s and volume V suspended in a fluid with density ρ_f. The net body force F_N on this sphere equals the difference between the mass of the sphere and the mass of the fluid it displaces times the gravitational constant. Thus, the net body force equals

$$(6.25) \qquad\qquad F_N = gV(\rho_s - \rho_f).$$

In turn, the drag force on a very small sphere with diameter d is given by Stokes's formula (see eq. [6.4]):

$$(6.26) \qquad\qquad D_f = -3\pi\, dv\rho_f U.$$

This formula holds true for Re $\leq$ 1.0. At equilibrium, $F_N + D_f = 0$, which leads to a formula for the terminal velocity,

$$(6.27) \qquad\qquad U_T = \frac{gV}{3\pi d}\frac{1}{v}\frac{(\rho_s - \rho_f)}{\rho_f}.$$

Notice that the sphere will ascend when $\rho_s < \rho_f$ and descend when $\rho_s > \rho_f$. Because $U_T = \mathrm{Re}(v/d)$ (eq. [6.3]), we can rearrange equation (6.27) to obtain

(6.28)
$$\mathrm{Re} = \frac{gV}{3\pi} \frac{1}{v^2} \frac{(\rho_s - \rho_f)}{\rho_f}.$$

Applying the condition $\mathrm{Re} \le 1$, we find an upper estimate for the volume of the sphere for which the settling velocity and the drag force are linearly related:

(6.29)
$$V \le 3\pi \frac{v^2}{g} \frac{\rho_f}{(\rho_s - \rho_f)}.$$

Note that for large spheres, the drag force is proportional to the square of the settling velocity.

Equation (6.29), together with knowledge of the kinematic viscosity v and the density of water or air ρ_f (see table 2.1), can be used to define what is meant by a "small" sphere. Assuming that a hypothetical sphere has a density of 1,200 kg/m³, we see that, in air, the volume of the sphere cannot exceed 0.217×10^{-12} m³, and the radius of the sphere r can be no larger than 37 μm. In the case of water, the calculation yields $V \le 4.75 \times 10^{-12}$ m³ and $r \le 100$ μm for the condition $\mathrm{Re} \le 1$ to be true. Clearly, the word "small" has a relative meaning (relative to the physical properties of the fluid). Nevertheless, objects that reside in the size range 37–100 μm are small by most macroscopic standards.

Formulas to compute the drag forces on nonspherical geometries for $\mathrm{Re} \le 1$ have been developed and can be used to estimate terminal settling velocities for some pollen grains, spores, and plankton. For example,

(6.30a)
$$D_F = 16\, v\rho_f r\, U$$

and

(6.30b)
$$D_F = 10.67\, v\rho_f r\, U$$

for a flat disk with radius r aligned against and parallel to the oncoming fluid, respectively. Likewise, for *very long* cylinders with length L and diameter D aligned with a fluid's motion,

(6.31a) $D_F = 2\pi v \rho_f\, UL/[\ln(2L/D) - 0.807]$ with $L \gg D$,

and for a spheroid with length L and diameter D aligned with a fluid's motion

(6.31b) $D_F = 2\,\pi v \rho_f\, U\, L/[\ln(2L/D) - 0.5]$ with $L \gg D$

(see Cox 1970).

These formulas show that shape distortion can have a significant effect on terminal velocities because a greater surface area (at an equivalent volume and density) results in a greater volume-specific frictional surface (i.e., an increase in *form resistance*) that reduces U_T relative to that predicted by Stokes's formula. For example, in the case of prolate and oblate spheroids, the extent to which observed and predicted values for U_T deviate is correlated with the quotient of the axis of the geometry aligned with the object's motion (here denoted by x) and the square root of the product of the other two axes (denoted by y and z); that is, $x/(yz)^{0.5}$. In the case of slender objects, spheroids with the greatest distortion in shape (very narrow prolate shapes and very flat oblate shapes) offer the greatest form resistance compared with spheres with the same volume. In addition, longer cylinders are expected to sink more slowly than their equivalent spheres, provided all Stokesian conditions are fulfilled (e.g., laminar flow at low Re).

Because the effects of shape distortion on terminal velocity have not been resolved analytically for many biological objects, some workers have turned to the more prosaic method of calculating a coefficient of form resistance, here denoted by φ. This parameter is the dimensionless quotient constructed by using equation (6.27) to calculate the terminal velocity for a sphere with the same volume and density as the object of biological interest, U_T^{calc}, and dividing this result by the observed settling velocity: $\varphi = U_T^{calc}/U_T^{obs}$. Statistical comparisons between the values of φ and specific morphological features can provide insights into how these features affect U_T^{obs}. This approach has merit, but it also presents some pitfalls. For example, Hutchinson (1967) examined the sinking rates of small chains of two, three, four, or eight glass beads that were glued together and, based on a linear regression, reported that $\varphi = 0.837 + 0.163\,n$, where n is the number of beads per chain. This formula is correct within the size range $1 < n \leq 8$, but it should not be extended beyond this range.

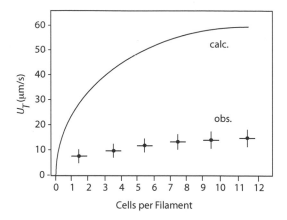

FIGURE 6.10. A bivariate plot of observed settling velocities for filaments (of different cell numbers) of the freshwater alga *Aulacoseira subarctica* and those calculated for spheres with equivalent volumes and densities. (Data from Reynolds 1984.)

A far more informative study regarding the sinking rates of chains of cells is that of Reynolds (2006), who measured the terminal settling velocities of filaments of cells of the freshwater diatom *Aulacoseira subarctica* and compared those settling velocities with those of spheres with equivalent volumes and densities. A direct comparison between the velocities observed for filaments and the velocities calculated for spheres shows that the filaments of this alga descend in still water much more slowly than if all the cells in each filament were grouped together to form a single cell (fig. 6.10). Thus, multicellularity confers an immediate advantage, at least in this particular case.

Some early workers suggested that shape distortions might have adaptive significance that has nothing to do with the benefits gained by reducing terminal velocity. For example, shape distortions might result in a preferential orientation as an object ascends or descends. However, fluid mechanics theory and experimental studies reveal that, in a truly vortex-free viscous medium, objects with uniform mass distributions will maintain their orientations once they achieve their terminal velocities. The stipulation of uniform mass distribution along the axis parallel to sinking is an important qualification and explains why teardrop-shaped objects travel with their blunt "heavy" ends facing forward, why pine pollen grains fall with their air-filled bladders (called sacci) trailing behind their cell-filled bulk, and why some unicellular algae continuously spiral downward as they sink in water.

6.11 Fluid dispersal of reproductive structures

Plants have evolved a variety of ways to discharge and disperse reproductive structures. Some plants use mechanical catapults to eject seeds or fruits (e.g., Howe and Smallwood 1982; Hayashi et al. 2010). Other species capitalize on the behavior of water or air to passively spread gametes, spores, pollen, seeds, and fruits. The size range occupied by these disseminules is remarkable, particularly in the aquatic habitat. For example, the nonflagellated spermatium of the red alga *Duresnaya crassa* measures about 3×10^{-6} m in diameter (and has an estimated terminal velocity of 10 μm/s), whereas the floating seeds of the coco-de-mer (*Lodoicea maldivica*) can exceed 0.38 m in length and travel hundreds of miles floating on water. The size range of wind-dispersed reproductive structures is less extensive, but still respectable. For example, the wind-dispersed seeds of orchids can measure 1×10^{-4} m in diameter and, because of their small mass ($\approx 5 \times 10^{-10}$ kg), achieve terminal velocities comparable to those of wind-dispersed pollen grains (i.e., ≈ 0.3 m/s), which can travel hundreds of kilometers (Robledo-Arnuncio 2011). Autogyroscopic seeds and fruits reside on the other end of the wind-dispersed size spectrum, ranging between 10^{-2} and 10^{-1} m in length and weighing as much as 0.05 kg. An additional and largely unexplored aspect of fluid dispersal is the complex methods that have evolved in many plants to trap suspended particulates such as pollen grains. An interesting example is the aerodynamic effects of flower structure on the motion and deposition of wind-borne pollen grains (fig. 6.11).

Regardless of which example of fluid dispersal we consider, it is clear that this phenomenon occurs in both the small and viscous world of low Reynolds numbers and the large and inertial world of very large Re. And, because plants sink less rapidly in water than in air, water can transport much larger organic structures over longer distances than can wind. But how far can an object travel passively in a fluid? Intuitively, the answer depends on how long the object remains suspended and the extent to which the fluid's horizontal and vertical velocity components vary.

In the case of a sinking object (spore, pollen grain, dust seed, etc.), the extent to which it remains in suspension depends on the quotient of its terminal velocity U_T and the extent to which the fluid's vertical velocity component u_V varies. It also depends on the height of release h. Likewise, the extent to which the object is horizontally transported depends on variation in the horizontal velocity component u_H. In perfectly still fluids,

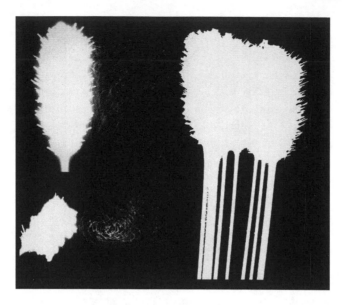

FIGURE 6.11. Stroboscopic photographs of pollen grains moving around leeward surfaces of the inflorescence of the grass *Setaria* oriented vertical to airflow (*top left*) and at a 45° angle to airflow (*bottom left*). An oscillating motion can be induced when an inflorescence begins to shed eddies of airflow (*shown at right*). Ambient airflow (at 5 m/s) is from left to right in each case.

$u_V = u_H = 0$, and the suspension time is governed by U_T and h. When vorticity occurs, however, u_V and u_H are expected to vary significantly such that their mean values, $\overline{u_V}$ and $\overline{u_H}$, affect both suspension time and horizontal displacement.

Under macroscopic turbulent conditions, we might expect $\overline{u_V}$ to be more influential in predicting the dispersal distance of objects with very slow descent velocities because their descent would be slow in comparison to the intensity of fluid upwelling that would delay their reaching the substrate. Indeed, Okubo and Levin (1989) demonstrate mathematically and empirically that the horizontal transport distance d_H of slowly descending objects (i.e., $U_T < 1.0$ m/s) can be approximated by the formula

$$(6.32a) \qquad\qquad d_H \sim h\frac{U_H}{u_V} \text{ for } \overline{u_V} \gg U_T,$$

where U_H is ambient horizontal fluid velocity (i.e., essentially the same as $\overline{u_H}$), whereas for objects with $U_T > 1.0$ m/s,

(6.32b) $$d_H \sim h\frac{U_H}{U_T} \text{ for } \overline{u_V} \ll U_T.$$

Based on a more detailed consideration of the effects of advection and other aspects of fluid behavior, Okubo and Levin (1989) adduced the generalized relationship

(6.32c) $$d_H = h\frac{U_H}{\left(U_T + \overline{u_V}\right)} \text{ for } U_T + \overline{u_V} \gg 0.$$

Notice, however, that these formulas do not account for the pronounced non-Gaussian dispersal curves frequently observed for spores, pollen, and small, passively dispersed seeds and fruits. Numerous workers have investigated these curves empirically, and many have tried to model them mathematically (e.g., Frampton et al. 1942; Gregory 1968; Cremer 1977; Green 1980; Augspurger 1986). Two of the most frequently encountered formulas used to approximate these curves are

(6.33a) $$p(d) = ad^{-b}$$

and

(6.33b) $$p(d) = ae^{-bd},$$

where $p(d)$ is the probability density for the dispersal, d is the distance from the source, and a and b are constants. The inverse power model (eq. [6.33a]) has the advantage of transforming to a straight line on a log-log plot, which allows the numerical values of a and b to be estimated (a is the log-log y-intercept slope and b is the slope). The negative exponential model (eq. [6.33b]) transforms to a linear relationship on a semilog plot, and the probability function remains a finite number as the distance from the source converges on zero. The disadvantage of both models is that neither gives theoretical insights, so there is no clear way of determining when to use one versus the other. Gregory (1968) evaluated 124 empirical dispersal curves using both models and found that 59 were better approximated by equation (6.33a) and that 65 were better approximated using equation (6.33b). In passing, it should be noted that spores, pollen grains, seeds, and fruits manifest differences in their shape, geometry, and bulk density even when harvested from the same sporangia, carpels, or plants. These differences can result in corresponding differences in the terminal velocities of these disseminules and may thus contribute to the leptokurtic shapes of dispersal curves. However, it is clear from the literature that empirical investigations of dispersal distances and patterns are extremely difficult and that some standard models are problematic (Sánchez et al. 2011).

Space precludes a detailed treatment of the aerodynamic behavior of autogyroscopic seeds and fruits. This topic has been discussed extensively by Ward-Smith (1984), who provides a useful formula for their terminal velocity:

$$(6.34) \qquad U_T = \left[\frac{2W}{\rho k(2-k)A_D} \right]^{0.5},$$

where W is the weight of the disseminule, ρ is air density, k is a proportionality constant that depends on the wake velocity generated downstream of a disk that has an area equivalent to the conical section of the rotating disseminule, and A_D is the area of the actuator disk (i.e., the area of the conical section's base). When $k = 1$, U_T is minimal and equals $2^{0.5}(W/\rho A_D)^{0.5} \approx 1.29(W/A_D)^{0.5}$ for air at 20°C (if ρ is expressed in units of kg/m³). However, at $0.5 \leq k \leq 1.0$, equation (6.34) gives values for U_T that vary no more than 15% from the minimum U_T. Finally, it is worth noting that autogyroscopic propagules accelerate before they reach their U_T. Their unsteady motion during acceleration is very complex (see Norberg 1973). However, measurements indicate that these propagules must accelerate to a factor of six times their final autorotational descent speed before achieving autogyratory motion.

Literature Cited

Augspurger, C. K. 1986. Morphology and dispersal potential of wind-dispersed diasporas of Neotropical trees. *Amer. J. Bot.* 73:353–63.

Biehle, G., T. Speck, and H.-C. Spatz. 1998. Hydrodynamics and biomechanics of a submerged water moss *Fontinalis antipyretica*—a comparison of specimens from habitats with different flow velocities. *Bot. Acta* 111:42–50.

Bussinger, J. A. 1975. Aerodynamics of vegetated surfaces. In *Heat and mass transport in the biosphere*, edited by D. A. DeVries and N. H. Afgan. New York: John Wiley.

Cox, R. G. 1970. The motion of long, slender bodies in a viscous fluid. Part 1. General theory. *J. Fluid Mech.* 44:791–810.

Cremer, K. W. 1977. Distance of seed dispersal in eucalypts estimated from seed weights. *Austral. For. Res.* 7:225–28.

De Langre, E. 2008. Effects of wind on plants. *Annu. Rev. Fluid Mech.* 40:141–68.

Feynman, R. P., R. B. Leighton, and M. Sands. 1967. *The Feynman lectures on physics.* Vol. 2. Reading, MA: Addison-Wesley.

Frampton, V. L., M. B. Linn, and E. D. Hansing. 1942. The spread of virus diseases of the yellow type under field conditions. *Phytopathology* 32:799–808.

Green, D. S. 1980. The terminal velocity and dispersal of spinning samaras. *Amer. J. Bot.* 67:1218–24.

Gregory, P. H. 1968. Interpreting plant disease dispersal gradients. *Annu. Rev. Phytopathol.* 6:189–212.

Hayashi, M., S. P. Gerry, and D. J. Ellerby. 2010. The seed dispersal catapult of *Cardamine parviflora* (Brassicaceae) is efficient but unreliable. *Amer. J. Bot.* 97:1595–1601.

Howe, H. F., and J. Smallwood. 1982. Ecology of seed dispersal. *Annu. Rev. Ecol. Syst.* 13:201–28.

Hutchinson, G. E. 1967. *A treatise on limnology.* Vol. 2. New York: Wiley and Sons.

Koehl, M. A. R. 1999. Ecological biomechanics of benthic organisms: Life history, mechanical design and temporal patterns of mechanical stress. *J. Exp. Biol.* 202:3469–76.

Koehl, M. A. R., and S. A. Wainwright. 1977. Mechanical adaptations of a giant kelp. *Limnol. Oceanogr.* 22:1067–71.

Longuet-Higgins, M. S. 1953. Mass transport in water waves. *Proc. Roy. Soc. London,* ser. A, 245:535–81.

Mayhead, G. J. 1973. Some drag coefficients for British forest trees derived from wind tunnel studies. *Agr. Meteor.* 12:123–30.

Milne, R. 1995. Modelling mechanical stresses in living Sitka spruce stems. In *Wind and trees,* edited by M. P. Coutts and J. Grace. Cambridge: Cambridge University Press.

Niklas, K. J., and H.-C. Spatz. 2000. Wind-induced stresses in cherry trees: Evidence against the hypothesis of constant stress levels. *Trees Struct. Funct.* 14:230–37.

Norberg, R. A. 1973. Autorotation, self-stability, and structure of single-winged fruits and seeds (samaras) with comparative remarks on animal flight. *Biol. Rev.* 48:561–96.

Okubo, A., and S. A. Levin. 1989. A theoretical framework for data analysis of wind dispersal of seeds and pollen. *Ecology* 70:329–38.

Peltola, H. M. 2006. Mechanical stability of trees under static loads. *Amer. J. Bot.* 93:1501–11.

Peltola, H. M., and S. Kellomäki. 1993. A mechanistic model for calculating windthrow and stem breakage at stand edge. *Silva Fenn.* 27:99–111.

Purcell, E. M. 1977. Life at low Reynolds number. *Amer. J. Phys.* 45:3–11.

Reynolds, C. 1984. Phytoplankton periodicity: The interaction of form, function, and environmental variability. *Freshwater Biol.* 14:112–42.

———. 2006. *Ecology of phytoplankton.* Cambridge: Cambridge University Press.

Robledo-Arnuncio, J. J. 2011. Wind pollination over mesoscale distances: An investigation with Scots pine. *New Phytol.* 190:222–33.

Sánchez, J. M. C., D. F. Green, and M. Quesada. 2011. A field test of inverse modeling of seed dispersal. *Amer. J. Bot.* 98:698–703.

Schindler, D., R. Vogt, H. Fugmann, M. Rodriguez, J. Schonborn, and H. Mayer. 2010. Vibration behavior of plantation-grown Scots pine trees in response to wind excitation. *Agr. For. Meteorol.* 150:984–93.

Sinn, G. 2003. *Baumstatik, Stand- und Bruchsicherheit von Bäumen an Straßen, in Parks und der freien Landschaft.* Braunschweig: Thalacker Medien.

Spatz, H.-C., and F. Brüchert. 2000. Basic biomechanics of self-supporting plants: Wind loads and gravitational loads on a Norway spruce tree. *For. Ecol. Mgmt.* 135:33–44.

Tritton, D. J. 1977. *Physical fluid dynamics.* Berkshire, UK: Van Nostrand Reinhold.

Vogel, S. 1981. *Life in moving fluids: The physical biology of flow.* Boston: Willard Grant Press.

———. 1984. Drag and flexibility in sessile organisms. *Amer. Zool.* 24:37–44.

———. 1988. *Life's devices: The physical world of animals and plants.* Princeton, NJ: Princeton University Press.

Ward-Smith, A. J. 1984. *Biophysical aerodynamics and the natural environment.* Chichester, UK: John Wiley.

White, F. M. 1979. *Fluid mechanics.* New York: McGraw-Hill.

CHAPTER SEVEN

Plant Electrophysiology

Good fences make good neighbors.—Seventeenth-century proverb

The goal of this chapter is to describe electric potentials occurring in plant cells as well as in whole plants. Electrical phenomena in plants were described by l'Abbé Pierre Bertholon de Saint-Lazare in 1783, even earlier than the famous studies of animal electricity by Galvani in 1791, but the success of studies in the electrophysiology of neurons overshadowed plant electrophysiology. Only recently have parallel aspects in the two fields become apparent (Davies 2006), such that the first symposium of its kind on "plant neurobiology" was organized (Baluska et al. 2006). In this chapter, we introduce the basic concepts of plant electrophysiology.

7.1 The principle of electroneutrality

One of the symmetry laws in physics is the principle of electroneutrality. It implies that in a finite volume, the number of positive charges equals the number of negative charges. In the context of this chapter, electroneutrality implies that in a biological cell, the number of positive charges carried by cations will be equal, or nearly equal, to the number of negative charges carried by anions. As we will see, even small deviations from equality lead to sizeable electric potential differences $\Delta \psi$ across the cell membrane, which can be viewed as a somewhat leaky insulator between the cell's inside and outside.

To calculate $\Delta \psi$, we must introduce the concept of electrical capacitance C, defined as the relationship between net charge Q and the potential difference $\Delta \psi$,

(7.1) $Q = \Delta\psi / C,$

where $\Delta\psi$ has units of volts (V), Q has units of coulombs (ampere-seconds), and C is given in farads. The net charge Q can be expressed as the concentration of net charges Δc, expressed in mol/m³, times the volume of the cell. To convert Q into coulombs, this number has to be multiplied by the Faraday constant $F = 96{,}485$ coulomb/mol. For most biological membranes, the capacitance per unit area C_{Area} is found to be approximately 0.01 farad/m². For a spherical cell with radius r, equation (7.1) can therefore be expressed as

(7.2a) $\dfrac{4\pi}{3} r^3 \, \Delta c \, F = \Delta\psi \, C_{Area} 4\pi r^2,$

which simplifies to

(7.2b) $\Delta\psi = \dfrac{1}{3} r \, \Delta c \dfrac{F}{C_{Area}} = \dfrac{9.65 \cdot 10^6}{3} r \, \Delta c.$

For a cell with a radius $r = 30$ µm, we find a potential difference $\Delta\psi \approx 100$ mV at a difference in the concentration of positive and negative charges of $\Delta c = 10^{-3}$ mol/m³ = 1 µM. This compares to an intracellular ion concentration of approximately 100 mol/m³ = 100 mM. Put differently, a deviation from electroneutrality of only 1 in 10^5 charges leads to a potential difference of 100 mV across the cell membrane.

7.2 The Nernst-Planck equation

Before we take up the discussion of electric potentials across the plant cell membrane, we will consider the flow of ions in a solution. Irreversible thermodynamics states that the net flow J_i of a substance i through any cross section (which may or may not be that of a membrane) is proportional to the driving force, which in our case is the gradient of the chemical potential μ_i (compare eq. [3.6]). In its one-dimensional form, the flow is given by

(7.3) $J_i = -\omega_i \, c_i \dfrac{d\mu_i}{dx},$

where c_i is the concentration of the substance i and ω_i is the mobility of the substance i in units of m²/s × mol/joule. Thus, J_i has units of mol m^{-2} s^{-1}.

For dilute solutions, equation (7.3) takes the form of the Nernst-Planck equation,

$$(7.4) \qquad J_i = -\omega_i \, c_i \, RT \left[\frac{1}{c_i} \frac{dc_i}{dx} + \frac{\overline{V}}{RT} \frac{dP}{dx} + \frac{z_i F}{RT} \frac{d\psi}{dx} \right],$$

where RT is the thermal energy per mole, $\overline{V}$ the partial specific volume, P the pressure, z the number of charges of the ion (i.e., substance i), and F the Faraday constant. Unlike equation (3.6), the Nernst-Planck equation contains a term for the effect of an electric potential ψ, but it usually does not include terms for surface tension or for gravity effects.

In this section, we will introduce two special simplifications. The simplest case is diffusion in the absence of a pressure gradient as well as a gradient of electric potential. For this case, the Nernst-Planck equation takes the form

$$(7.5) \qquad J_i = -\omega_i \, RT \frac{dc_i}{dx}.$$

Notice that this equation was introduced earlier in our discussion of environmental biophysics as Fick's first law (see eq. [2.1], which gives eq. [7.5] in terms of diffusivity), wherein

$$(7.6) \qquad D_i = \omega_i \, RT.$$

Another simplified form of equation (7.4) describes diffusion potentials. These potentials arise if the electrolyte concentrations on the two sides of a cross section anywhere in the solution are different (fig. 7.1). For simplicity, we assume that the pressure is uniform throughout the solution. In addition, we will treat the case of monovalent electrolytes with $z_C = +1$ and $z_A = -1$, where the subscripts C and A refer to cations and anions, respectively. Recall that the principle of electroneutrality implies that the concentrations of cations and anions on either side of the cross section must be the same, or very nearly the same; that is, $c_C = c_A = c$ (see section 7.1). It also implies that the net flow of cations and anions must be the same; that is, $J_C = J_A$. Eliminating the term from equation (7.4) that describes the contribution of hydrostatic pressure differences, we obtain

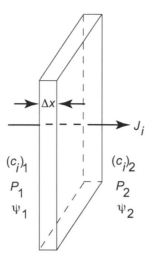

FIGURE 7.1. Two compartments in a solution are separated by an element with thickness Δx. Generally, if the concentrations of substance i, the pressures P, and the electrical potentials ψ are different on either side of the element, a flux J_i of substance i will occur through the element. The element may or may not be a membrane.

(7.7a)
$$0 = J_C - J_A = -\omega_C\, c\, RT\left[\frac{1}{c}\frac{dc}{dx} + \frac{F}{RT}\frac{d\psi}{dx}\right] +$$
$$\omega_A\, c\, RT\left[\frac{1}{c}\frac{dc}{dx} - \frac{F}{RT}\frac{d\psi}{dx}\right],$$

which leads to

(7.7b)
$$\frac{d\psi}{dx} = \frac{RT}{F}\frac{\omega_A - \omega_C}{\omega_A + \omega_C}\frac{1}{c}\frac{dc}{dx}$$

or

(7.7c)
$$\Delta\psi = \frac{RT}{F}\frac{\omega_A - \omega_C}{\omega_A + \omega_C}\ln\frac{c_1}{c_2},$$

where ln is the natural logarithm and the subscripts 1 and 2 refer to the two sides of the cross section.

If, for example, the cross section is a membrane, and only the transfer of cations is allowed through the membrane (i.e., $\omega_C \neq 0$ and $\omega_A = 0$), equation (7.7c) gives the well-known expression for the Nernst potential,

(7.8)
$$\Delta\psi = -\frac{RT}{F}\ln\frac{c_1}{c_2}.$$

If only anions are transferred (i.e., $\omega_c = 0$ and $\omega_c \neq 0$), the negative sign in equation (7.8) has to be replaced by a positive sign.

7.3 Membrane potentials

Potential differences across a membrane are typically on the order of 100 mV. It is important to note that even at pressure differences as high as 1.0 MPa, the term in the Nernst-Planck equation describing the contribution of hydrostatic pressure is much smaller than the term describing the contribution of an electric potential difference of 100 mV. Considering the flow of ions across a membrane, we see that it is justified to simplify equation (7.4) to

(7.9)
$$J_i = -\omega_i\, c_i\, RT\left[\frac{1}{c_i}\frac{dc_i}{dx} + \frac{z_i F}{RT}\frac{d\psi}{dx}\right].$$

Unfortunately, the mobility of ions within or through a membrane is a membrane property, which is not easily measured. In addition, the concentration of ions in the membrane may not be the same as that in the adjacent solution. Accordingly, we have to simplify equation (7.9) by introducing a partitioning coefficient β_i, which essentially delineates the concentration difference between membrane and solutions (fig. 7.2). Furthermore, we will assume that the concentration gradient in the membrane is proportional to the concentration difference of the solutions on either side of the membrane, divided by the thickness Δx of the membrane, which in itself may be difficult to determine exactly. Approximating the actual concentration gradient in the membrane by

(7.10)
$$\frac{dc_i}{dx} \approx \beta_i\left(\frac{\Delta c_i}{\Delta x}\right)_{solution},$$

it can be seen that, in the absence of a potential difference across the membrane (or for uncharged molecules), the net flow can be expressed as

(7.11)
$$J_i \approx -\omega_i\, RT\,\beta_i\left(\frac{\Delta c_i}{\Delta x}\right)_{solution} = -P_i\,(\Delta c_i)_{solution},$$

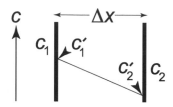

FIGURE 7.2. The partitioning coefficient is defined as the ratio of the concentrations of a substance in two adjoining phases at equilibrium. In the nonequilibrium situation of different solutions on either side of a membrane, the ratio of the concentrations of a substance at the interfaces between the membrane and the solutions is given by the partitioning coefficient such that $c_1' = \beta c_1$ and $c_2' = \beta c_2$.

where the three unknown variables ω_i, β_i, and Δx are lumped together into a single quantity p_i, which is called permeability. Importantly, this quantity can be easily measured in tracer experiments.

A special case of a diffusion potential is the Donnan potential. This potential arises if one of the compartments contains, in addition to other ions, immobile charges or polyelectrolytes, which cannot cross the barrier between the two compartments. If the polyelectrolyte carries n negative charges, electroneutrality requires that

(7.12) $(c_C)_1 = (c_A)_1 + n(c_P)_1$ and $(c_C)_2 = (c_A)_2,$

where subscripts C, A, and P refer to monovalent cations, monovalent anions, and polyelectrolytes, respectively.

At equilibrium, the net fluxes are zero, such that $J_C = J_A = 0$. It therefore follows from equation (7.9) that the electric potential across the membrane is given by

(7.13) $$\Delta \psi = \frac{RT}{F} \ln \frac{(c_C)_1}{(c_C)_2} = \frac{RT}{F} \ln \frac{(c_A)_2}{(c_A)_1},$$

which leads to the relationship called the Donnan equilibrium,

(7.14) $(c_C)_1 \cdot (c_A)_1 = (c_C)_2 \cdot (c_A)_2,$

where once again the subscripts 1 and 2 refer to the two compartments (see fig. 7.1).

BOX 7.1 **The Goldman equation**

The Goldman equation is a special solution of the Nernst-Planck equation. The essential assumption, that the potential gradient in the membrane is constant, can be written in the form

(7.1.1)
$$\frac{d\psi}{dx} = \frac{\Delta\psi}{\Delta x} = \text{constant},$$

where Δx is the thickness of the membrane. With this assumption, the Nernst-Planck equation (in the form of eq. [7.9]) can be simplified to

(7.1.2)
$$J_i = -\omega_i RT\left[\frac{dc_i}{dx} + c_i\frac{z_i F}{RT}\frac{\Delta\psi}{\Delta x}\right].$$

Equation (7.1.2) can be written in the form of a differential equation,

(7.1.3)
$$\frac{dc_i}{dx} = -az_ic_i - b \text{ with } a = \frac{F}{RT}\frac{\Delta\psi}{\Delta x} \text{ and } b = \frac{J_i}{\omega_i RT}.$$

This differential equation has the following solution:

(7.1.4)
$$c_i = \alpha \, \exp(-az_i \, x) - \frac{b}{a},$$

with a as an integration constant.

The boundary conditions can be written as

(7.1.5)
$$c_i = (c_i)_1 \text{ for } x = 0 \text{ and } c_i = (c_i)_2 \text{ for } x = \Delta x.$$

This leads to

(7.1.6)
$$(c_i)_1 = \alpha - \frac{b}{a} \text{ and } (c_i)_2 = \alpha \, \exp(-az_i \, \Delta x) - \frac{b}{a}.$$

After some algebraic manipulation, the integration constant α can be eliminated:

(7.1.7)
$$(c_i)_1 - (c_i)_2 \, \exp(az_i \, \Delta x) = \frac{b}{a}[\exp(az_i\Delta x) - 1].$$

Reinserting the expression for b gives

BOX 7.1 **(Continued)**

(7.1.8)
$$J_i = -\omega_i\, z_i\, F\frac{\Delta\psi}{\Delta x}\, \frac{(c_i)_1 - (c_i)_2 \cdot \exp(az_i\Delta x)}{1 - \exp(az_i\Delta x)}.$$

For each of the ions that can be transported through the membrane, an equation of this type is valid. The principle of electroneutrality requires that the net fluxes of all cations equal the net fluxes of all anions. This requirement can be stated as

(7.1.9)
$$\Sigma J_C + \Sigma J_A = 0.$$

For monovalent ions, the last two equations lead to an easily manageable expression. As an example, we will consider conditions in which only the cations Na^+ and K^+ and the anion Cl^- have finite mobilities through the membrane. With $z_{Na} = +1$, $z_K = +1$, and $z_{Cl} = -1$, equation (7.1.8) takes the form

(7.1.10a)
$$J_{Na} = -\omega_{Na}\, F\frac{\Delta\psi}{\Delta x}\, \frac{(c_{Na})_1 - (c_{Na})_2 \cdot \exp(a\Delta x)}{1 - \exp(a\Delta x)},$$

(7.1.10b)
$$J_K = -\omega_K\, F\frac{\Delta\psi}{\Delta x}\, \frac{(c_K)_1 - (c_K)_2 \cdot \exp(a\Delta x)}{1 - \exp(a\Delta x)},$$

(7.1.10c)
$$J_{Cl} = +\omega_{Cl}\, F\frac{\Delta\psi}{\Delta x}\, \frac{(c_{Cl})_1 - (c_{Cl})_2 \cdot \exp(-a\Delta x)}{1 - \exp(-a\Delta x)}.$$

Equation (7.1.10c) can also be written as

(7.1.10d)
$$J_{Cl} = -\omega_{Cl}\, F\frac{\Delta\psi}{\Delta x}\, \frac{(c_{Cl})_1 \cdot \exp(a\Delta x) - (c_{Cl})_2}{1 - \exp(a\Delta x)}.$$

Since electroneutrality requires that $J_{Na} + J_K - J_{Cl} = 0$, equation (7.1.10) can be summarized as

(7.1.11)
$$\begin{aligned}
&\omega_{Na}\,[(c_{Na})_1 - (c_{Na})_2 \cdot \exp(a\Delta x)] \\
&+\omega_K\,[(c_K)_1 - (c_K)_2 \cdot \exp(a\Delta x)] \\
&-\omega_{Cl}\,[(c_{Cl})_1 \cdot \exp(a\Delta x) - (c_{Cl})_2] = 0
\end{aligned}$$

BOX 7.1 **(Continued)**

Rearrangement leads to

(7.1.12a) $$\exp(a\Delta x) = \frac{\omega_{Na}(c_{Na})_1 + \omega_K(c_K)_1 + \omega_{Cl}(c_{Cl})_2}{\omega_{Na}(c_{Na})_2 + \omega_K(c_K)_2 + \omega_{Cl}(c_{Cl})_1}$$

or

(7.1.12b) $$a\Delta x = \ln\frac{\omega_{Na}(c_{Na})_1 + \omega_K(c_K)_1 + \omega_{Cl}(c_{Cl})_2}{\omega_{Na}(c_{Na})_2 + \omega_K(c_K)_2 + \omega_{Cl}(c_{Cl})_1}.$$

Reinserting the expression for a (eq. [7.1.3]) leads to the Goldman equation,

(7.1.13) $$\Delta\psi = \frac{RT}{F}\ln\frac{\omega_{Na}(c_{Na})_1 + \omega_K(c_K)_1 + \omega_{Cl}(c_{Cl})_2}{\omega_{Na}(c_{Na})_2 + \omega_K(c_K)_2 + \omega_{Cl}(c_{Cl})_1}.$$

So far we have not taken partitioning coefficients between membrane and solution into account. Furthermore, as discussed in section 7.3, the mobility of ions in the membrane may not be directly accessible. It is therefore useful to express the Goldman equation in terms of permeabilities:

(7.1.14) $$\Delta\psi = \frac{RT}{F}\ln\frac{p_{Na}(c_{Na})_1 + p_K(c_K)_1 + p_{Cl}(c_{Cl})_2}{p_{Na}(c_{Na})_2 + p_K(c_K)_2 + p_{Cl}(c_{Cl})_1}.$$

We derived equation (7.1.14) by considering independent passive movements of Na^+, K^+, and Cl^- ions through the membrane. If the flux of divalent ions, such as Ca^{2+}, is also taken into account, the equation becomes considerably more complicated. However, in biological membranes, the fluxes of these ions are usually much smaller than those of Na^+, K^+, and Cl^- ions, so that for most cases plant membrane potentials are adequately described by this equation (Nobel 2005).

The cell membrane usually has a number of specific ion channels, notably for sodium (Na^+), potassium (K^+), calcium (Ca^{2+}), and chloride (Cl^-) ions, so that, as previously mentioned, it can be considered a leaky insulator allowing for a limited flow of several ions into and out of the cell. In this regard, the solution of the Nernst-Planck equation (see eq. [7.9]) becomes more difficult if more than one species of positive or negative ion passes through the membrane. Using the assumption of a constant gradient of the electric potential across the membrane, Goldman (1943) has provided a solution known as the Goldman equation, or the constant field equation. If K^+, Na^+, and Cl^- are the main contributors to the ion flow, this equation can be written in the form

$$(7.15) \qquad \Delta \psi = \frac{RT}{F} \ln \frac{p_K(c_K)_1 + p_{Na}(c_{Na})_1 + p_{Cl}(c_{Cl})_2}{p_K(c_K)_2 + p_{Na}(c_{Na})_2 + p_{Cl}(c_{Cl})_1},$$

where p denotes the permeability of the ion specified by the subscript through the membrane. The Goldman equation has proved to be very successful in describing resting potentials as well as action potentials, particularly in neurons, and it is important for our understanding of membrane potentials in plant cells as well. The somewhat cumbersome derivation of the formula is given in box 7.1.

It should be noted that the Goldman equation is not compatible with the formula derived for diffusion potentials (see eq. [7.7]) because the assumption of a constant gradient of the electric potential across the membrane violates the principle of electroneutrality. The agreement of calculated and experimentally determined values shows that the Goldman equation, nevertheless, serves as a very good approximation of the membrane potential.

7.4 Ion channels and ion pumps

Phospholipids, which are a major constituent of the cell membrane, provide for a low dielectric constant and, consequently, a distinctly electrophobic milieu. Transporting charged molecules through a phospholipid bilayer would require very high activation energies. Therefore, ion transport must be facilitated by transporters for active transport and channels for passive transport. It is not surprising, therefore, that a number of specialized transporter proteins are integrated into cell membranes.

As reviewed by Hedrich and Schroeder (1989) and Ward et al. (2009), X-ray crystallography, genomics, mutant analysis, and patch clamp techniques (see section 9.7) have greatly contributed to our understanding of the form and function of ion channels. For example, it is now known that Na^+, K^+, and Ca^{2+} channels consist of four transmembrane domains connected by side chains extending into the intra- and extracellular space. Each domain consists of six α-helices and a β-sheet loop. The four domains are spatially arranged to form a hydrophilic pore in between. The less well studied Cl^- channels contain ten to twelve transmembrane helices.

Some of the genes for plant channel proteins have been identified, in part by their sequence homology with animal channel protein genes. A good example is the plant Shaker-like K^+ channel gene family, which is homologous to the *Drosophila Shaker* gene, which has been shown to code for a K^+ channel (Yan and Yan 1992). On the other hand, Na^+ channels have not been found in plant cells, although they occur in animal cells. Calcium channels are present in plant cells. Genes for Ca^{2+} channels have been identified in *Arabidopsis* root cells (White et al. 2002), and a gene for an ubiquitously expressed Ca^{2+} channel hast been found in rice (Kurusu et al. 2004).

Different channels are characterized by their selectivity for monovalent or divalent ions, for cations or anions, and for different sizes of (hydrated) ions. These properties result from differences in the size of the pore and the spatial arrangement of charged amino acids within the pore. Another important characteristic is the mode by which the channels are activated. Patch clamp experiments on single channels (Neher and Sakmann 1976) have shown that the phenomenological term "activity" can be understood as the probability that a channel is either open or closed. Some channels are activated by Ca^{2+} ions, others by small organic ligands such as cAMP or glutamate. Mechanosensitive channels open under pressure, shear stress, or strain. Particularly important for the generation of action potentials are voltage-gated channels, whose activation (or inactivation) is under the control of membrane potentials.

The relationship between the membrane potential and the probability that a channel is open is often nonlinear. Some channels have a high probability of being open at positive membrane potentials and remain closed under negative membrane potentials; other channels are closed at positive potentials and open under negative potentials. These channels, therefore, have rectifying characteristics. If channels are specific for a particular ion species, and if the electrochemical potential of that ion is different on the inside and on the outside of a cell, channels are described as being either inward or outward rectifying channels. Some channels display a combination of these properties; that is, they have a high probability of being open both at elevated positive and negative potentials, but remain closed in an intermediate range of membrane potentials.

Patch clamp recordings are far more difficult to make when dealing with plant cells than with animal cells. Nevertheless, they have led to the characterization of many ion channels in plant cell membranes (Hedrich and Schroeder 1989). An alternative approach to the identification of channel proteins from plants is an analysis of the expression of putative channel protein genes in model systems such as yeast or *Xenopus* oocytes, in which recordings pose fewer difficulties (Maathuis et al. 1997). Care has to be taken, however, because the properties of the channel proteins expressed in different systems may be very different. For example, the voltage-dependent gating of the rectifying K^+ channel AKT2, but not its ion selectivity, is affected by the expression system (Latz et al. 2007).

Since the concentration difference of potassium is the main determinant of the membrane potential, the rectifying characteristics of K^+ channels are important for membrane potential regulation. A number of physiological processes involve the participation of K^+ channels. Examples include the uptake of potassium in root cells as well as in various plant cells during cell expansion, movement, and growth. An outward rectifying Shaker-like K^+ channel in guard cells contributes to stomatal closure (see section 3.1). A cyclic nucleotide–gated cation channel activates a pathogen defense response (Yoshioka et al. 2006). Nonselective, voltage-dependent cation channels, permeable to Ca^{2+}, have been found in the vacuolar membrane. They are similar to TPC channels in animal cells and seem to be involved in the release of Ca^{2+} into the cytosol upon depolarization.

The functions of anion channels in land plants, as well as in algae, include turgor regulation and osmoregulation, anion homeostasis in root cells, and the regulation of passive salt loading into the xylem. Aluminum tolerance is mediated by chelating Al^{3+} with malate extruded from root cells through anion channels (Schroeder 1995). In guard cells of *Vicia faba*, both R-type and S-type anion channels contribute to the process of stomatal movement and thus to the regulation of transpirational water loss by plants (Schroeder and Keller 1992). The R-type anion channel is rapidly activated by depolarization, inactivated during prolonged stimulation, and rapidly deactivated at hyperpolarized membrane potentials. The slow S-type channel is selective for Cl^- and regulated by Ca^{2+}. Chloride channels were also recorded from the large internodal cells of the green alga *Chara* (Coleman 1986). Chloride efflux in *Chara* is important for cytosolic pH homeostasis, as it facilitates proton efflux from the cytosol (Johannes et al. 1998). A voltage-dependent chloride channel regulated by Ca^{2+} was identified by Okihara et al. (1991). Thiel et al. (1993) accumulated evidence that two chloride channels with transient activity are not directly voltage dependent, but indirectly activated by an increase of cytosolic Ca^{2+} upon stimulation. A possible significance of these findings for the generation and propagation of action potentials will be discussed in the next section.

Active transport is defined as transport against a gradient of the chemi cal potential and therefore requires energy. That energy may be supplied either in the form of ATP or by a concentration gradient of another ion, most commonly a proton gradient across the cell membrane. Ion pumps are essential to establish and maintain these concentration gradients. Two different classes of pumps can be distinguished: neutral pumps and electrogenic pumps. Neutral pumps, which include K^+/H^+ and Na^+/H^+ antiporters as well as K^+/Cl^- symporters, do not change the number of charges transported. If a neutral pump is inactivated by a blocking agent, the membrane potential is not affected, provided that the agent does not influence the ion channels mediating passive transport. However, membrane potentials are affected by active electrogenic pumps, such as the Cl^- pump in the green alga *Acetabularia* or the light-activated, ATP-dependent proton pump in the charophycean alga *Nitella translucens* (Spanswick 2006). In the aquatic vascular plant *Elodea densa*, an electrogenic K^+ pump was identified by the observation that the membrane potential was more negative than the Nernst potential for K^+ ions.

The interpretation of experiments with agents that block specific pumps is not always straightforward, since these chemicals may also affect other integral membrane proteins, or even properties of the membrane itself. Fortunately, there are other ways to distinguish between active and passive ion transport. One way is to measure ion fluxes into and out of a cell separately in experiments using radioactive tracers under otherwise identical conditions—with temperatures, membrane potentials $\Delta\psi$, and all component concentrations inside and outside the cell identical except for the tracer. If ions are only passively transported through the cell membrane, the ratio of inward to outward flux is given by the Ussing-Teorell equation (Teorell 1949; Ussing 1949):

$$(7.16) \qquad \frac{J_i^{inward}}{J_i^{outward}} = -\frac{(c_i)_{soulution}}{(c_i)_{cell} \cdot \exp\left(\dfrac{F}{RT} z_i \, \Delta\psi\right)}.$$

This equation, derived in box 7.2, is valid for all passively transported ion species, provided that they are transported independently of one another

BOX 7.2 **The Ussing-Teorell equation**

Like the Goldman equation, the Ussing-Teorell equation for passive transport can be derived from the Nernst-Planck equation. We can start from equation (7.1.8) derived in box 7.1:

(7.2.1a)
$$J_i = -\omega_i\, z_i\, F\frac{\Delta\psi}{\Delta x}\frac{(c_i)_1 - (c_i)_2 \cdot \exp(az_i\Delta x)}{1 - \exp(az_i\Delta x)},$$

where, as in equation (7.1.3),

(7.2.1b)
$$a = \frac{F}{RT}\frac{\Delta\psi}{\Delta x}.$$

The inward flux can be measured by extrapolating the flux into cells, initially not containing tracers, to time zero after the radioactive tracer is added to the solution. The outward flux can be measured in the same way after the cells, loaded with radioactive tracers, are placed in a solution not containing tracers. If the experiments are carried out in this way, we can rewrite equation (7.2.1) as

(7.2.2a)
$$J_i^{inward} = -\omega_i\, z_i\, F\frac{\Delta\psi}{\Delta x}\frac{(c_i)_{solution}}{1 - \exp(az_i\Delta x)}$$

and

(7.2.2b)
$$J_i^{outward} = +\omega_i\, z_i\, F\frac{\Delta\psi}{\Delta x}\frac{(c_i)_{cell} \cdot \exp(az_i\Delta x)}{1 - \exp(az_i\Delta x)},$$

where J_i refers to the flux of the tracer at time zero and c_i to its initial concentration. The ratio of inward and outward fluxes under otherwise identical conditions is given by the expression

(7.2.3)
$$\frac{J_i^{inward}}{J_i^{outward}} = -\frac{(c_i)_{solution}}{(c_i)_{cell} \cdot \exp(az_i\Delta x)}.$$

Reinserting the constant a from equation (7.2.1b) leads to the Ussing-Teorell equation,

BOX 7.2 **(Continued)**

(7.2.4) $$\frac{J_i^{inward}}{J_i^{outward}} = -\frac{(c_i)_{solution}}{(c_i)_{cell} \cdot \exp\left(\dfrac{F}{RT} z_i \, \Delta\psi\right)}.$$

This derivation can also be delineated taking partition coefficients into account (Nobel 2005). However, since they cancel out at the end, we have not made use of this extension.

(an assumption that is implicit in the Nernst-Planck equation). Deviations from the Ussing-Teorell equation can therefore be interpreted in two ways: either the ions are actively transported into or out of the cell, such that active pumps contribute to the fluxes, or the ions are *not* transported independently of one another.

7.5 Electrical currents and gravisensitivity

Scott and Martin (1962) were the first to report endogenous currents in the roots of bean seedlings and their relationship to salt accumulation in roots. Such currents have been subsequently observed by direct current measurements in more than twenty other species. A comparison among all studied species reveals that the current densities and patterns are uniform, although variations are observed among species and as a result of the media used to submerge roots for observation. Taking vertically growing *Lepidium sativum* L. roots as an example (data reported by Behrens et al. 1982; Weisenseel et al. 1992), measurements with vibrating electrodes reveal a $0.7 \, \mu A/cm^2$ current entering or leaving the apical meristem. In the meristematic zone (approximately 0.3 ± 0.6 mm behind the root cap), a 1.6 $\mu A/cm^2$ current enters the root most of the time. A $2.2 \, \mu A/cm^2$ inward current enters the elongation zone (0.7 ± 1.5 mm behind the cap), while at the beginning of the root hair zone (2 ± 4 mm behind the cap), a $1.0 \, \mu A/cm^2$ current leaves the root (fig. 7.3).

This configuration changes significantly when *Lepidium* roots are placed horizontally. The current direction changes along the "dorsal" side of the root cap and in the immediate region of the apical meristem, while

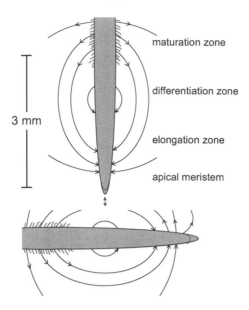

maturation zone

differentiation zone

3 mm

elongation zone

apical meristem

FIGURE 7.3. Schematic of a *Lepidium sativum* root and its electrical current pattern. When the root is growing vertically, a current source occurs near the proximal elongation and root hair zones, while a current sink occurs in the distal elongation and apical meristem zones. When the root is rotated 90° from the vertical, a new current source appears on the dorsal side toward the growing tip within several minutes. (Adapted from Weisenseel and Meyer 1997.)

the direction of the current remains unchanged along the more proximal regions and along the entire lower surface. The current change at the upper portion of the root tip causes an asymmetrical current pattern at the root cap and in the meristematic region. This change occurs within a few minutes after the reorientation of the root, and thus reflects the effects of gravitropism. No change in current pattern and no gravitropism occur in roots growing in media with only sub-micromolar Ca^{2+} concentrations. Similar results are reported from studies using corn roots (*Zea mays* L.) as the model system (Collings et al. 1992).

7.6 Action potentials

The electric potentials recorded on the leaves of the Venus flytrap (*Dionea muscipula*) upon mechanical stimulation (Burdon-Sanderson 1873) are the first reported examples of action potentials in plants. When specialized trichomes on the dorsal leaf blade are mechanically stimulated, a series of electrical and hydrostatic events leads to the rapid closure of the leaf and the subsequent release of digestive fluids (fig. 7.4).

Other examples of electrical excitation in higher plants have been extensively studied because they involve similarly rapid movements

FIGURE 7.4. Time-lapse images of a resting Venus flytrap leaf and its closure upon stimulation by an insect. In the upper image, three specialized trichomes (i.e., epidermal cellular outgrowths) are visible on the dorsal surfaces of each half of the leaf. When two or more trichomes are mechanically stimulated, the leaflets rapidly close in on each other. The time interval between the middle and lower frame is approximately 1 second.

(Umrath 1929); examples include the action potentials in the leaves of *Mimosa pudica* (Haberlandt 1890) and the sundew *Drosera intermedia* (Williams and Pickard 1972). Action potentials have also been investigated in plants not possessing motor activity, such as *Salix viminalis* (Fromm and Spanswick 1993) and *Solanum tuberosum* (Fisahn et al. 2004), and in the large internodal cells of the green algae *Nitella mucronata* (Umrath 1930) and *Chara corallina* (Wacke and Thiel 2001).

Two types of electric potentials have been found in plants: variation potentials and action potentials. A variation potential, otherwise known as a slow wave potential, can be generated by wounding, cutting, or burning. It consists of a slowly propagating membrane depolarization, which varies in proportion to the intensity of the stimulus and decreases with increasing distance from the injured site. Most important, it depends on xylem pressure. The changes in potential as a function of distance from the injury or trauma are believed to be the result of hydrostatic pressure waves in the xylem and their effect on mechanosensory channels in adjacent living cells.

Action potentials can be elicited by mechanical, chemical, or electrical stimuli, by light-dark transitions, or by brief cooling. In contrast to variation potentials, action potentials are self-propagating, all-or-nothing signals transmitted at a constant amplitude and velocity (Davies 2006). Self-propagation requires the participation of ion channels that respond directly or indirectly to changes in membrane potentials. General agreement exists that action potentials are triggered by the activation of Ca^{2+}-dependent Cl^- channels in the plasma membrane, which causes a depolarization of the (negative) membrane potential. The membrane potential is subsequently restored to resting levels by the opening of outward rectifying K^+ channels (Thiel 1995; Davies 2006).

A model that describes the typical kinetics, and particularly the all-or-none character, of the transient increase of cytoplasmic Ca^{2+} accompanying an action potential (Fisahn et al. 2004) has been developed by Wacke et al. (2003). This model is an extension of a model for animal cells (Othmer 1977). Upon electrical stimulation, a second messenger is formed, most likely IP_3, which is assumed to bind to an IP_3-sensitive Ca^{2+} channel. This channel is thought to be activated upon the binding of IP_3 and one Ca^{2+} ion. The opening of the channel leads to a release of Ca^{2+} from an as yet unspecified cytoplasmic compartment. This aspect of the model accounts for the autocatalytic character of the Ca^{2+} elevation. The Ca^{2+} channel is thought to close upon binding a second Ca^{2+} ion. As Ca^{2+} is pumped back to replenish the cytoplasmic compartment, Ca^{2+} levels in the cytosol return to normal levels. This mechanism accounts for the transitory character of the Ca^{2+} pulse. With the proper choice of rate constants for this reaction scheme, the model provides a good representation of the experimental findings.

In vascular plants, the propagation of action potentials seems to occur predominantly in the phloem (Fromm 2006). Propagation velocities in the range of 0.01–0.2 m/s have been found. In the green algae *Chara* and *Nitella*, the action potential is transmitted from one internodal cell to its neighboring internodal cell. Velocities between 0.03 and 0.05 m/s have been recorded in single internodal cells of *Chara corallina* (Trontelj et al. 2006).

7.7 Electrical signaling in plants

In a situation that is reminiscent of the debate in the nineteenth century about the nature of the signal in the nervous systems of animals, the question as to whether an action potential in a plant is a primary signal or an epiphenomenon remains undecided. If the latter, an action potential merely reflects a unidirectional propagating string of chemical events, such as the transient opening of Ca^{2+}-dependent Ca^{2+} channels. If the former, an action potential is a true signal and is therefore analogous to what happens in animal neurons. The notion that an action potential is a true signal is perpetuated by the observation that rapid leaf movements in *Mimosa pudica* (see fig. 2.4) and in the Venus flytrap (see fig. 7.4) are accompanied by action potentials. In vascular plants, the phloem can be seen as analogue to the axon of a neuron (Fromm and Lautner 2007). Furthermore, neurotransmitter homologues that may translate an electrical signal into a chemical event have been found in plants (Brenner et al. 2006).

Although direct proof is still missing, rapid communication in a whole plant, or between different organs or cells via electrical signals, remains an attractive hypothesis. Since action potentials are closely associated with increased cytoplasmic concentrations of the second messenger Ca^{2+} or putative transmitter molecules, various effects on biochemical reactions in a plant cell are to be expected, including gene expression, biosynthesis (Fisahn et al. 2004), and posttranslational modification of enzymes, in part via phosphorylation and dephosphorylation (Davies 1993). Action potentials consequently should have pronounced effects on physiological responses of plant organs such as turgor regulation, phloem transport, respiration, and photosynthesis (Volkov 2006).

Future work is required to settle this issue. Regardless of its resolution, the study of plant electrophysiology is showing that plants are extremely sophisticated organisms that have evolved systems very similar to those of animals when examined at the level of cells and cell-to-cell communication.

Literature Cited

Baluska, F., S. Mancuso, and D. Volkmann. 2006. *Communication in plants: Neuronal aspects of plant life.* Heidelberg: Springer Verlag.

Behrens, H. M., M. H. Weisenseel, and A. Sievers. 1982. Rapid changes in the pattern of electric current around the root tip of *Lepidium sativum* L. following gravistimulation. *Plant Physiol.* 70:1079–83.

Bertholon, P. 1783. *De l'électricité des végétaux.* Paris: Didot éditeur.

Brenner, E., R. Stahlberg, S. Mancuso, J. Vivanco, F. Baluska, and E. Van Volkenburgh. 2006. Plant neurobiology: An integrated view of plant signaling. *Trends Plant Sci.* 8:413–19.

Burdon-Sanderson, J. 1873. Note on the electrical phenomena which accompany stimulation of the leaf of *Dionea muscipula. Proc. Roy. Soc. London* 21: 495–96.

Coleman, H. A. 1986. Chloride currents in *Chara*—a patch clamp study. *J. Membr. Biol.* 93:55–61.

Collings, D. A., R. G. White, and R. L. Overall. 1992. Ionic current changes associated with the gravity-induced bending response in roots of *Zea mays* L. *Plant Physiol.* 100:1417–26.

Davies, E. 1993. Intercellular and intracellular signals in plants and their transduction via the membrane-cytoskeleton interface. *Semin. Cell Biol.* 4:139–47.

———. 2006. Electrical signals in plants: Facts and hypotheses. In *Plant electrophysiology*, edited by A. G. Volkov. Berlin: Springer Verlag.

Fisahn, J., O. Herde, L. Willmitzer, and H. Peña-Cortés. 2004. Analysis of the transient increase in cytosolic Ca^{2+} during action potential of higher plants with high temporal resolution: Requirement of the Ca^{2+} transients for the induction of jasmonic acid biosynthesis and the *PINII* gene expression. *Plant Cell Physiol.* 45:456–59.

Fromm, J. 2006. Long-distance electrical signaling and physiological functions in higher plants. In *Plant electrophysiology*, edited by A. G. Volkov. Berlin: Springer Verlag.

Fromm, J., and S. Lautner. 2007. Electrical signals and their physiological significance in plants. *Plant Cell Env.* 30:249–57.

Fromm, J., and R. Spanswick. 1993. Characteristics of action potentials in willow (*Salix viminalis* L.). *J. Exp. Bot.* 44:1119–25.

Galvani, G. 1791. *De viribus electricitatis in motu musculari commentaries*. Bologna: Bononiae Instituti Scientarum.

Goldman, D. E. 1943. Potential, impedance, and rectification in membranes. *J. Gen. Physiol.* 27:37–60.

Haberlandt, G. 1890. *Das reizleitende Gewebesystem der Sinnpflanze*. Leipzig: Engelmann.

Hedrich, R., and J. I. Schroeder. 1989. The physiology of ion channels and electrogenic pumps in higher plants. *Annu. Rev. Plant Physiol.* 40:539–69.

Johannes, E., A. Crofts, and D. Sanders. 1998. Control of Cl⁻ efflux in *Chara corallina* by cytosolic pH, free Ca^{2+}, and phosphorylation indicates a role of plasma membrane anion channels in cytosolic pH regulation. *Plant Physiol.* 118:173–81.

Kurusu, T., Y. Sakurai, A. Miyao, H. Hirohiko, and K. Kuchitsu. 2004. Identification of a putative voltage-gated Ca^{2+}-permeable channel (OsTPC1) involved in Ca^{2+} influx and regulation in growth and development in rice. *Plant Cell Physiol.* 45:693–702.

Latz, A., N. Ivashikina, S. Fischer, P. Ache, T. Sano, D. Becker, R. Deeken, and R. Hedrich. 2007. In planta AKT2 subunits constitute a pH- and Ca^{2+}-sensitive inward rectifying K^+ channel. *Planta* 225:1179–91.

Maathuis, F. J. M., A. M. Ichida, D. Sanders, and J. I. Schroeder. 1997. Roles of higher plant K^+ channels. *Plant Physiol.* 114:1141–49.

Neher, E., and B. Sakmann. 1976. Single-channel currents recorded from membrane of denervated frog muscle fibers. *Nature* 260:779–802.

Nobel, P. S. 2005. *Physicochemical and environmental plant physiology*. 3rd ed. Amsterdam: Elsevier.

Okihara, K., T. Ohkawa, I. Tsutsui, and M. Kasai. 1991. A Ca^{2+}- and voltage-dependent Cl⁻ sensitive anion channel in the *Chara* plasmalemma: A patch-clamp study. *Plant Cell Physiol.* 32:593–601.

Othmer, H. G. 1977. Signal transduction and second messenger systems. In *Case studies in mathematical modeling—ecology, physiology and cell biology*, edited by H. G. Othmer, F. R. Adler, M. A. Lewis, and J. Dallon. Englewood Cliffs, NJ: Prentice Hall.

Schroeder, J. I. 1995. Anion channels as central mechanisms for signal transduction in guard cells and putative functions in roots for plant-soil interactions. *Plant Mol. Biol.* 28:353–61.

Schroeder, J. I., and B. U. Keller. 1992. Two types of anion channel currents in guard cells with distinct voltage regulation. *Proc. Nat. Acad. Sci. USA* 89:5025–29.

Scott, B. I. H., and D. W. Martin. 1962. Bioelectric fields of bean roots and their relation to salt accumulations. *Austral. J. Biol. Sci.* 15:83–100.

Spanswick, R. 2006. Electrogenic pumps. In *Plant electrophysiology*, edited by A. G. Volkov. Berlin: Springer Verlag.

Teorell, T. 1949. Membrane electrophoresis in relation to bio-electrical polarization effects. *Arch. Sci. Physiol.* 3:205–10.

Thiel, G. 1995. Dynamics of chloride and potassium currents during action potential in *Chara* studied with action potential clamp. *Eur. Biophys. J.* 24:85–92.

Thiel, G., U. Homann, and D. Gradmann. 1993. Microscopic elements of electrical excitation in *Chara*: Transient activity of Cl⁻ channels in the plasma membrane. *J. Membr. Biol.* 134:53–66.

Trontelj, Z., G. Thiel, and V. Jazbinsek. 2006. Magnetic measurements in plant electrophysiology. In *Plant electrophysiology*, edited by A. G. Volkov. Berlin: Springer Verlag.

Umrath, K. 1929. Über die Erregungsleitung bei höheren Pflanzen. *Planta* 7: 174–207.

———. 1930. Untersuchungen über Plasma und Plasmaströmungen in Characeen. IV. Potentialmessungen an *Nitella mucronata* mit besonderer Berücksichtigung der Erregungserscheinungen. *Protoplasma* 9:576–97.

Ussing, H. H. 1949. The distinction by means of tracers between active transport and diffusion. *Acta Physiol. Scand.* 19:43–56.

Volkov, A. G., ed. 2006. *Plant electrophysiology*. Berlin: Springer Verlag.

Wacke, M., and G. Thiel. 2001. Electrically triggered all-or-none Ca²⁺ liberation during action potential in the giant alga *Chara*. *J. Gen. Physiol.* 118:11–21.

Wacke, M., G. Thiel, and M.-T. Hütt. 2003. Ca²⁺ dynamics during membrane excitation of green alga *Chara*: Model simulations and experimental data. *J. Membr. Biol.* 191:179–92.

Ward, J. M., P. Mäser, and J. I. Schroeder. 2009. Plant ion channels: Gene families, physiology, and functional genomics analyses. *Annu. Rev. Physiol.* 71: 12.1–12.23.

Weisenseel, M. H., H. F. Becker, and J. G. Ehlgötz. 1992. Growth, gravitropism, and endogenous ion currents of cress roots (*Lepidium sativum* L.). *Plant Physiol.* 100:16–25.

Weisenseel, M. H., and A. J. Meyer. 1997. Bioelectricity, gravity and plants. *Planta* 203:S98–S106.

White, P. J., H. C. Bowen, V. Demidchik, C. Nichols, and J. M. Davies. 2002. Genes for calcium-permeable channels in the plasma membrane of plant root cells. *Biochim. Biophys. Acta* 1564:299–309.

Williams, S. E., and B. G. Pickard. 1972. Properties of action potentials in *Drosera* tentacles. *Planta* 103:193–221.

Yan, L. Y., and Y. N. Yan. 1992. Structural elements involved in specific K⁺ channel functions. *Annu. Rev. Plant Physiol.* 54:535–55.

Yoshioka, K., W. Moeder, H. G. Kang, P. Kachroo, and K. Masmoudi. 2006. The chimeric *Arabidopsis* cyclic nucleotide–gated ion channel 11/12 activates multiple pathogen resistance responses. *Plant Cell* 18:747–63.

A Synthesis:
The Properties of Selected
Plant Materials, Cells, and Tissues

All organic beings have been formed on two great laws—Unity of Type, and the Conditions of Existence. By unity of type is meant that fundamental agreement in structure, which we see in organic beings of the same class, and which is quite independent of their habits of life. ... The expression of conditions of existence ... is fully embraced by the principle of natural selection [which] acts by either now adapting the varying parts of each being to its organic and inorganic conditions of life; or by having adapted them in the long-past periods of time. —Charles Darwin, *On the Origin of Species*

The purpose of this chapter is to provide an overview of the properties of biologically important plant materials by discussing a variety of cell and tissue types. The organization of this chapter is from the "outside" toward the "inside" of the vascular plant body, beginning with the plant cuticle, which extends over leaves, stems, and even portions of roots, and ending with the heartwood that lies at the inner core of older woody stems and roots. The treatment presented here is neither comprehensive nor inclusive. Some materials and structures have received comparatively little attention, in part because technological advances have only recently permitted their direct biophysical study (e.g., the cuticular membrane), while other materials have been extensively examined for many decades by a wide variety of workers interested in using and learning more about the different properties of specific tissues (e.g., wood).

Throughout this chapter (as well as in other parts of the book), the word "material" is used purely for linguistic convenience, which raises an important caveat. As generally defined, a true material is a homogeneous

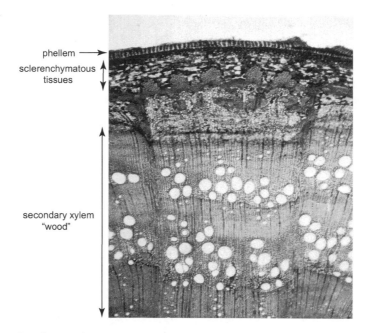

phellem →
sclerenchymatous
tissues

secondary xylem
"wood"

FIGURE 8.1. Cross section of a five-year-old twig from an oak (*Quercus*), showing the arrangement of the outer bark (phellem), subtending regions of sclerenchymatous tissues (derived from the cortex and phloem), and internal secondary xylem (wood). Inspection of any of these regions reveals that each is of composed of different cell types.

substance such that, once quantified, the properties of any representative sample allow us to predict the physical behavior of any specimen made of the same material as a whole (provided that the sample had no defects). In general, this condition does not hold true for plant materials because cells, tissues, and organs are typically complex composites or microstructures that are chemically and structurally heterogeneous (fig. 8.1). As a consequence of this heterogeneity, plant materials are extremely anisotropic, and far worse for most experimental purposes, they are typically nonlinear viscoelastic materials. This does not mean that plant materials cannot be quantified rigorously. It just means that we have to accept a certain level of uncertainty if we want to generalize from the results of any particular set of experimental measurements. Provided that the physical properties of a sufficiently large and representative sample of any plant material are determined, it is possible to extrapolate to some degree the behavior of the whole structure from which that material was removed. Naturally, this statement raises the question of what constitutes a "large and representative sample." As will be seen throughout this chapter, it is unwise to underestimate the complexity of even the seemingly simplest plant "material."

Another concern emerges when we examine the numerical values of the properties reported for the same kind of plant material. Consider the tissue called sclerenchyma, which can provide considerable mechanical support, particularly if this tissue is placed at or near the perimeter of an organ, where mechanical stresses reach their maximum magnitudes. Sclerenchyma is composed of thick-walled, generally (but not invariably) lignified cells. Although any tissue that manifests these anatomical features can be referred to as "sclerenchyma," it is obvious that the same material can have very different material properties. As shown in table 8.1, the elastic modulus for this tissue measured in tension varies over a large range (i.e., $E = 3.08$–4.5 GN/m^2). This variation is expected, given the anatomical and chemical differences observed for the same tissue taken from different plants or even from different parts of the same plant. Yet the oversimplification that results when the same botanical term is used to describe a cell or tissue type manifesting variable chemical and anatomical attributes is often ignored or neglected in the rush to make general statements about the mechanical behavior of plants.

Even with these caveats in mind, the mechanical and physical properties of many plant materials compare remarkably well with those observed for many engineered materials, particularly when we consider the specific weight γ of a material; that is, its density ρ multiplied by the acceleration of gravity ($g = 9.8067$ m/s^2). Normalizing properties shows us how much a material contributes to the ability of a structure to cope with mechanical loads versus the mechanical load the material itself contributes to the load the structure must support. Consider cellulose microfibrils, which are the principal tensile load–bearing members in the primary cell walls of the land plants. Crystalline cellulose microfibrils are reported to have a tensile elastic modulus E that ranges between 100 and 140 GN/m^2, depending on the extent to which they are hydrated (see table 8.1). The density of microcrystalline cellulose is on the order of 1,500 kg/m^3; thus, $\gamma = 14{,}700$ N/m^3 and $E/\gamma = 6.8 \times 10^6$ m – 9.5×10^6 m. In comparison, the E/γ of nylon thread ranges between 0.18×10^6 m and 0.32×10^6 m (E ranges between 2.0 and 3.6 GN/m^2 and $\gamma = 11{,}280$ N/m^3), while that of Kevlar ranges between 4.2×10^6 m and 8.9×10^6 m. Indeed, cellulose compares well with many carbon-based composite fibers, which have an E/γ between 13×10^6 m and 22×10^6 m. These numbers tell us that cellulose is comparable to modern engineered materials in terms of the weight it contributes to the structures

TABLE 8.1 **Elastic moduli of some plant and engineered materials**

Source (type of material)	Young's modulus, E (MN/m^2)		Shear modulus, G (MN/m^2)
	In tension	In bending	In torsion
Isolated plant materials			
Cellulose microfibrils[a]	100,000–140,000	—	—
Phloem fibers (kiln dried)[a]	90,000	—	—
Tomato fruit cuticle[b]	25–50	—	—
Vascular plants			
Apium graveolus (collenchyma)[a]	20–25		
Aristolochia macrophylla (sclerenchyma)[c]	3,080 ± 300	—	—
Arundo donax (lignified parenchyma)[d]	8,000	—	320
Lolium perenne (leaf sclerenchyma)[a]	22,600	—	—
Phormium tenax (leaf fibers)[e]	31,400	—	—
Phormium tenax (leaf cell walls)[f]	71,400	—	—
Secale ancestrale (sclerenchyma)[f]	14,500	—	—
Solanum tuberosa (stem parenchyma)[g]	19–40	3.6	—
Solanum tuberosa (stem periderm)[g]	—	18	—
Stenocereus gummosa (stem cortex)[g]	—	3.5–4.8	—
Triticum aestivum (sclerenchyma)[h]	3,100	—	—
Algae			
Chondracanthus exasperatus (stipe tissues)[i]	1.3–3.9	—	—
Durvillaea antarctica (stipe)[j]	3.7 ± 0.5	8.5 ± 1.5	0.3 ± 0.1
Durvillaea antarctica (cortical tissues)[j]	6.5 ± 3.5	—	—
Durvillaea antarctica (medulla tissues)[j]	2.9 ± 3.0	—	—
Durvillaea willana (stipe)[j]	4.8 ± 1.3	12.2 ± 8.0	0.3 ± 0.1
Durvillaea willana (cortical tissues)[j]	6.5 ± 3.5	—	—
Durvillaea willana (medulla tissues)[j]	4.8 ± 2.7	—	—
Laminaria digitata (stipe)[j]	13.2 ± 8.6	84 ± 85	0.7 ± 0.5

TABLE 8.1 (continued)

Source (type of material)	Young's modulus, E (MN/m²)		Shear modulus, G (MN/m²)
	In tension	In bending	In torsion
Algae			
Laminaria digitata (cortical tissues)[j]	18.5 ± 11.8	—	—
Laminaria digitata (medulla tissues)[j]	8.7 ± 3.6	—	—
Nereocystis luetkeana (stipe)[j]	16.6 ± 3.2	—	—
Nitella opaca (internode cell wall)[k]	500–4,000	—	—
Postelsia palmaeformis (stipe)[l]	14.5 ± 8.1	6.4 ± 2.4	—
Engineered materials			
Glass-filled epoxy (35%)	25,000	—	—
Kevlar epoxy (53%)	63,600	—	—
Nylon (6/6)	2,000–3,600	—	1,200–1,400
Polypropylene	1,400	—	—

Sources: [a]Niklas, unpublished; Ishikawa et al. (1997). [b]Matas et al. (2004). [c]Köhler et al. (2000). [d]Spatz et al. (1997). [e]Spatz et al. (1998). [f]King and Vincent (1996). [g]Niklas et al. (2003). [h]Spatz et al. (1993). [i]Harder et al. (2006). [j]Johnson and Koehl (1994). [k]Probine and Preston (1962). [l]Holbrook et al. (1991).

it reinforces. By the same token, across a broad spectrum of angiosperm woods, the average maximum compressive strength S_C and average specific weight are 36 MN/m² and 7,900 N/m³ (table 8.2). Thus, across these same species, $S_C/\gamma \approx 4.5 \times 10^3$ m, which is numerically larger than that of hard-fired brick (and nearly twice that of ordinary limestone). Conifer woods are equally impressive. With an average compressive strength of 22 MN/m² and an average specific weight of 5,800 N/m³, their S_C/γ is about 3.8×10^3 m. These numbers show that cellulose-reinforced cell walls (even in green, hydrated wood) are comparatively light and remarkably strong, both in tension and in compression, which helps to explain why trees are the most massive and tallest forms of terrestrial life.

8.1 The plant cuticle

We begin our discussion of plant materials by considering the cuticle. The epidermis of the aerial organs of all vascular land plants is covered with a

TABLE 8.2 **Average densities and mechanical properties of green (50% moisture content) wood species measured along the grain**

Genus, species	Density, ρ (kg/m^3)	Modulus of rupture, M_R (MN/m^2)	Young's modulus, E (MN/m^2)	Maximum shear strength, S_G (MN/m^2)	Maximum compressive strength, S_C (MN/m^2)
Conifers					
Abies alba	545	41	6,700	6.4	22.4
Abies alba	577	43	8,100	5.8	22.0
Abies balsama	529	43	8,100	5.7	20.8
Abies grandis	449	35	5,700	4.6	17.0
Abies procera	465	34	5,700	4.9	16.0
Agathis vitiensis	673	59	9,700	7.5	27.8
Araucaria angustifolia	689	52	8,700	7.6	28.8
Chaemaecyparis lawsoniana	497	34	3,900	5.2	14.3
Larix decidua	673	53	7,900	6.9	24.3
Larix eurolepis	577	43	5,900	5.9	19.2
Larix kaempferi	609	48	6,800	6.1	21.9
Picea abies	497	36	6,300	4.8	17.0
Picea abies	497	36	6,100	5.4	19.6
Picea alba	529	39	7,400	5.8	18.2
Picea omorika	497	35	6,400	4.6	16.5
Picea sitchensis	481	34	5,900	4.5	16.1
Picea sitchensis	529	39	8,800	5.0	18.3
Pinus brutia	—	—	—	—	49.9
Pinus caribaea	977	66	10,400	9.2	33.0
Pinus caribaea	977	83	11,400	9.9	37.4
Pinus contorta	593	41	6,400	5.6	18.8
Pinus holfordiana	513	—	—	—	18.5
Pinus nigra	609	42	7,000	5.6	19.4
Pinus nigra	705	63	10,400	—	25.5
Pinus pinaster	609	36	6,600	5.0	17.4
Pinus ponderosa	561	38	6,000	4.8	17.7
Pinus radiata	577	38	6,800	5.0	16.9
Pinus radiata	641	41	7,700	5.8	19.1
Pinus strobus	433	28	4,800	4.8	13.9
Pinus sylvestris	625	—	7,300	5.9	21.9
Pinus sylvestris	625	44	7,700	5.9	21.0
Podocarpus sp.	641	48	6,100	6.3	22.1
Podocarpus guatemalensis	657	63	8,100	8.4	30.9
Pseudotsuga menziesii	625	53	8,300	6.8	24.6
Pseudotsuga menziesii	673	54	10,400	7.2	25.9
Thuja plicata	465	38	5,400	4.9	18.3
Tsuga heterophylla	545	41	6,800	6.1	19.7
Tsuga heterophylla	593	49	8,700	6.6	24.1
Tsuga heterophylla	609	52	8,900	7.0	25.4
mean ± SE	593 ± 18.4	45.1 ± 1.8	7,326 ± 277	6.09 ± 0.22	22.1 ± 1.1

TABLE 8.2 (continued)

Genus, species	Density, ρ (kg/m³)	Modulus of rupture, M_R (MN/m²)	Young's modulus, E (MN/m²)	Maximum shear strength, S_G (MN/m²)	Maximum compressive strength, S_C (MN/m²)
Dicots					
Aesculus hippocastanum	657	41	5,300	—	17.4
Acacia mollissima	897	80	11,200	11.2	35.6
Acer pseudoplatanus	721	66	8,400	10	27.5
Afzelia quanzensis	1,137	89	8,700	13.4	47
Alnus glutinosa	625	49	7,600	6.3	21.7
Alstonia boonei	497	36	6,400	5.3	20.1
Anthocephalus chinensis	567	54	7,600	6.2	27.7
Aspidosperma sp.	993	79	8,300	14.7	39.8
Autranella congolensis	1,144	109	14,200	13.6	55.8
Berlinia confusa	849	72	9,100	9.4	34
Betula sp.	801	63	9,900	7.7	26.3
Brachystegia nigerica	865	79	8,800	11.9	39.4
Brachylaena hutchinsii	1,153	92	8,600	17.4	53.6
Byrsonima coriacea	865	84	10,800	11.8	39.8
Calophyllum brasiliense	817	77	10,100	10.2	37.9
Canarium schweinfurthii	593	41	6,200	6.4	21.6
Carpinus betulus	865	66	9,700	9.9	27
Cassispourea malosana	897	63	7,900	10	30.9
Castanea sativa	657	52	7,200	7.5	24.2
Cedrela odorata	433	37	5,200	4.9	18.8
Celtis sp.	961	90	12,800	—	44.9
Ceratopetalum apetalum	753	59	11,000	7.9	28.3
Chlorophora excelsa	817	74	8,300	10.3	35.3
Cordia millenii	545	54	6,100	7.1	26.3
Cullenia ceylanica	769	74	11,000	8.3	35.4
Cylicodiscus gabunensis	1,185	101	12,800	16.6	56.7
Cynometra alexandri	1,121	94	9,900	15.1	48.5
Dipterocarpus sp.	929	83	16,000	—	43.1
Dipterocarpus acutangulus	913	82	11,800	8.1	39.2
Dipterocarpus caudiferus	753	70	12,800	7.2	35.4
Dipterocarpus zeylanicus	977	90	13,500	9.9	47.9
Dryobalanops beccarii	865	81	10,900	8.5	41.2
Dryobalanops keithii	902	81	12,400	8.8	42.9
Dryobalanops lanceolata	865	88	11,000	8.1	42.9
Entandrophragma angolense	705	52	6,900	7.4	25.4
Entandrophragma cylindricum	833	74	9,600	9.8	36
Entandrophragma utile	833	79	9,600	10.8	38.2
Eperua sp.	1,169	104	15,000	—	57.8
Erythrophleum sp.	1,362	124	13,300	16.7	71.2
Eucalyptus diversicolor	1,041	77	13,400	10.4	37.6
Eucalyptus marginata	1,009	72	9,600	10.3	37.2
Eucalyptus microcorys	1,234	110	16,500	14.1	63.8
Eucalyptus paniculata	1,346	121	18,100	17.2	67.6
Eucalyptus piluaris	897	87	12,800	12.2	45.2
Eusideroxylon zwageri	1,282	143	17,700	15.4	79.9

TABLE 8.2 (*continued*)

Genus, species	Density, ρ (kg/m³)	Modulus of rupture, M_R (MN/m²)	Young's modulus, E (MN/m²)	Maximum shear strength, S_G (MN/m²)	Maximum compressive strength, S_C (MN/m²)
Fagus sylvatica	833	65	9,800	9.4	27.6
Fraxinus excelsior	801	66	9,500	9	27.2
Gmelina arborea	625	54	5,900	8.5	25.6
Gonystylus macrophyllum	85	71	10,100	7.8	38.7
Gossweilerodendron balsamiferum	641	52	6,000	7.7	24.3
Guarca excelsa	689	66	8,400	10	31.7
Guarca thompsonii	817	85	10,600	10.5	43.2
Heritiera simplicifolia	753	81	10,700	8.5	41.3
Heritiera simplicifolia	801	81	11,700	9.7	39.8
Hevea brasiliensis	865	62	8,200	9.8	30.8
Hopea sengal	817	90	10,800	9.4	46.5
Khaya anthotheca	657	53	7,400	8.3	25.4
Khaya anthotheca	657	60	7,700	8.8	28.4
Khaya grandiflora	817	69	9,100	12.1	35.9
Khaya ivorensis	641	54	7,400	7.3	26.8
Khaya nyascia	705	63	7,300	10.4	33.2
Koordersiodendron pinnatum	1,089	108	13,100	12.5	56.1
Loneiocarpus castillo	1,169	128	15,400	16.3	65.5
Lophira alata	1,292	118	13,600	15.2	65.6
Lovoa trichilioides	673	57	7,300	—	28.8
Maesopsis eminii	609	55	8,100	7.4	28.5
Mansonia altissima	801	90	9,700	12.5	44.1
Mitraguna sp.	689	54	8,100	—	49.4
Mora excelsa	1,137	94	14,800	12.5	27.3
Nauclea diderrichii	945	94	11,900	13.1	51.6
Nectrandra sp.	689	63	9,400	8.1	31.1
Newtonia buchaneni	705	68	9,000	9.4	32.8
Nothofagus sprocera	561	43	5,000	6.2	16.8
Ocotea rodiaei	1,250	144	15,900	15	72.5
Ocotea usambarensis	769	59	8,100	8.7	30.6
Octomeles sumatrana	481	37	5,300	4.3	21.8
Olea hochstetteri	1,121	105	13,700	16.5	48.8
Oxystigma oxyphyllum	801	81	10,500	10.2	39.2
Parashorea sp.	705	62	10,400	—	31.6
Parashorea malaanonan	641	61	9,100	7.1	31.9
Parashorea tomentelia	577	57	7,700	6.3	29.6
Peltogyne sp.	1,105	105	14,000	14.9	56.5
Pericopsis elata	977	108	11,400	13.1	53.8
Pipradeniostrum africanum	849	76	9,900	11.2	36.7
Platanus hybrida	785	54	6,400	9.7	24.2
Populus canadensis	529	41	6,800	5.9	19.3
Populus × *canescens*	577	44	7,200	5.9	20.1
Protium decendrum	801	76	10,100	10.5	36.4
Prunus avium	753	64	8,300	9.2	27.8
Pseudosindora palustris	833	81	10,100	10.3	39.2
Pterocarpus angolensis	881	85	7,600	12.5	40.6
Pterygota bequaertii	849	73	8,800	6.8	35.4

TABLE 8.2 (continued)

Genus, species	Density, ρ (kg/m³)	Modulus of rupture, M_R (MN/m²)	Young's modulus, E (MN/m²)	Maximum shear strength, S_G (MN/m²)	Maximum compressive strength, S_C (MN/m²)
Pterygota macrocarpa	705	57	7,400	6.14	26.7
Qualea sp.	897	81	12,800	10.4	43.1
Quercus sp.	833	59	8,300	9.1	27.6
Quercus cerris	929	70	10,100	11.2	28.5
Quercus rubra	865	72	10,500	9.2	28.7
Ricinodendron rautanenii	224	14	1,800	—	5.9
Salix alba	529	36	4,800	5.2	14.7
Salix alba var. coerulea	513	31	5,600	4.4	13.6
Salix fragilis	529	35	5,600	4.8	14.8
Sclerocarpa sp.	657	28	2,400	6	12.4
Scottellia coriacea	849	83	11,300	10	38.6
Shorea acuminatissima	609	57	7,400	8.3	27.6
Shorea dasyphylla	609	63	9,700	—	32
Shorea faguetiana	673	66	9,700	8.4	32.1
Shorea gibbosa	625	67	9,000	8.3	32
Shorea guiso	993	105	14,100	11.6	53.2
Shorea hopeifolia	689	58	8,400	8.3	27.9
Shorea leptoclados	545	50	7,000	5.9	27
Shorea macrophylla	449	43	7,500	4.9	22.7
Shorea parviflora	513	48	6,700	5.6	23.9
Shorea pauciflora	689	68	9,700	8.7	33.9
Shorea smithiana	513	50	7,200	4.8	25.3
Shorea superba	945	96	11,700	11.6	50.1
Shorea superba	1,057	111	14,300	11.5	58.3
Shorea waltonii	529	52	8,100	7.1	26.4
Staudtia stipitata	1,139	123	14,200	16.6	66.6
Sterculia oblonga	913	81	10,300	7.2	38.7
Sterculia rhinopetala	961	87	10,800	9.6	42.5
Swartzia leiocalycina	1,298	148	17,100	16.7	72.6
Symphonia globulifera	881	86	13,000	10.2	41.6
Syncarpia globulifera	1,025	77	10,600	10.5	41.2
Tarrietia utilis	817	70	8,400	10	36.5
Tectona grandis	801	90	8,900	11.5	41.2
Tectona grandis	801	83	8,900	12.5	37.9
Tectona grandis	817	84	8,800	10.1	42.8
Terminalia amazonica	961	94	13,000	12.5	46.1
Tieghemelia heckerii	801	75	8,200	10.7	36.5
Tilia vulgaris	657	54	9,200	7.3	26.1
Triplochiton scleroxylon	465	37	4,600	5.2	18.5
Ulmus glabra	753	68	9,400	8.3	30.4
Ulmus hollandica	641	44	5,400	7.6	18.7
Ulmus procera	641	40	5,200	7.9	16.9
Virola koschnyi	657	45	9,500	5.2	21.9
Vochysia sp.	657	55	8,600	6.4	26.4
Vochysia hondurensis	577	43	6,900	5.5	22.1
mean ± SE	809 ± 18.3	72.8 ± 2.1	9,707 ± 254	9.72 ± 0.28	36.3 ± 1.2

Source: Niklas and Spatz (2010).
Note: Duplicate entries for species are for samples drawn from trees growing under different conditions.

waxy cuticle that provides an important barrier between the plant body and the air that surrounds it. The cuticle restricts water loss by passive diffusion through the epidermis, impedes microbial attack, and provides protection against UV-induced cellular damage. It also prevents neighboring developing organs from fusing together early in their ontogeny. Furthermore, recent investigations have shown that the cuticle can contribute substantially to the mechanical support of self-supporting plant organs when those organs are turgid (Wiedemann and Neinhuis 1998; Matas et al. 2004). For all these reasons, the cuticle is one of the most fundamental of the adaptations that allowed plants to colonize the land and elevate their body parts into the air.

Following the terminology of Jeffree (1996) and others (fig. 8.2), the cuticle typically consists of an external layer of epicuticular waxes that covers a comparatively thin layer of saponifiable lipids (the cuticle proper, CP), which in turn covers an inner layer of waxes and fibrous polysaccharides embedded in a cutin matrix (the cuticular layer, CL). The CP and CL constitute the cuticular membrane (CM), which, *sensu stricto*, develops within primary cell walls (Esau 1977). The structure of the CM and the extent to which it extends beneath the epidermis vary among, and sometimes within, species (Jeffree 1996). For example, although it is frequently confined to the outer periclinal and anticlinal walls of epidermal cells, the CM may develop in subepidermal cell walls (figs. 8.2–8.3). Significant ultrastructural and chemical differences in the CM also exist across and within species (see Kolattukudy 1980, 1996; Jeffree 1996; Wiedemann and Neinhuis 1998). These differences can affect the physiological performance of the CM as well as its mechanical function (Petracek and Bukovac 1995; Wiedemann and Neinhuis 1998).

Because the walls of the epidermis and adjoining cells are placed in tension when primary organs are turgid (Kutschera 1989; Niklas and Paolillo 1997, 1998; see section 8.3), the CM is in an ideal location to function as a tensile "skin." In one of the most comprehensive and detailed studies of cuticular biomechanics, Matas et al. (2004) examined the mechanical and anatomical relationships between the isolated CM and intact samples of the outer fruit wall (= epidermal and subepidermal peels) of different cultivars of tomato (*Solanum lycopersicum*), employing a spectrum of mechanical tests to quantify the mechanical behavior of these structures when placed in tension (to mimic the tensile forces exerted by a turgid fruit on its outermost elastic "skin"). Their tests indicated that the outer fruit wall and the

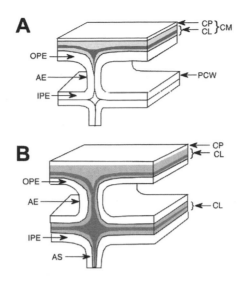

FIGURE 8.2. Schematic representations of two different patterns of cuticularized epidermal and subepidermal cells. In both cases, the cuticular membrane (CM) consists of the cuticle proper (CP) and the cuticular layer (CL). (A) In one pattern, the CM develops in the outer periclinal walls of the epidermis (OPE) and in the epidermal anticlinal walls (AE). (B) In the other pattern, the CM develops in the inner periclinal walls of the epidermis (IPE) and the anticlinal walls of subepidermal cells (AS). Darker areas in the CL represent pectin-rich regions.

CM of tomatoes are isotropic, viscoelastic, and to different degrees, strain-hardening structures. In addition, the average work of fracture for the CM was comparable to that of some polyesters and epoxy resins (i.e., $W \sim$ 100 J/m^2). The strain hardening observed was attributed to realignment of CM–epidermal cell wall polymeric fibrils in the direction of applied tensile forces, which in theory should increase the effective tensile modulus of the entire structure. This hypothesis has received empirical support from subsequent work examining the biomechanical behavior and quantitative contributions of cutin and polysaccharides. By isolating CM from mature green and ripe red tomato fruits and treating it with anhydrous hydrogen fluoride in pyridine (which selectively eliminates polysaccharides attached to or within the cutin matrix), López-Casado et al. (2007) demonstrated that cutin samples treated in this manner showed a drastic decrease in elastic modulus and stiffness (up to 92%) compared with untreated CM. Thus, polysaccharides incorporated into the cutin matrix are probably responsible for the high tensile elastic modulus, stiffness, and linear elastic behavior of the cuticle as a whole (López-Casado et al. 2007).

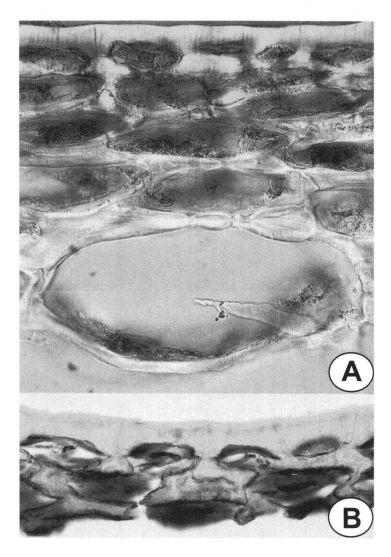

FIGURE 8.3. Peel and isolated cuticles from the ripened fruit of Inbred 10, a tomato cultivar. (A) Peel showing the cuticle extending 2–3 cell layers into the interior, interfacing with collenchyma. Unstained. (B) Isolated cuticle stained with chlor-zinc-iodine. Primary cell walls are stained dark gray. (From Matas et al. 2004.)

TABLE 8.3 **Thickness, tensile elastic modulus, and breaking stress and strains for hydrated plant cuticles**

Taxon	Thickness (μm)	Tensile E (GN/m²)	Breaking stress (GN/m²)	Breaking strain (%)
Cuticles from leaves				
Agave americana	9.5	0.6	>13	≈4
Agave stricta	8.5	0.55	>15	≈5
Aloe buhrii	11.5	0.72	>14	≈9
Clusia fluinensis*	2.4	0.7	>15	≈3
Hedera helix*	3.7	0.29	>15	≈14
Nerium oleander*	10.2	0.37	>15	≈8
Sansevieria hahnii	10	0.65	>14	≈3
Yucca aloifolia*	12	0.73	>15	≈3
Cuticles from fruits				
Malus sp.	15	0.04	>5	≈5
Solanum esculentum*	13.4	0.06	>2	≈3
Vitis vinifera	14	0.05	>3	≈3

Source: Data denoted by * are from Wiedemann and Neinhuis (1998). All other data are from Niklas and Spatz (unpublished).

The rheological behavior of the CM reported by Matas et al. (2004) is consistent with other studies. For example, Wiedemann and Neinhuis (1998) showed that the Young's modulus and breaking stress of hydrated tomato CM are on the order of 60 MN/m² and 2 MN/m², respectively (considerably lower than those reported for cuticle isolated from leaves; table 8.3). Likewise, the viscoelasticity of the CM has been observed previously, whereas inspection of the data from sequential loading-unloading cycle tests graphed by Thompson (2001) indicates that the CM rate of creep decreases with successive cycles, which indicates strain hardening. Petracek and Bukovac (1995), who also showed that the CM is viscoelastic, report that elastic strains typically exceed plastic strains when the hydrated CM is modestly extended. Thus, a general mechanical phenomenology can be adduced for the CM, although significant variation exists regarding the manifestation of this phenomenology and the absolute magnitudes of material properties.

Finally, the data reported by Matas et al. (2004) show that the epidermis and subepidermal layers in peels of tomato fruits play an important role in allowing the CM to sustain tensile mechanical forces. Unfortunately, the modulus of elasticity E of tomato fruit epidermal and subepidermal cell layers (which are collenchymatous) can only be inferred because these layers are difficult to isolate from the CM or each other. However, a Voigt model (see section 4.16) indicates that these tissues have an E on the order of 24 MN/m^2, which is significantly less than that of the CM of tomato fruits and numerically consistent with the E reported for collenchyma isolated from the leaves of celery (*Apium graveolens*) and lovage (*Levisticum officinale*) (i.e., 22–23 MN/m^2).

In passing, it should be mentioned that the plant cuticle experiences biaxial tensile stress, whereas, with very rare exceptions, published accounts deal with samples placed in uniaxial tension. This discrepancy does not pose a serious concern because for isotropic materials, one can measure the elastic moduli in uniaxial tension and use the relationships $\varepsilon_1 = (\sigma_1 - v\sigma_2)/E$ and $\varepsilon_2 = (\sigma_2 - v\sigma_1)/E$ to infer the biaxial behavior. The subscripts in these relationships refer to the stresses and strains measured in two orthogonal directions (e.g., width and length). A reasonable estimate for the Poisson's ratio is 0.3. If the material is anisotropic, the strains resulting from the same stresses measured in orthogonal directions can be determined to calculate E in each of the two directions.

8.2 A brief introduction to the primary cell wall

With rare exceptions (e.g., gametes), plant cells are enveloped by a rigid or semirigid wall that is produced by the protoplast and deposited outside the plasma membrane. It is a heterogeneous and dynamic polymeric structure composed of a three-dimensionally interwoven network of cellulose microfibrils embedded in a complex matrix. Cellulose is an unbranched polysaccharide composed of 1,4-linked β-D-glucopyranose units. This polymer is roughly 0.8 nm in its maximum width and 0.33 nm^2 in cross section. It is one of the most abundant polymers in living organisms, making up between 30% and 60% of dehydrated plant cell walls. Cellulose microfibrils are approximately 5 nm by 9 nm in cross section. Each consists of an inner core of roughly 50 parallel chains of cellulose molecules organized in a crystalline array surrounded by roughly the same number of cellulose molecules arranged in a paracrystalline array. The spaces between microfibrils are occupied by the constituents of the cell wall matrix: pectins, heterogeneous hemicelluloses (e.g., xylans, mannans, xyloglucans), and a variety of structural proteins. By weight, the main constituent of the primary cell wall is water, which can contribute as much as 80% of the fresh weight of walls.

The exact nature of the interactions and bonds that hold the wall together is still uncertain. However, the cell wall must meet a variety of mechanical as well as physiological requirements: for example, it must sustain large tensile stresses, and it must permit reversible (elastic) deformation when mature, yet it must extend plastically to allow for cell growth in size (Schopfer 2006). Thus, in mature cells, the wall is a rigid mechanical barrier against sudden protoplasmic distension resulting from a rapid influx of water caused by steep osmotic gradients. Accordingly, the cell wall mechanically sets the limits for the size and shape of the non-growing protoplast (just as it largely defines the texture and mechanical properties of mature tissues and organs).

Changes in cell shape or permanent growth in cell size require irreversible plastic extension followed by stiffening. This reduction and subsequent increase in stiffness are consequences of enzymatic activity within the cell wall. The osmotic gradient is the driving force for protoplasmic expansion in volume. The hydrostatic pressure exerted by the protoplast on the innermost layer of the cell wall extends and, in the absence of cell wall deposition, thins the wall. During episodes of growth, enzymatic activity decreases the yield stress of the cell wall. This phenomenon is sometimes referred to as cell wall loosening or cell wall stress relaxation. The mechanical properties of the walls of living cells, therefore, can change from those of a fluidlike, extensible material to those of a stiff and rigid material.

Although the chemical and mechanical properties of the primary cell wall have been extensively studied in a variety of species, the mechanisms responsible for cell wall loosening are not completely understood. It is generally agreed that primary cell walls consist of a chemically heterogeneous (sometimes erroneously referred to as an "amorphous") matrix in which a variety of long-chain polymers are embedded. Cellulose is considered by most workers to be among the most important of these polymers because it chemically dominates the essentially fibrous infrastructure of most primary cell walls. Like most polymers, cellulose has a higher tensile elastic modulus and a greater tensile strength measured along its molecular chain length than when measured normal to its length. In cell walls in which cellulose fibrils are aligned principally parallel to cell length, the tensile elastic modulus in the longitudinal direction can be over three times the modulus measured in the transverse direction (i.e., for some *Chara* internodal cells, $E_L = 3.5$ GN/m^2 and $E_T = 1.0$ GN/m^2). Enzymatically induced changes in the chemical composition of the matrix and in the chemical cross-links that bind cellulose molecules into microfibrils allow cell wall polymers to slide past one another or become "locked" in place.

In this sense, changes in the wall yield stress are changes in the *shear* yield stress.

As noted, any change in the cell wall yield stress requires molecular modifications. Stress relaxation might result from scission of stress-bearing cross-links or from slippage of those cross-links along a scaffold. Either of these phenomena could result in a reduction in the wall yield stress without a substantial change in the original mass of the wall. Subsequent enlargement of the wall occurs secondarily, as a consequence of cellular water uptake in response to the turgor relaxation that inevitably accompanies wall stress relaxation. Four molecular candidates have been proposed as wall-loosening agents: expansins, xyloglucan endotransglycolase/hydrolases, endo-(1,4)-β-D-glucanases, and hydroxyl radicals (Cosgrove 2005).

- Expansins are believed to be wall-loosening proteins that induce wall stress relaxation and irreversible wall extension in a pH-dependent manner. However, they appear not to hydrolyze wall polymers.
- Xyloglucan endotransglycolase/hydrolases carry out various functions, including wall loosening and rigidification, integration of new xyloglucans into cell walls, trimming of xyloglucan strands that are not tightly linked to the surface of cellulose, hydrolyzing of xyloglucans, and fruit softening.
- Endo-(1,4)-β-D-glucanases, which are sometimes called "cellulases," belong to the glycoside hydrolase family of enzymes. These are membrane-bound endoglucanases that appear to be involved in cellulose formation. Potential substrates for these enzymes include cellulose and xyloglucan. It has been suggested that their mechanism of action is the digestion of noncrystalline regions of cellulose in the cell wall or the release of xyloglucans trapped in cellulose microfibrils.
- Hydroxyl radicals are the most reactive of all oxygen species. They consist of a free hydroxyl group in which oxygen is missing an electron in its outermost shell. These oxidants can "steal" electrons from—and thereby damage and weaken—polysaccharides, such as cellulose and hemicellulose.

The scope of this chapter precludes a detailed discussion of each of these candidate mechanisms. However, it is useful to note that they are not mutually exclusive. Some or all may act in concert to alter the yield stress of cell walls. Indeed, it is useful to distinguish between two classes of wall-loosening agents, which can be referred to as "primary" and "secondary" mechanisms. Primary mechanisms catalyze stress relaxation directly, which can be demonstrated when the yield stress of isolated cell walls is reduced by the mechanism. Secondary mechanisms modify the wall to amplify the physical effects of primary mechanisms. Both classes of loosening mechanisms might be important for the control of wall enlargement in plants.

The anisotropy of cell wall polymers such as cellulose also helps to explain how cells can achieve different shapes as they grow in size. If microfibrils are deposited more or less randomly in the cell wall, the wall will exhibit isotropic mechanical behavior when hydrostatically stressed from within by its expanding protoplast. If microfibrils are deposited preferentially in certain directions, however, the wall will exhibit anisotropic mechanical properties (e.g., different shear yield stresses in different directions) that will prefigure the manner and extent to which a cell can elongate or expand in girth. The extent of cell wall anisotropy is revealed by simple shell theory, which can be used to predict the differences in the lines of strain required to manifest cylindrical cell shapes when an internal pressure P with equivalent directional lines of force is exerted from within. Note that multidirectional but nonequivalent strain lines within the "shell" (cell wall) are not necessary for cells to grow in length because turgor pressure has equidirectional magnitude. For a spherical thin shell with radius r and thickness t, the stress exerted by an internal pressure P_i is given by the formula $\sigma = rP_i/2t$, which can be used to estimate the stresses that develop in cell walls (box 8.1). For a cylindrical thin shell with radius r and thickness t, the stress in the longitudinal direction exerted by P also equals $rP_i/2t$. However, the circumferential stress (sometimes referred to as the "hoop stress") equals rP_i/t (see Stephens 1970), a relationship that is sometimes called Laplace's law. Thus, the ratio of the longitudinal to the circumferential stresses in a cylindrical shell is 2:1, from which it follows that a cylindrical wall must be anisotropic in its ability to resist deformation if elongation (rather than expansion) is to occur. Cellulose microfibrils that are preferentially oriented in the transverse direction in the youngest cell wall layer provide one way to deal with this 2:1 stress ratio because they can resist the circumferential expansion of a cell while permitting cell elongation. As new layers are added to the cell wall, these microfibrils will passively reorient as the cell wall layers in which they reside are pushed outward and stretched longitudinally. Given sufficient growth in cell length and continued cell wall layer deposition, some of the oldest transverse microfibrils may become nearly longitudinally aligned.

BOX 8.1 **Cell wall stress and expansion resulting from turgor**

As noted in the text, the stress that develops in the wall of a spherical thin shell with radius r, thickness t, and internal pressure P_i is given by the formula $\sigma = rP_i/2t$. Here, we use this formula to estimate the stresses that develop in the primary cell wall resulting from turgor pressure and the extent to which turgor pressure can drive cell expansion. For this purpose, we will consider a cell with a diameter of 30 μm and a wall thickness of 2 μm. Accordingly, $r/t = 15$ μm/2 μm = 7.5. A reasonable estimate for the turgor pressure in a living parenchyma cell is 0.5 MPa. Therefore, the stress in the wall is given by $\sigma = 7.5(0.5$ MPa)/2 ~ 1.88 MPa. Because the modulus of elasticity equals stress divided by strain—that is, $E = \sigma/(\Delta r/r)$, we see that, for a cell wall modulus on the order of 500 MPa, the change in cell radius $\Delta r/r$ equals 1.88 MPa/500 MPa = 0.0038, which is less than one-half of one percent. Likewise, when we calculate the change in cell volume that accompanies this change in radius, we see that it is less than 11 μm^3 for a cell that had an original volume of 14,138 μm^3 (which amounts to a 7.78 $\times$ 10^{-4} percent change in volume).

These calculations show that the primary cell wall is remarkably stiff and that its elastic properties must be considered when analyzing cell expansion. This last point is easily shown by considering the volumetric elastic modulus, denoted here by E_V, which is the quotient of the change in turgor pressure and the change in cell volume: $E_V = \Delta P/(\Delta V/V)$. This formula shows that the change in volume equals the change in turgor pressure divided by the volumetric elastic modulus. E_V typically ranges between 1.0 and 50 MPa as cells change in volume between 0.2% and 10% for each 0.1 MPa change in pressure resulting from the influx of water.

8.3 The plasmalemma and cell wall deposition

The inner layer of the cell wall is in immediate contact with the outermost cell membrane, the plasmalemma, which assists in the synthesis of cellulose microfibrils and the cell wall matrix. The plasmalemma defines the outer boundary of the living cytoplasm, where it regulates the influx and efflux of substances facilitated by various transport proteins. Other membranes delimit the boundaries of organelles as well as the functional units within double membrane–bound organelles.

According to the fluid mosaic model, all biological membranes share the same basic molecular organization: membranes consist of a double

layer of phospholipids (or, in the case of the chloroplast, glycosylglycer-
ides) and a variety of embedded proteins. The fluidity of the membrane
bilayer is conferred by virtue of *cis* double bonds, which prevent tight
packing of the phospholipids. Membrane fluidity is temperature dependent, decreasing at lower temperatures. The presence of unsaturated fats
in the bilayer increases membrane fluidity.

With the possible exception of the plasmalemma, biological membranes
typically do not experience hydrostatic pressures. For example, the water
potential on the two sides of the tonoplast, in the cytosol and in the plant
cell vacuole, must be in equilibrium such that there is no net flow between
these two cellular compartments. If it were not, a hydrostatic pressure difference would cause a folded tonoplast to expand outward or a taut cell
vacuole to burst. Indeed, the yield stress of most biological membranes
appears to be on the order of 0.2–1.0 MN/m^2. By inference, the membrane breaking stresses are also low. This observation emphasizes the mechanical importance of the cell wall, which provides a stiff bulwark against
which the plasmalemma is pushed by a turgid protoplast. Nevertheless,
data on the mechanical properties of the plasmalemma are sparse, and it is
not wise to base inferences about those properties on measurements using
animal systems, especially those, such as sea urchin eggs or red blood cells,
that tend to be atypically rigid (e.g., the red blood cell membrane is closely
attached to a rigid cytoskeleton). In their study of protoplasts isolated
from the leaves of rye (*Secale cereale*), Wolfe and Steponkus (1981) report
that the protoplast plasmalemma has an area elastic modulus of 230 mN/m^2
when hydrostatic stresses are applied over a few seconds. Over longer periods, the membrane surface area increases as a result of its stretching
and growth. The stress-strain relationship therefore involves two kinds
of responses. Over short periods (seconds), the plasmalemma expands
and its mass is conserved (it thins out); over longer periods (minutes),
the surface area of the plasmalemma expands beyond its original size by
the addition of new membrane constituents (the deformation follows a
surface energy law). Based on their measurements, Wolfe and Steponkus
conclude that an intrinsic elastic expansion of more than 2% causes the
protoplast membrane to break in tension.

As noted, the plasmalemma is intimately involved in the deposition of
cellulose microfibrils and the matrix in which they are embedded (Cosgrove 2005; Driouich and Baskin 2008). Cellulose is deposited by transmembrane complexes called terminal complexes (TCs), which contain

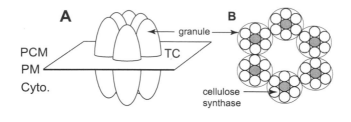

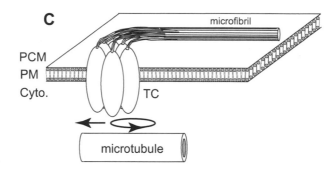

FIGURE 8.4. Graphic representation of the transmembrane terminal complex (TC) in relation to its inferred substructure and relationship to the cell cytoskeleton. (A) Each granule
of the TC spans the plasmalemma (PM) and thus contacts the primary cell wall (PCM) and
cytoplasm (Cyto.) at either end. (B) Among land plants, each TC consists of six granules, each
of which is posited to contain six functionally nonredundant cellulose synthase proteins (their
precise arrangement is unknown). (C) Each of the six granules contributes to the construction
of a cellulose microfibril that is formed external to the plasmalemma and is thus in contact
with the cell wall. The TC is believed to be tethered to a cortical microtubule within the cytoplasm. It is unclear whether the microtubule orientation directly dictates that of the TC or
whether the direction of microfibrillar extrusion influences the subsequent orientation of the
TC. (Adapted from Niklas 2004.)

cellulose synthase proteins encoded by the CesA gene family (Delmer
1999; Richmond and Somerville 2000; Nobles et al. 2001). Among all land
plants and some of their close algal relatives, TCs, as seen from the exterior surface of the plasmalemma, are solitary rosettes, each consisting
of six granules, each of which is made up of six CesA cellulose synthase
proteins (fig. 8.4). Although the precise arrangement of different cellulose
synthases within an individual granule is currently unknown, research indicates that some (designated CesAi proteins) initiate glucan chain formation by accepting glucose residues from a sugar nucleotide donor (most
likely uridine 5′-diphosphate, or UDP). Other proteins (designated Ce-
sAe proteins) in the TC further extend individual glucan chains. Several

investigators have also suggested that dimers of cellulose synthase are required in order to provide a cooperative cellobiose-generating system.

The mechanical behavior of microfibrils is governed in part by their dimensions (width, thickness, and length). Microfibrillar width and thickness are determined by the arrangement of TCs and the number of cellulose synthase units per TC, both of which tend to be highly conserved for individual species. For example, cellulose-producing eubacteria and all phaeophytes and rhodophytes (e.g., *Pelvetia* and *Ceramium*, respectively) typically have single linear TC arrays similar to those reported for the tunicate *Metandrocarpa uedai*. Likewise, many xanthophyte and chlorophyte algae have stacked linear TC arrays (e.g., *Vaucheria* and *Oocystis*, respectively). In contrast, all embryophytes and some charophyte algae (e.g., *Nitella*) have solitary TCs in the shape of hexagonal rosettes (when viewed from the exterior plasmalemma surface; see fig. 8.4B). Among embryophytes, as noted above, each granule contains six cellulose synthase proteins, each of which produces one glucan chain. The typical embryophyte microfibril is thus described as consisting of 36 glucan chains.

Although TC arrangement and cellulose synthase number per TC dictate microfibrillar width and thickness (and possibly cellulose allomorphism), little is known about the factors influencing microfibrillar length, which is also an important determinant of cell wall strength. Across many different natural and artificial sources of cellulose, the degree of cellulose crystallinity is correlated with glucan chain length (Delmer 1999). The siphonous green alga *Boergesenia forbesii* holds the current record in this regard (i.e., 23,000 units per chain). However, microfibrillar length often exceeds the glucan chain length predicted on the basis of cellulose crystallinity, indicating that chain initiation and termination probably occur multiple times during the fabrication of individual microfibrils. Although various hypotheses have been proposed to explain glucan chain termination (e.g., chain cleavage to relax localized tension resulting from "out-of-step" synthases in adjoining granules), none satisfactorily explains how entire TCs are initiated or terminated.

Much remains to be learned about how TCs are globally coordinated to fabricate the cellulosic infrastructure of an entire cell wall. The most prevalent hypothesis is that TCs are guided by cortical microtubules (see fig. 8.4C). If this is true, then cytoskeletal architecture controls microfibrillar deposition and cell wall texture, and a change in that architecture ought to produce different depositional patterns and wall textures. This hypothesis is consistent with the parallel alignment of microfibrils and microtubules that has been frequently reported for recently synthesized parts of plant cell walls (Wymer and Lloyd 1996; Taiz and Zeiger 2002).

Although the mechanics of cell wall expansion and isotropic growth are comparatively well understood, recent work indicates that microtubule and microfibril organization is only part of a much more complex story. For example, the cells of the temperature-sensitive *morl-1* mutant of *Arabidopsis thaliana* lose their orderly cortical microtubule arrangements at their restrictive culture temperatures. Although they also lose their capacity for anisotropic growth (which suggests a random arrangement of microtubules), these cells nevertheless retain parallel microtubule arrangements. Using the same mutant, Himmelspach et al. (2003) investigated whether well-ordered, preexisting microfibrils or cortical microtubules are essential for the resumption of normal (longitudinally aligned) microfibrils. Their protocol involved the transient disruption of microfibril organization with a brief treatment of the cellulose synthesis inhibitor 2,6-dichlorobenzonitrile and the subsequent examination of the alignment of newly formed microfibrils as cellulose synthesis was recovered at the mutant's permissive culture-temperature requirements. Despite the presence of disordered microtubules (and the initially random cell wall texture of the microfibrils), new microfibrils formed in transverse and longitudinal patterns.

These and other experiments indicate that preexisting microtubule or microfibril templates may not be required for the resumption of prior microfibrillar organization. Nevertheless, these results do not preclude the possibility that microtubules influence the direction of cellulose extrusion, which in turn dictates subsequent microfibrillar orientation. Given that recently formed portions of microfibrils are anchored in the cell wall, it is possible that TCs are dynamically propelled in the same direction in which they initially extrude microfibrils. If so, an initial parallel alignment of microtubules and microfibrils may be a transient phenomenon.

8.4 The epidermis and the tissue tension hypothesis

The tissue tension hypothesis posits that the epidermis and subepidermal peripheral layers of cells in fully turgid or actively growing organs are held in biaxial tension as they are stretched and pulled as a result of the influx of water (Kutschera 1989; Kutschera and Niklas 2007). Although the phenomenon of "tissue tension" was described qualitatively by eighteenth-

and nineteenth-century botanists, it was Hofmeister's (1859, 1860) now classical experiments that first shed quantitative light on the physiological and mechanical phenomena operating within the outer cell layers of growing organs (stems and petioles) of higher plants (for a recent review, see Kutschera and Niklas 2007). The results of two of the most important of his experiments led to the tissue tension hypothesis: (1) the immediate outward recurvature of median slices through sunflower (*Helianthus*) internodes when dissected, which is even more evident when internodes are placed in water, and (2) the rapid extension of the pith (exposed by removing peripheral cell layers with a cork borer) when the internodes were submerged in water. From the results of his first experiment, Hofmeister concluded that the epidermal and subepidermal peripheral cell layers are held in tension as internal tissues expand when hydrated. In his second experiment, Hofmeister quantitatively estimated the physical stresses exerted by the turgid pith against the epidermal and subepidermal cell layers. After adding water, he noted that the pith rapidly elongated by up to 20% of internodal length. He concluded that the hydrated pith is held in compression by the epidermal and subepidermal cell layers, which in turn are held in tension.

Müller (1880) subsequently constructed an apparatus that could quantitatively estimate (albeit crudely) the longitudinal stresses exerted by the hydrated pith of the *Helianthus* stem against the outer cell layers. He showed that a column of excised pith slices, split longitudinally and confined within a glass cylinder, exerted a pressure of 13.5 atm (1.35 MPa) against a weight applied to the top of the tissue mass. Much more recently, Peters and Tomos (2000) showed that tissue elongation is insensitive to cyanide (KCN), an inhibitor of oxidative phosphorylation and hence of cellular ATP production, for at least 2 hours after its application. Thus, Heinich's (1908) observation that the pith elongates when exposed to water, independently of the metabolism of the cells, was confirmed.

Based on detailed histological and developmental observations on the internodes of sweetgum (*Liquidambar styraciflua*), Brown et al. (1995a,b) concluded that cell division and expansion/extension in the developing pith provide the "driving force" for internodal elongation, and that it is an oversimplification to say that "the passive extension of peripheral tissues, especially the epidermis, controls the rate of growth in axial organs" (Brown et al. 1995a, 776). These authors also concluded that the rapid production and differentiation of vascular tissues in elongating internodes "contribute significantly to the development of tensile forces ascribed only to outer peripheral tissues in herbaceous plants" (Brown et al. 1995b, 781). The first of these conclusions was based in large part on two observations: (1) as internodes attain their maximum rates of elongation, the highest rates of cell division occur in the pith and cortex (where cell division continues until the cessation of internode growth) and (2) concomitant reduced rates of cell division in peripheral internode tissues are associated with a significant increase in cell elongation rates in the epidermis and subepidermis (cortical collenchyma) (Brown et al. 1995a). Both of these observations are consistent with the theoretical expectation that the subepidermal tissues of internodes experience compressive stresses as the peripheral tissues are placed in longitudinal (and circumferential) tension, and neither observation provides a basis for inferring a priori which of these two features constrains or "drives" internode growth in length. Nevertheless, Brown et al. (1995a,b) are entirely justified in arguing for a more synoptic mechanical view of axial organ growth, one that involves a complex stress-strain dynamic among different adjoining tissues in which all tissues and cell types participate. More recently, Passioura and Boyer (2003) used the concept of tissue tension in the epidermis to develop a time- and position-dependent model of stem growth incorporating both water uptake and turgor-driven cell expansion. Their model, which was based on the physics of tissue tension as described in the *Nitella* internode through corresponding experiments, correctly predicted the measured organ dynamics.

8.5 Hydrostatic tissues

The epidermis and the ground tissues of the land plants operate mechanically as cellular hydrostats. Their effective stiffness and strength is a function of the turgor pressure exerted by the living cytoplasmic component (the symplast) against cell walls (the apoplast). Collectively, the cell walls in these tissues mechanically reinforce one another as they "push" against the walls of neighboring cells. The bending stiffness of these tissues is maximized at full turgor, but decreases as water is withdrawn and turgor pressure declines. This phenomenon is most obviously expressed when organs composed mainly of hydrostatic tissues lose water and gradually wilt.

Hydrostats are remarkably cost efficient, as is revealed by considering the biaxial stresses within a thin-walled sphere with an outer radius r_o, a wall thickness t, and an internal pressure P_i. For reasons of symmetry, all four normal stresses on any small surface element on the wall must be identical. In addition, there can be no shear stress. Since the cell wall is at equilibrium for any P_i, it must satisfy Newton's first law of motion; that is, the stress around the wall must have a net resultant to balance the internal pressure across the cross section of the hollow sphere. Thus, as already mentioned in section 8.2, the stresses are given by the relationship $\sigma = P_i(r/2t)$. However, as P_i increases and the wall inflates and deforms, the change in the stresses developing in the wall will comply with the formula

$$(8.1) \qquad \Delta\sigma = P_i\, \Delta\left(\frac{r}{2t}\right) + \Delta\, P_i\left(\frac{r}{2t}\right).$$

Note that the expression $\Delta(r/2t)$ equals $(r/2t)[(\Delta r/r) - (\Delta t/t)]$, or simply $(r/2t)(1 + v)\varepsilon$, where v is the Poisson ratio and ε is the tensile strain in the cell wall. Therefore, equation (8.1) can be written in the form

$$(8.2) \qquad \Delta\sigma = P_i\left(\frac{r}{2t}\right)(1+v)\,\varepsilon + \Delta P_i\left(\frac{r}{2t}\right).$$

It can be further noted that ΔP and ε are related by the formula

$$(8.3) \qquad \varepsilon = \frac{\Delta r}{r} = \frac{(1-v)}{E_w}\frac{r}{2t}\Delta P_i.$$

where E_w is the elastic modulus of the cell wall when placed in tension. These foregoing relationships lead to the formula

$$(8.4) \qquad \Delta\sigma = P_i\left(\frac{r}{2t}\right)(1+v)\,\varepsilon + \frac{E_w}{1-v}\,\varepsilon.$$

The effective elastic modulus of the cell wall E_w^{eff} (i.e., the stiffness of the wall) when the wall is placed in tension as a result of internal (turgor) pressure is given by the formula

$$(8.5) \qquad E_w^{eff} = \frac{\Delta\sigma}{\varepsilon} = P_i\left(\frac{r}{2t}\right)(1+v) + \frac{E_w}{1-v}.$$

Equation (8.5) reveals one of the important properties of a hydrostatic cellular construction: the effective stiffness of the cell wall depends on the cell's internal pressure P_i, which in turn depends on the availability of water. When P_i decreases, E_w^{eff} decreases.

High internal pressures pose their own risk because every material, no matter how strong, has a maximum tensile yield stress (box 8.2). The risk of bursting is evident when we consider the circumferential stresses that develop within the walls of short, cylindrically shaped hydrostats experiencing inner and outer pressures P_i and P_o, respectively. For aquatic plants with long cylindrical cells, such as the green algae *Nitella* and *Chara*, the normal physiological condition is given by $P_i > P_o$, where the internal pressure P_i can be as much as 10 times the outside pressure P_o. Under these circumstances, equation (8.2.4) reveals that the maximum circumferential stress (which occurs on the innermost layer of the wall) increases dramatically either as wall thickness decreases or as cell girth increases with a constant wall thickness. This phenomenon may help to explain why most unicellular algae are comparatively small, while cell wall thickness remains fairly constant across phyletically very different species. Indeed, according to Laplace's law (i.e., $\sigma_c = rP_i/t$; see section 8.2), turgid eukaryotic cells, regardless of cell size, ought to devote the same fraction of their volume to cell wall material (for any particular cell wall material, turgor pressure, and mechanical safety factor). Whether this scaling relationship also applies to the turgor-resisting peptidoglycan layer of cyanobacterial cell walls is problematic (see Hoiczyk and Hansel 2000).

BOX 8.2 **Stresses in thick-walled cylinders**

For a thick-walled cylinder, the circumferential stress is given by the formula

(8.2.1)
$$\sigma^{circ} = \frac{P_i r_i^2 - P_o r_o^2}{r_o^2 - r_i^2} - \frac{r_i^2 r_o^2 (P_o - P_i)}{r^2 (r_o^2 - r_i^2)},$$

where r_o and r_i are the outer and inner radii of the vessel wall and r is the radius to any point within the wall (Timoshenko and Goodier 1970). The circumferential stress on the outside of the hollow cylinder (i.e., for $r = r_o$) is given by

(8.2.2)
$$\sigma_o^{circ} = \frac{P_i r_i^2 - P_o r_o^2}{(r_o^2 - r_i^2)} - \frac{r_i^2(P_o - P_i)}{(r_o^2 - r_i^2)},$$

or simply

(8.2.3a)
$$\sigma_o^{circ} = \frac{2P_i r_i^2 - P_o(r_o^2 + r_i^2)}{r_o^2 - r_i^2}.$$

For the inside of the hollow cylinder where $r = r_i$, the circumferential stress is given by

(8.2.3b)
$$\sigma_i^{circ} = \frac{P_i(r_o^2 + r_i^2) - 2P_o r_o^2}{(r_o^2 - r_i^2)}.$$

Note that the stress in the radial direction at the outer and inner radius is given by

$$\sigma_o^{radial} = -P_o \quad \text{and} \quad \sigma_i^{radial} = -P_i.$$

Equations (8.2.3a–c) can be simplified further by means of the dimensionless expression $r_i/r_o = \gamma$. Inserting this expression into equations (8.2.3a–c) gives

(8.2.4a)
$$\sigma_o^{circ} = \frac{2P_i\gamma^2 - P_o(1+\gamma^2)}{(1-\gamma^2)}$$

and

(8.2.4b)
$$\sigma_i^{circ} = \frac{P_i(1+\gamma^2) - 2P_o}{(1-\gamma^2)}.$$

In fully turgid tissues, the condition $P = P_i = P_o$ must be true such that equation (8.2.1) takes the form $\sigma_r^{circ} = -P$ independently of r. Likewise, $\sigma_r^{radial} = -P$, such that under these conditions, the cell wall is under uniform stress.

Although materials that are exceptionally strong when placed in tension can confer an advantage, materials with low elastic moduli can also be used to advantage if large deformations are functionally desirable. Indeed, in some biological circumstances, stiff structures are not always the best. For example, Lee (1981) examined ash (*Fraxinus americana*) and calculated the elastic modulus of the cell walls in phloem (the tissue that is specialized to conduct cell sap). The elastic modulus of these cell walls has a range of 5.6–7.4 MN/m^2, which is very low compared with that of most fabricated materials and many other plant materials (see table 8.1). But these low values permit the cell walls to undergo very large circumferential deformations as a function of internal pressure. Indeed, a pressure change of 0.3–0.4 MN/m^2 would result in a 10% change in the radius of the cell, which, under the right conditions, could increase its capacity to contain and conduct sap. This potential for large deformations has an important bearing on mathematical models (which typically rely on measurements of cell diameters made on dead, essentially unpressurized tissues) treating the flow of cell sap in trees, since the rate of flow through a phloem cell would depend in part on the radius of the cell (see box 6.2).

As noted, we expect the effective stiffness of hydrostatic tissues to increase as a function of their water content. This expectation was first investigated empirically by Virgin (1955), who found that the apparent elastic modulus of parenchyma isolated from the pith of potato tubers was directly proportional to the tissue turgor pressure. This finding was subsequently verified and elaborated on by Falk et al. (1958). The maximum elastic modulus reported by these authors for tissue samples with a turgor pressure of 0.67 MPa was 19 MN/m^2 (which falls in the lower portion of the range of E reported for actively growing meristematic tissues, i.e., 19–40 MN/m^2), whereas, at a turgor pressure of 0.31 MPa, the maximum elastic modulus was roughly 8 MN/m^2.

So far we have treated the expansion of spherical cells upon uptake of water and a corresponding increase in turgor pressure. A different situation is encountered if a turgid cell is subjected to a deformation—for example, by compression between two plates. Provided that no water is extruded, uniaxial compression of a spherical cell increases the internal pressure. This increase in pressure is due to the elastic expansion of the cell wall, since the deformation of the cell leads to an increased surface area (box 8.3). Extrapolating this behavior to that of a strip of parenchyma composed of turgid spherical cells leads to an expression for the elastic modulus of the tissue for small strains:

(8.6)
$$E_{tissue} = \frac{dP}{d\varepsilon} \approx 2P_o + \frac{3t}{r}\frac{E_w}{(1-v)}\varepsilon^2.$$

The first term in equation (8.6) expresses the contribution to the effective elastic modulus of the tissue resulting from turgor pressure; the second term expresses the influence of cell wall material properties and geometry within the tissue. Subsequent work has confirmed the attributes and predictions of equation (8.6), particularly the influence of the second term; for example, the apparent stiffness of potato tuber parenchyma samples increases as a function of the number of cells in the transverse dimension because the volume fraction of cell walls in samples is a function of cell number (Niklas 1988). However, great care should be exercised when using equations like equation (8.6) that assume that anatomically "simple" tissues are isotropic and uniform, because it is clear that this is not necessarily a valid assumption. For example, working with the parenchyma in apple fruits, Vincent (1989) has demonstrated clearly that even within the same fruit, considerable variation exists in tissue density and that at any given density, parenchyma isolated from the inner part of the fruit is less stiff than parenchyma located near the epidermis.

BOX 8.3 **Compression of spherical turgid cells**

A spherical cell with a radius r filled with an incompressible liquid such as water may be compressed between two flat plates. While the area of contact between the plates and the cell wall will be flat, the spherical cell will be distorted into a shape that, for small compressive strains, can be approximated by an ellipsoid of rotation. If we define the pressure P^* as the force applied divided by the area of contact, P^* must be equal but opposite to the radial pressure inside the cell. For very thin-walled cells, P^* is equal to the turgor pressure P_0. For cell walls with a finite thickness t, an additional pressure increment ΔP results if the cell wall is strained due to a change in the surface area upon compression. Provided that the thickness of the cell wall is finite but small compared with the radius r, the relationship between the internal pressure P and the stress σ in the cell wall is given by the formula $\sigma = P(r/2t)$. The pressure needed to compress the cell can be expressed as

(8.3.1)
$$P^* = P_0 + \Delta P = P_0 + \frac{2t}{r}\Delta\sigma.$$

BOX 8.3 **(Continued)**

Analogous to the volumetric modulus (eq. [4.14]), we can introduce an aerial modulus C with $\Delta\sigma = C\,\Delta A/A_0$, with A_0 as the surface area of the spherical cell and the modulus of elasticity of the cell wall E_w,

(8.3.2)
$$E_w = 2C(1-v),$$

where v is the Poisson's ratio.

With this relationship, equation (8.3.1) takes the form

(8.3.3)
$$P^* = P_0 + \frac{2t}{r}\frac{E_w}{2(1-v)}\frac{\Delta A}{A_0}.$$

The remaining problem is the calculation of the change in surface area upon deformation of the cell. Assuming that the cell is filled with an incompressible liquid, and disregarding any change in the volume of the cell wall, the total volume of the cell upon deformation remains constant. This can be expressed by

(8.3.4)
$$\frac{4\pi}{3}r_0^3 = \frac{4\pi}{3}a^2 b,$$

where a is the major semiaxis and b is minor semiaxis of the ellipsoid of rotation. Noting that compression is in the direction of b, we see that $b = r_0(1 - \varepsilon)$ and $a = r_0(1 - \varepsilon)^{-\frac{1}{2}}$, where ε is the strain in the direction of the compressive force. The surface area of an ellipsoid of rotation is given by

(8.3.5)
$$A = 2\pi a\left[a + \frac{b^2}{\sqrt{a^2-b^2}}\operatorname{arcsin}h\left(\frac{\sqrt{a^2-b^2}}{b^2}\right)\right].$$

With the expressions for a and b, the relative change in the surface area can be approximated to $\Delta A/A_0 \approx 0.5\varepsilon^2$ for $\varepsilon \leq 0.12$. Inserting this into equation (8.3.3), we obtain for small compressive strains

(8.3.6)
$$P^* \approx P_0 + \frac{2t}{r}\frac{E_w}{2(1-v)}\frac{\varepsilon^2}{2}.$$

Note that for small compressive strains, t and r can be considered constant.

BOX 8.3 **(Continued)**

If tissues consisting of spherical cells are compressed, it is more appropriate to define the pressure applied as the external force divided by the cross-sectional area of the cells under compression. With this definition, the pressure P can then be related by geometric considerations to the pressure $P*$ as defined above,

(8.3.7) $$P = P*(2\varepsilon - \varepsilon^2) \approx 2P*\varepsilon \qquad \text{for } \varepsilon \leq 0.1.$$

In combination with equation (8.3.6), and neglecting higher terms, we obtain

(8.3.8) $$P \approx 2P_0\varepsilon + \frac{t}{r}\frac{E_w}{(1-v)}\varepsilon^3.$$

From the foregoing, we see that the modulus of elasticity of the tissue is given by

(8.3.9) $$E_{tissue} = \frac{dP}{d\varepsilon} \approx 2P_0 + \frac{3t}{r}\frac{E_w}{(1-v)}\varepsilon^2 \qquad \text{for } \varepsilon \leq 0.1,$$

where the first term describes the influence of the turgor pressure and the second term the elastic resistance of the cell wall against expansion. It should be noted, however, that the derivation of equation (8.3.9) assumes that the cells are mechanically independent of one another, which is not a very realistic assumption (see main text).

Nevertheless, equation (8.6) draws attention to some general aspects of plant tissue mechanical behavior. First, the elastic modulus of tissues composed of cells with thin walls, such as parenchyma or collenchyma, can vary significantly as a function of tissue water content (this is true even for dead tissues such as heartwood). Second, the structure of a plant tissue—the geometry of its cell wall infrastructure as well as the material properties of its cell wall constituents—plays a major role in dictating its mechanical behavior. Both of these aspects of plant tissue behavior reinforce what has been said many times in this book: plant "materials" behave more as complex structures than as materials *sensu stricto*.

Indeed, the mechanical behavior of hydrostatic tissues is much more complex than formally expressed by most mathematical treatments because most hydrostatic tissues exhibit nonlinear elastic behavior, short-term elastic recovery, long-term plasticity, stress relaxation, and creep (see chap. 4). In addition, the rate at which cell fluids (gases or liquids) are expelled from protoplasts or intercellular spaces when tissue samples are mechanically stressed can influence the mechanical properties of the tissues. Thus, the slope of the stress-strain diagram increases with the strain rate because higher strain rates give fluids less time to cross the plasma membrane or leave the confines of intercellular spaces. With higher strain rates, hydrostatic tissues demonstrate less compressibility, and their apparent E increases. In turn, stress relaxation is the result of pressure stabilization as turgor pressure within cells increases. Loading followed by unloading produces recoverable cell wall elastic deformations and unrecoverable plastic deformations owing to the loss of fluids from cells. Creep is also the result of fluid evacuation.

Tissue failure under compression occurs when the applied stress equals or exceeds the rupture stress of cell walls. In addition, the mechanical behavior of the tissues depends on the material properties of the pectinaceous middle lamella that binds adjoining cell walls together (see Lin and Pitt 1986). This last point is frequently neglected, but is absolutely vital to understanding plant biomechanics. Specifically, tissue failure can result either by cells separating from one another at their middle lamella or by cell walls rupturing. Thus, the mechanical behavior of parenchyma and collenchyma depends on the rate of loading, cell turgor, plasma membrane permeability, cell wall stiffness, and the volume fraction of expressible fluids, all of which can vary ontogenetically.

8.6 Nonhydrostatic cells and tissues

Unlike those of hydrostatic cell and tissue types, the mechanical properties of cells possessing thick secondary walls are comparatively insensitive to changes in turgor pressure because these cells are generally dead when mature. Nevertheless, the water content of secondary cell walls can vary, with consequent effects on the elasticity of cell walls and therefore on the mechanical behavior of tissues, as is evident when we compare the mechanical properties of dry and fresh ("green") wood. For example, the relationship between the elastic modulus E_0 measured at M_0% moisture content and the elastic modulus E_M adjusted to a higher moisture content $M_H\%$ is given by the formula

(8.7) $E_M = E_0[1 - H(M_H - M_0)]$,

where H is the coefficient of moisture effect; that is, the change in the elastic modulus with a 1.0% change in moisture content. For most wood species, H equals 0.02, although this numerical value varies nonlinearly for very high moisture contents. This empirically determined formula predicts that the stiffness of tissues in which secondary walls dominate decreases with increasing moisture content. For example, at 50% moisture content, the elastic modulus of Sitka spruce (*Picea sitchensis*) is reported to be between 5.9 and 8.8 GN/m^2 (see table 8.2). But, at 60% moisture content, the elastic modulus is reduced to between 4.64 and 7.04 GN/m^2—a reduction of nearly 27%. Data from isolated primary phloem fibers indicate a similar dependency of tissue stiffness on cell wall hydration. For example, the elastic modulus of air-dried primary phloem fibers equals 19 GN/m^2. But when kiln dried, the same cells have an elastic modulus of 90 GN/m^2 (see table 8.1). These examples caution against using the mechanical properties of dried wood samples to infer the mechanical properties of the green wood in living trees (see Niklas and Spatz 2010).

The secondary cell wall is a prominent feature of some cell types found in primary tissues and the majority of cells found in secondary tissues. Primary phloem and xylem fibers and primary xylary elements, such as vessel members and tracheids, have secondary walls comparable in thickness to those observed in wood fibers. Regardless of whether it develops in primary or secondary tissues, the secondary wall is produced by a living protoplast and deposited against the primary cell wall (fig. 8.5). The initial deposition of the secondary wall may or may not coincide with the terminal phase of cell expansion and elongation, but the bulk of the secondary wall is typically deposited after cell elongation and expansion have stopped. The sequential wall layers deposited in the secondary wall are traditionally designated S1, S2, and S3. The numbers in this series denote the temporal sequence of deposition. The S1 layer is the outermost (oldest) secondary wall layer, while the S3 layer in the innermost (youngest) layer. In some cells, such as those in tension wood, a gelatinous layer, called the G-layer, is deposited between the S3 layer and the plasmalemma (see section 8.9).

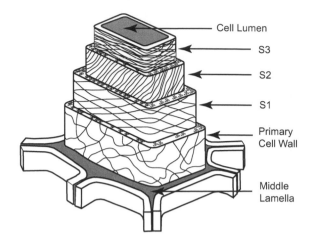

FIGURE 8.5. Graphic representation of the relationships between the primary and secondary cell wall and among the various layers in the secondary cell wall (S1, S2, and S3) The three S layers differ in the average orientation of their cellulose microfibrils (as indicated by the patterns shown on their exposed surfaces). The primary cells walls of adjoining cells are held together by their middle lamellae.

The secondary wall in wood has received the greatest attention because wood's numerous commercial applications rely on the mechanical effects of the cellulose microfibrillar angle (MFA), which changes from one cell wall layer to another (see fig. 8.5). The MFA refers to the angle made between microfibrils in the secondary wall and the longitudinal axis of the cell. All other things being equal, the smaller the MFA, the stronger the cell will be (see box 5.5). For example, Cave (1966) showed that a reduction in MFA from 40° to 10° can increase cell wall stiffness nearly fivefold. The following discussion of the secondary cell wall and MFA is therefore based principally on studies devoted to wood anatomy, although it has considerable application to understanding the secondary wall in general.

The secondary wall in tracheids, vessel members, fibers, and other cell types with secondary walls has a helicoidal structure because, as noted, the S1, S2, and S3 layers have different, albeit intergrading, microfibrillar orientations. The microfibrils in the S1 and S3 layers tend to be more perpendicular to the longitudinal axis of the cell than those in the S2 layer. In turn, each S layer is composed of many thinner layers, which vary in their MFA. Intervening layers between the S layers also exist. The MFAs in these intervening layers are reported to intergrade with those of the two S layers that flank them. The secondary cell wall thus has a helicoidal microfibrillar architecture that is continuous with that of the primary cell wall. However, the different MFAs observed in the S layers cannot be attributed to the passive realignment of fibrils induced by cell growth (as in the case for the primary cell wall) because the bulk of the secondary cell wall is deposited after cell elongation.

Because the thickness of the S2 layer is generally much greater than that of the S1 or S3 layer, measurements of MFA involving the whole wall of a single cell, or which average MFAs across a sample of wood several cells thick, will approximate the angle in the S2 layer. In turn, the MFA in the S2 layer is a main determinant of the stiffness and strength of wood.

A review of the older literature reveals that the MFA decreases from higher values in the innermost growth layers of wood to lower values in the outermost layers at all heights in the trunks of tree species. The MFA is also reported to decrease with tree height for any particular growth layer of wood until approximately 30% to 50% of tree height, at which point it increases again within the same growth layer (see Barnett and Bonham 2004). These trends must be approached with caution because considerable variation exists across species. However, they shed considerable light on many aspects of tree ecology because changes in MFA from the central to the more peripheral growth layers of wood, or from the base to the top of a tree, can result in significant changes in stem stiffness and strength. For example, the tracheids and fibers close to the center of a tree trunk are produced when the tree is young. Because these cells have high MFAs, this arrangement confers a low Young's modulus, and thus a high flexibility, on young trees, which allows them to bend and sway in the wind. As trees increase in size, however, the wood in the outer growth layers develops secondary cell walls with lower MFAs, which confers a higher tensile modulus and thus greater stiffness.

Even if microfibrillar orientation in the S layers is ignored, the secondary wall is neither chemically nor spatially homogeneous. For example, the S2 layer is typically thicker than the S1 layer in late wood produced late in the growth season, while the S3 layer may be thin or entirely absent. In contrast, the S2 layer is much thinner than either the S1 or S3 layer in wood produced in the early spring. Secondary wall deposition may also retroactively alter the composition of the primary cell wall. Although lignin is deposited in the cell wall when the S layers are formed, it is typically found in higher concentrations in the compound middle lamella. Lignin may also be concentrated at the corners of cell walls, in radial walls as opposed to tangential walls, or in concentric lignin-rich layers alternating with cellulose-rich layers in the cell wall. Lignification occurs in sites re-moved from physical contact with the plasmalemma and involves chemi-cal "impregnation" throughout the wall matrix that does not disrupt the architecture of the cellulose microfibrils. In this regard, delignification produces little or no change in the X-ray diffraction patterns of cell walls, which are dictated by the crystalline cellulose network. Lignin, which has little resistance to tensile stresses, appears to function mechanically in two important ways. First, it serves as a "bulking agent" that fills up cell wall interstices and thus increases the density and the compressive strength of walls. Second, it reduces the extent to which cellulose microfibrils can become hydrated. Dry cellulose is much stiffer and stronger than wet cel-lulose. Therefore, lignification increases the tensile strength of the cell wall indirectly and the compressive strength of the wall directly.

Although there may not be a direct causal relationship, it is neverthe-less not surprising that many mechanical properties of wood correlate with its degree of lignification and density. For example, conifer wood species are, on average, significantly less dense than dicot wood species. Conifer wood is also significantly less stiff in bending and weaker in both shearing and compression (see table 8.2). These and other important mechanical properties increase with increasing wood density across species as well. In-deed, there is a nearly one-to-one correlation between increases in density and increases in many biomechanical properties (table 8.4). In passing, it is worth noting that the strongest correlation among the properties listed in table 8.2 is that between the maximum compressive strength and the modulus of rupture (r^2 = 0.969 across all species). The negative value of $\log \beta$ in the allometric representation of the data (see table 8.4) highlights the fact that most wood species fail in compression before they fail in ten-sion.

Because the volume fraction of cell wall materials in wood contrib-utes significantly to the strength and stiffness of wood, the late wood in woody stems is generally stronger than the early wood because late wood tends to have a greater volume fraction of cell wall materials. Variation in the width of growth layers also affects the strength of wood, but dif-ferently when the wood of gymnosperms is compared with the wood of angiosperms. Among conifer species, a reduction in the width of growth layers lowers the proportion of cells with thin walls and large diameters, which are characteristics of early wood. Therefore, up to a limit (and as-suming all other relationships hold), conifer wood with narrow growth layers tends to be stronger and stiffer than wood with wide growth layers.

TABLE 8.4 **Summary statistics of ordinary least squares regression analyses of $\log_{10}$-transformed data for wood mechanical properties**

$\log Y_o$ vs. $\log Y_a$	n	α	$\log \beta$	r^2
Conifer tree species				
M_R vs. density	37	1.14	−1.52	0.802
E vs. density	38	1.00	1.08	0.569
S_G vs. density	37	0.96	−1.88	0.711
S_C vs. density	39	1.10	−1.73	0.745
M_R vs. E	37	0.76	−1.27	0.623
S_C vs. E	38	0.77	−1.66	0.644
S_G vs. S_C	37	0.83	−0.32	0.880
S_C vs. M_R	37	0.97	−0.27	0.929
Dicot tree species				
M_R vs. density	137	1.18	−1.57	0.859
E vs. density	137	1.02	1.02	0.737
S_G vs. density	128	1.15	−2.36	0.812
S_C vs. density	137	1.28	−2.17	0.812
M_R vs. E	137	0.97	−2.03	0.827
S_C vs. E	127	1.08	−2.75	0.814
S_G vs. S_C	128	0.78	−0.23	0.745
S_C vs. M_R	137	1.09	−0.48	0.963
All tree species				
M_R vs. density	174	1.24	−1.76	0.877
E vs. density	175	1.00	1.09	0.745
S_G vs. density	165	1.19	−2.50	0.843
S_C vs. density	176	1.33	−2.33	0.840
M_R vs. E	174	1.03	−2.26	0.805
S_C vs. E	175	1.13	−2.96	0.800
S_G vs. S_C	165	0.81	−0.28	0.820
S_C vs. M_R	174	1.08	−0.45	0.969

Note: See table 8.2 for raw data. Regression formula takes the form $\log Y_o = \log \beta + \alpha \log Y_a$. E = elastic modulus, M_R = modulus of rupture, S_C = maximum compressive strength, and S_G = maximum shear strength.

In contrast, a reduction in the width of growth layers in the stems of angiosperm species results mainly at the expense of the late wood. Therefore, for these species, stems with wide growth layers tend to be stronger and stiffer than stems with narrow growth layers.

Other factors also contribute to the mechanical properties of wood. Wood that contains a large volume fraction of libriform fibers and fiber-tracheids tends to be stronger than wood lacking these cell types. Ring-porous wood with aggregates of large vessels tends to be weaker than diffuse-porous wood, as does wood containing large parenchymatous rays (although this weakness can be counterbalanced by the fact that wood species with a large volume fraction of rays also tend to have wood with abundant thick-walled fibers). The ontogenetic transition between sapwood and heartwood does not appear to alter the strength or stiffness of wood, which suggests that the materials that occlude the lumens of the cells in heartwood are not especially stiff or strong.

8.7 Cellular solids

The mechanical behavior of nonhydrostatic plant tissues, such as sclerenchyma and wood, conforms in many respects to that of engineered materials classified as cellular solids (Gibson et al. 2010). These solids are characterized by having both a solid phase (comparable to secondary cell walls) and a fluid phase (comparable to the air- or liquid-filled spaces found in plant tissues). An example of a liquid-filled plant cellular solid is sapwood; an example of a gas-filled plant cellular solid is cork (i.e., phellem).

A cellular solid may be defined as any material consisting of open-walled (spongelike) or closed-walled (foamlike) compartments whose bulk density ρ is less than the density ρ_S of the material making up the cellular walls; in other words, its relative density, ρ/ρ_S, is less than one. (Here, ρ is to be understood as the density without the fluid enclosed.) The relative density of a cellular solid largely dictates the mechanical attributes of this kind of material. When a fabricated cellular solid is subjected to compressive stresses, its stress-strain diagram shows three responses: a linear elastic response under low stress magnitudes; an elastic or plastic deformation response under constant moderate stress levels; and a densification response associated with the crushing of the solid phase and the expulsion of the fluid phase (fig. 8.6). Each of these three phases can be explained by means of elementary strength of materials theory because the "cell walls" or "struts" in a cellular solid can be modeled as rigid cantilevered beams operating on flexible hinges.

In much the same way that the architecture of the solid phase in a cellular solid influences its material properties, such as the rupture modulus or the maximum compressive strength, the anatomy of nonhydrostatic plant tissues profoundly influences the extent to which those tissues manifest each of the three responses to increasing stress levels. The influence of anatomy on mechanical behavior is complex, however, because the cellular geometry within a tissue can vary significantly as a function of the plane of sectioning, which in turn influences the extent to which a tissue

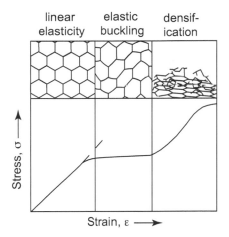

FIGURE 8.6. Geometric distortions within a honeycomb cellular solid subjected to compressive stresses (*top*) and the corresponding stress-strain relationships (*bottom*). The stress-strain relationship shows three distinct phases: a linear elastic response, an elastic buckling phase, and finally, the collapse of the solid's infrastructure and the densification of its solid material component. (Adapted from Niklas 1989.)

responds to an applied mechanical force. Many plants capitalize on the resulting mechanical anisotropy in terms of how the geometries of their tissues are oriented with respect to normally occurring types of loading. Thus, the layers of cork in the outer bark of a tree can sustain very high impact loadings directed toward the center of a stem, but substantially less vertical compression along the length of the same stem. Wood can sustain very high compressive stresses applied along the length of its grain, but it is typically very much weaker when stressed across the grain.

As noted, the relation between stress and strain for a cellular solid is not constant over the entire range of stress. When densification occurs, many cellular solids get stiffer, deforming proportionally less as the stress level increases. This response confers advantages under many conditions, as when the wood in a stem is placed in bending. The cellular solid tissues within such a stem can resist deformation even at stress levels that exceed the proportional limits of their cell walls.

Gibson and Ashby (1997) have provided theoretical treatments of the mechanics of three-dimensional cellular solids and compared the results of their theory with experimental data. A detailed quantitative treatment of the mechanical behavior of cellular solids is outside the scope of this book. A highly redacted treatment is useful, however, because it sheds

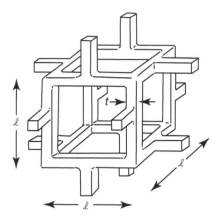

FIGURE 8.7. Sketch of a cubic cell within an open-walled cellular solid. The cell has edge length l and edge thickness t. (Adapted from Gibson and Ashby 1997.)

light on a number of plant functional traits. For example, the geometry of an open-walled cellular solid, such as aerenchyma, consists of many beamlike struts interconnected end to end. In its simplest case, the unit cell of a three-dimensional open-walled cellular solid can be represented by a cubic array of members of length ℓ and edge thickness t (fig. 8.7). Note that the relative density ρ/ρ_S is proportional to $(t/\ell)^2$.

During mechanical loading, the cells' edges bend, and their nonvertical sides undergo deflections δ. These changes distort the unit cell. According to equation (5.2.3), the overall strain ε will be proportional to the applied force F as

(8.8)
$$\varepsilon \propto \frac{\delta}{\ell} \propto \frac{F\ell^2}{E_S I_S}.$$

E_S is the modulus of elasticity of the material making up the cell walls, and I_S the second moment of area, which is proportional to t^4. The overall stress σ is proportional to F/ℓ^2. The proportionality of the elastic modulus E of the cellular solid to E_S can therefore be expressed as

(8.9)
$$\frac{E}{E_S} \propto \frac{t^4}{\ell^4} \propto \left(\frac{\rho}{\rho_S}\right)^2,$$

where the proportionality factor depends on the particular geometry of the unit cell.

The linear elastic range of behavior is seen in response to small strains (about 5% in compression and between 5% and 8% in tension). Beyond this range, nonlinear deformations occur, principally in the form of elastic buckling, for which Euler's buckling formula (see section 5.10) provides reasonable approximations. Thus, for an open-walled cellular solid, the stress at which buckling occurs σ_B is given by the proportional relationship

(8.10)
$$\sigma_B \propto E_S \left(\frac{\rho}{\rho_S} \right)^2 .$$

This proportionality works for cellular solids with relative densities less than 0.3. At higher relative densities, the fluid-filled spaces are too sparsely distributed to permit Euler buckling in the solid phase. If the solid phase in the cellular solid is composed of a plastic material, compression or tension can induce plastic buckling in the solid phase. The yield plastic stress at which this occurs σ_P is proportional to the plastic moment of the struts within the cellular solid (and inversely proportional to strut length):

(8.11)
$$\sigma_P \propto \sigma_Y \left(\frac{\rho}{\rho_S} \right)^{3/2} ,$$

where σ_Y is the yield stress of the solid phase.

At higher stress levels, compression or tension results in the densification of the solid phase. The crushing or rupture stress σ_R is linearly related to the rupture stress of the solid phase σ_M and the ½ power of the cellular solid's relative density:

(8.12)
$$\sigma_R \propto \sigma_M \left(\frac{\rho}{\rho_S} \right)^{3/2} .$$

The foregoing proportional relationships have been verified empirically for commercially fabricated open-walled cellular solids. Niklas (1991) examined the mechanical properties of aerenchyma by bending the cylindrical leaves of the common rush (*Juncus effusus*) and found that this tissue behaves in a manner predicted by theory for an isotropic beam-and-strut cellular solid with a relative density near 0.3.

The mechanics of closed-wall cellular solids are more complicated. For the simplest case of a cubic cell, where the cell faces have the same thickness

t as the cell edges, the relative elastic modulus of the cellular solid is given by the formula

$$(8.13) \qquad \frac{E}{E_S} \approx \frac{V_S}{V} = \frac{\rho}{\rho_S},$$

where E, as before, is the elastic modulus of the cellular solid and E_s is the modulus of the material from which the cellular solid is fabricated. If air or any other gas at pressure $P - P_0$ is enclosed in the cell, a second term has to be added to this expression such that

$$(8.14) \qquad \frac{E}{E_S} \approx \frac{\rho}{\rho_S} + \frac{(P - P_0)(1 - 2v)}{E_S (1 - \rho/\rho_S)},$$

where P_0 is the ambient pressure and $v \approx 0.3$, which is quite independent of ρ/ρ_S. As mentioned before, the bulk density ρ is to be understood as the density without the gas enclosed.

A material consisting of closed cylindrical cells with length ℓ much greater than their radius r cannot be isotropic simply for geometric reasons. Provided that cells do not experience Euler buckling, equation (8.9) describes the mechanical properties in the longitudinal direction. In the orthogonal direction, stresses will lead to ovalization of the circular cross sections of the cylindrical cells. Provided that the cells are sufficiently long that end-wall effects can be neglected, the force that causes the compression of the circular cross section of a cylindrical thin-walled cell to an elliptical cross section is given by

$$(8.15) \qquad F^{oval} = 2 E_S^{trans} I_S \frac{1}{b} \left(\frac{1}{r} - \frac{a}{b^2} \right),$$

where E_S^{trans} is the elastic modulus transverse to the longitudinal direction, I_S is the second moment of area, r is the mean between the outer and inner radii (r_o and r_i, respectively) of the cell before compression, and a and b are the short and long axes of the ellipse defined at the midpoint of the cell wall. E_S^{trans} is taken as a constant, and I_S is given by the formula

$$(8.16) \qquad I_S = \ell \int_{r_i}^{r_o} (\varsigma - (r_o + r_i)/2)^2 \, d\varsigma = \frac{\ell}{12} (r_o - r_i)^3 = \frac{\ell t^3}{12},$$

where $t = r_o - r_i$.

For small degrees of ovalization, equation (8.15) can be simplified to yield

(8.17)
$$F^{oval} = \frac{1}{2} E_S^{trans} \frac{\ell t^3}{r^2} \varepsilon,$$

where $\varepsilon = (r - a)/r$ is the strain. The stress is given by $\sigma = F^{oval}/(\ell r)$, and the modulus of elasticity $E^{rad} = \sigma/\varepsilon$. Introducing this notation into equation (8.17) leads to

(8.18)
$$E^{rad} = \frac{1}{2} E_S^{trans} \frac{t^3}{r^3}.$$

Since $\dfrac{t}{r} \propto \dfrac{V_S}{V_{cell}} = \dfrac{\rho_{cell}}{\rho_S}$, equation (8.18) can also be expressed as

(8.19)
$$E^{rad} \propto E_S^{trans} \left(\frac{\rho_{cell}}{\rho_S} \right)^3.$$

Similar considerations can be made for different geometries, such as hexagonal cross sections.

Wood, with its highly elongated fibers and tracheids or vessel members, can be considered a closed-walled cellular solid with cylindrical unit cells. Of particular interest is the dependence of the modulus of elasticity on the density of dried wood for a large number of species (fig. 8.8). In the longitudinal direction, a linear relation is found between E^{rad}/E_S^{trans} and ρ_{cell}/ρ_S, while in the radial and tangential directions, the modulus of elasticity is found to follow a dependence closer to a second than to a third order and less than predicted by equation (8.19), which indicates that the model used to derive this equation is an oversimplification.

According to figure 8.8, the anisotropy of wood is a steep function of its relative density. For low-density balsa wood, the quotient of the modulus of elasticity measured in the axial and tangential directions is approximately 80, while that of red oak is 12. Easterling et al. (1982) examined the elastic modulus and crushing strength of balsa wood (*Ochroma lagopus*), which is one of the strongest wood species for its bulk density. They examined balsa specimens in all three principal planes with respect to the wood grain (radial, tangential, and longitudinal) and found that the mechanical properties of this wood species depend on the material properties of the solid phase (secondary cell walls) and on the cellular geometry as seen in

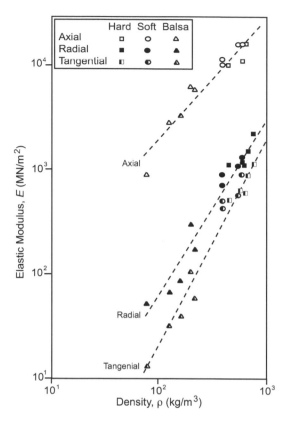

FIGURE 8.8. The modulus of elasticity of wood plotted against its density in the dried state. (Adapted from Easterling et al. 1982.)

the plain of the applied force. A comparison between the predictions of theory and the observed behavior of balsa wood left little doubt that this species behaves as a closed-wall cellular solid with marked mechanical anisotropy.

Like wood, cork can be considered a cellular solid with closed cells. The cellular structure of this tissue, which comes from the cork oak (*Quercus suber*), was studied by the engineer and biomechanicist Robert Hooke (1635–1703). The roughly prismatic cells in cork have more or less equivalent radial and tangential dimensions, but are almost twice as long in the direction of their longitudinal axes, which are aligned orthogonally (radially) to the lengths of branches and trunks. Thus, when they are pulled or compressed, they exhibit similar material properties in two of their three dimensions. In the longitudinal direction, cell walls of cork tend to be corrugated, such that they have a structure comparable to that of bellows. This explains why, upon compression in the longitudinal direction, cork expands only a little in the radial direction (see section 4.5).

The relative density ρ/ρ_s of cork is approximately 0.15, and the elastic properties of this tissue can be described by three moduli: $E_1 = 20$ MPa in the longitudinal direction and $E_2 = E_3 = 13.5$ MPa in the directions orthogonal to the longitudinal axis. The Poisson's ratios as defined in the same coordinate system are $v_{12} = v_{21} = v_{13} = v_{31} = 0.0\text{--}0.1$ and $v_{23} = v_{32} = 0.25\text{--}0.5$. The mechanical behavior of cork beyond the linear elastic range is presented by Gibson and Ashby (1997). In addition to its excellent thermal insulation capacity, cork has the capacity to dissipate mechanical energy at high strain rates, which gives it high damping and sound-absorbing qualities.

Plant tissues composed of living cells, such as parenchyma or collenchyma, can be considered cellular solids with cells enclosing an incompressible fluid (Niklas 1989). The elastic properties of these tissues are empirically known to depend on the geometry of cells and the mechanical properties of cell walls. In addition, we know that the stiffness of living tissues increases to a limit with increasing turgor pressure, particularly in the case of parenchyma, which has very thin walls in comparison to other plant tissues.

8.8 Tissue stresses and growth stresses

The production of new cells and the growth in size of individual cells within the plant body result in stresses that must be accommodated as tissue systems and organs increase or decrease in size. These stresses can be either transient or permanent. Transient stresses can result as cells, tissues, or organs expand or contract as the result of the influx or efflux of water. Technically, these stresses are not "growth stresses," since "growth" implies a permanent change in the size of an organic structure. Therefore, it is more appropriate to call these "tissue stresses" and to reserve the term "growth stresses" for cases where cells change in size permanently.

Perhaps the best-known growth stresses are those that occur in wood. Jacobs (1938, 1945) showed that planks cut from the median longitudinal section through *Eucalyptus* tree trunks often split explosively along the line of the pith, with the resulting halves bending away from the pith. This rapid longitudinal shearing is accompanied by an increase in the length of

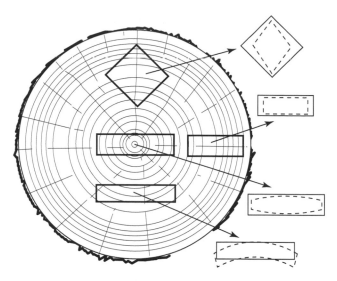

FIGURE 8.9. Illustration of the deformations in blocks of wood removed from a freshly cut (moist) transverse section of a tree trunk. (Adapted from Niklas 1992.)

the wood toward the center of a trunk and a decrease in its length at the perimeter of the trunk, which indicates that the wood toward the center of the trunk is placed in compression by subsequently formed growth layers of secondary xylem, while the wood toward the periphery of the trunk experiences tensile stresses, phenomena that can be seen even in transverse sections through a trunk (fig. 8.9).

At the cellular level, newly differentiated xylary elements (e.g., fibers and vessel members) undergo longitudinal contraction that places older elements toward the center under compressive loading as a tree grows and expands laterally. When planks of wood are cut, the tensile stresses toward the perimeter of the trunk are eliminated, and the cells toward the center are relieved of their compressive loading, which results in the potential for a violent mechanical response. This effect is mechanically similar to the curvature of split stem segments lacking secondary growth, in which the epidermis is mechanically analogous to the newly differentiated xylary elements in wood (see section 8.4). It is also analogous to what is seen when a turgid onion bulk is cut transversely and then cut in half. The concentric segments of the leaf bases are tightly appressed to one another. The outer epidermis of each leaf base is placed in tension, whereas the inner epidermis of the leaf base is placed in compression. When the transverse section is cut in half, these loadings are removed, and each leaf base section curves outward to varying degrees.

Although the growth stresses that can result within secondary tissues such as wood can theoretically reach levels high enough to cause tissue failure, Boyd (1950) has argued that these stresses are never fully achieved owing to the formation of minute compression failures in the form of slip planes and compression planes that locally relieve the strain energy stored within wood. Detailed theoretical treatments of growth stresses have been developed and are in reasonable agreement with empirical data (Gillis 1973). It appears, therefore, that localized mechanical failure in secondary tissues is part of the normal growth process of trees. The only significant difference between early formulations of the theory of growth stresses and experimental results is the magnitude of the curvatures occurring on strips of tissue cut from median longitudinal sections. The radial gradient of the longitudinal stress is generally less than would be expected based on theory. Gillis and Hsu (1979) modified early theoretical treatments to deal with this discrepancy by considering elastic-plastic (rather than pure elastic) behavior. Importantly, all treatments of this subject agree that the longitudinal growth stress distribution is uncoupled from both the radial and the circumferential stress distributions (Kübler 1959; Archer and Byrnes 1974; Gillis and Hsu 1979).

Slip planes and compression creases can develop naturally within wood that is compressionally loaded well below its breaking load. These deformations result from differential dehydration (when the surfaces of wood are exposed to air) as a result of the anisotropy in the mechanical properties of the tissue as a whole. When placed under compression by self-loading and wind-induced bending (as, for example, at the base of very tall and large tree trunks), the walls of tracheids and vessels buckle or concertina, while whole regions of wood densify (see fig. 8.6). Robinson (1920) was the first to report microscopic slip planes as tissue failure ensues. A slip plane is a shear zone running through any material. In wood, slip planes result from the buckling of microfibrils. They can be visualized with the aid of polarized light microscopy, under which they appear as bright lines running obliquely across cell walls at angles between 61° and 69.5° to the cell length axis. Note that if the walls were isotropic, the angle would be 45°; hence, the angles of the slip planes confirm the anisotropic nature of xylem cell walls. Slip planes caused by compression begin to form under applied loads that are roughly one-half the breaking load of wood, which can vary substantially depending on the species of wood (see table 8.2).

8.9 Secondary growth and reaction wood

Angiosperm species capable of producing wood and all conifer species thus far examined can reorient stems and roots that have been mechanically displaced from their original orientations as a result of wind-throw, soil movements, or the sustained application of an external force. The mechanism by which this reorientation occurs is the formation of reaction wood by the vascular cambium. Reaction wood functions to bend and reorient displaced parts of the plant body by either contracting or expanding at the cellular level. Among conifer tree species, reaction wood is called compression wood. This form of secondary xylem forms along the lower sides of displaced stems or roots. Compression wood is characterized by short, relatively rounded tracheids that have thick S2 layers with a high lignin content and high microfibril angles (Timell 1969). Many angiosperm species produce a type of reaction wood called tension wood as well as compression wood. Tension wood forms on the upper side of stems or roots that have been displaced. This type of reaction wood is characterized by the presence of only a few, small vessels. Approximately one-half of all woody flowering plant species thus far examined produce reaction wood containing fibers with an inner gelatinous cell wall layer (called the G-layer). This layer consists of almost pure cellulose arranged in microfibrils that run more or less parallel to the longitudinal axis of the cell.

It is perhaps confusing that the terminology of "tension" and "compression" wood gives the impression that tension wood is produced in regions that experience tensile stresses and that compression wood is produced in regions that experience compression. That is not the case. The terminology refers to the ways in which the two kinds of reaction wood operate mechanically to restore the orientation of a bent organ. This can be seen when a conifer branch is bent into a vertical loop (fig. 8.10). Compression wood forms on the lower surface of each part of the loop. From first principles (see sections 4.1 and 4.2), we know that the concave surface of the upper portion of a bent bar is placed in compression, while the convex surface of the lower portion is in tension. Accordingly, the locations where compression wood forms do not conform to the nature of the stresses experienced by the conifer branch.

Certain aspects of the development of reaction wood have been extensively explored, particularly for conifer species (Timell 1969, 1986). The prevailing hypothesis is that reaction wood is induced by gravistimulation

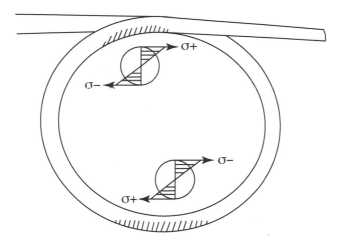

FIGURE 8.10. Location of compression wood formation (indicated by the two striated areas) in a conifer branch bent into a vertical loop, compared with the locations and magnitudes of compressive and tensile stresses ($\sigma-$ and $\sigma+$, respectively) in a representative cross section taken at each of the two locations.

rather than by mechanical stimulation. This hypothesis rests on data collected from a large number of bending experiments such as the one illustrated in figure 8.10. However, reaction wood can be induced by internal physiological mechanisms. For example, after the removal of the leader shoot at the top of a tree, one or more side shoots will begin to reorient and grow vertically. This reorientation is clearly not the result of any known form of gravistimulation. The molecular developmental mechanisms that result in the formation of reaction wood are not currently well understood. The formation of compression wood is often (but not always) accompanied by increased rates of cell division within the vascular cambium and by depressed rates of cell division along the opposite side of the displaced organ, such that the pith is displaced toward the upper side. Early in the development of the reaction wood in angiosperm species, xylem mother cell development into vessel members is inhibited. During the cell expansion phase, the shape of xylary elements is altered, and in late development, the chemistry and structure of secondary cell walls is modified significantly. In conifers, developing xylem cells can be reversibly induced during the whole differentiation process (see Timell 1986) such that there are xylem cells that exhibit only some of the characteristics typical of reaction wood.

8.10 Wood as an engineering material

Due to its availability and its relative ease to shape, even with simple tools, wood has been a preferred engineering material since ancient times. Its disadvantage is the danger of decay, particularly when it is wet. The tendency of wood to decay differs among tree species. Usually dry wood is used in large-scale construction work. Attempts have been made to use fresh wood and bend it into different shapes. The difficulty is that the wood will not necessarily maintain its shape after drying, but rather shows a tendency to twist and deform in an unpredictable way. On the other hand, heat bending of moist wood leads to reorganization of its cellulose fibers and thus to its permanent plastic deformation into a desired form.

As noted before, wood is a cellular composite characterized by pronounced material anisotropy (see section 4.7). For its density, it has superior mechanical properties, particularly its high modulus of elasticity in the longitudinal direction and its strength (Wood Handbook 1999; see table 8.1). However, the choice of wood species for a particular application is often not made exclusively on the basis of mechanical properties, since form stability and resistance to decay can be equally important. Furthermore, in the manufacture of furniture, the appearance of the wood grain pattern, its texture, and its color may be equally or more important than factors such as strength or stiffness.

In the case of wood for musical instruments, the situation is often very different, and mechanical properties may take precedence. Indeed, a scientific understanding of sound production requires an appreciation of a large number of material properties. Wegst (2006) has published illustrative graphs that show these properties for a large number of wood species together with their use in musical instruments. The properties of interest include the speed of sound in the material, the impedance of the material, the sound radiation coefficient, and the loss coefficient as a measure of the damping within the material. The first three properties depend on the modulus of elasticity E and density ρ, while the loss coefficient does not.

The speed of sound, $c = (E/\rho)^{0.5}$, is roughly invariant across wood species, but it can be significantly different in the tangential and the radial directions than in the longitudinal direction (with the grain). It decreases slightly with increasing frequency and amplitude of vibration. The impedance, $Z = (E\rho)^{0.5}$, is the property that, for example, determines the transmission of vibrations from the strings of a violin via the bridge to the soundboard. The sound radiation coefficient, $R = (E/\rho^3)^{0.5}$, describes how much of the vibration energy of a body is lost by radiation to the surroundings. Damping due to internal friction in the material is a complex process that depends on temperature and humidity and most notably affects the high frequencies. It is for this reason that violinists are concerned about the microclimate in the concert hall.

The soundboards of violins are usually made from softwood (*Abies alba*, *Picea abies*, and *Pinus sylvestris*) with the grain running parallel to the strings. The wood comes from trees grown at high altitudes and felled in winter. It is carefully selected for narrow and even annual rings, then dried for several years. Such wood has the highest radiation coefficient and low impedance, which is beneficial for sound transmission into the air. Sadly, the forests near Cremona, Italy, from which the famous violin makers took the wood for their soundboards, no longer exist; thus, it is not possible to create a true replica of a Stradivari violin. Maple wood (*Acer saccharinum* or *A. pseudoplatanus*) is used for the bridge, the ribs (side parts), and the back plate. Maple wood has nearly the same impedance as spruce wood. The curvature of the ribs is achieved by heat bending.

Pernambuco (*Guilandia echinata*) is the most valued wood for bows. It has a high modulus of elasticity, which gives strength to a slender bow to tighten the horsehair. Its sound velocity matches that of wood used for soundboards. Due to its high density, pernambuco has a low radiation coefficient. Its loss coefficient is roughly the same as that of wood used for soundboards, but less than that of maple wood, so damping in the wooden parts of the bow is low. In contrast, horsehair shows pronounced hysteresis in cyclic stress-strain experiments, equivalent to a loss of mechanical energy and therefore high damping. This property prevents unwanted oscillations of the bow.

Wind instruments (clarinet, oboe, bassoon, etc.) are made of dense wood selected for its dimensional stability, particularly at high levels of moisture. Today, tropical woods, such as African blackwood (*Dalbergia melanoxylon*), Brazilian rosewood (*D. nigra*), or Macassar ebony (*Dyospyros celebia*), are the preferred materials for clarinets and oboes, while bassoons are made from maple wood (*Acer platanoides*). In contrast to string instruments, the wooden tube is too thick to vibrate itself; however, the sound quality is influenced by the interaction of the material with the enclosed column of vibrating air. Damping due to air friction at the tube walls and turbulence generated at edges affect the tonal quality. Thus, the finish of the wooden parts of the instrument is important for its quality.

The reeds for clarinets, oboes, and bassoons are made from *Arundo donax*. This material is remarkable because it shows a decrease in the modulus of elasticity from the outside toward the inside of the hollow stem (Spatz et al. 1997) and, therefore, given the way the reeds are prepared, from the part where it is clamped toward the vibrating end.

For xylophones, a high loss coefficient is undesirable, as the sound of the bars should not decay rapidly. Typically, wood with a high density is chosen, such as Amazon rosewood (*Dalbergia spruceana*) or African padauk (*Pterocarpus soyouxii*). These woods provide for high side hardness, which is important if the wood has to sustain impact loads.

Literature Cited

Archer, R. R., and F. E. Byrnes. 1974. On the distribution of tree growth stresses. I. An anisotropic plane strain theory. *Wood Sci. Tech.* 8:184–96.

Barnett, J. R., and V. A. Bonham. 2004. Cellulose microfibril angle in the cell wall of wood fibres. *Biol. Rev.* 79:461–72.

Boyd, J. D. 1950. Tree growth stresses. 3. The origin of growth stresses. *Austral. J. Sci. Res.*, ser. B, 3:294–309.

Brown, C. L., H. E. Sommer, and L. V. Pienaar. 1995a. The predominant role of the pith in the growth and development of internodes in *Liquidambar styraciflua* (Hamamelidaceae). I. Histological basis of compressive and tensile stresses in developing primary tissues. *Amer. J. Bot.* 82:769–76.

———. 1995b. The predominant role of the pith in the growth and development of internodes in *Liquidambar styraciflua* (Hamamelidaceae). II. Pattern of tissue stress and response of different tissues to specific surgical procedures. *Amer. J. Bot.* 82:777–81.

Cave, I. D. 1966. X-ray measurement of microfibril angle. *For. Prod. J.* 16:37–42.

Cosgrove, D. J. 2005. Growth of the plant cell wall. *Nature* 6:850–61.

Delmer, D. P. 1999. Cellulose biosynthesis: Exciting times for a difficult field. *Annu. Rev. Plant Physiol. Plant Mol. Biol.* 50:245–76.

Driouich, A., and T. I. Baskin. 2008. Intercourse between cell wall and cytoplasm exemplified by arabinogalactan proteins and cortical microtubules. *Amer. J. Bot.* 95:1491–97.

Easterling, K. E., R. Harrysson, L. J. Gibson, and M. F. Ashby. 1982. On the mechanics of balsa and other woods. *Proc. Roy. Soc. London*, ser. A, 383: 31–41.

Esau, K. 1977. *Anatomy of seed plants.* 2nd ed. New York: John Wiley.

Falk, S., C. H. Hertz, and H. Virgin. 1958. On the relation between turgor pressure and tissue rigidity. I. Experiments on resonance frequency and tissue rigidity. *Physiol. Plant.* 11:802–17.

Gibson, L. J., and M. F. Ashby. 1997. *Cellular solids, structure and properties.* 2nd ed. Cambridge: Cambridge University Press.

Gibson, L. J., M. F. Ashby, and B. A. Harley. 2010. *Cellular materials in nature and medicine.* Cambridge: Cambridge University Press.

Gillis, P. P. 1973. Theory of growth stresses. *Holzforschung* 26:197–207.

Gillis, P. P., and C. H. Hsu. 1979. An elastic, plastic theory of longitudinal growth stresses. *Wood Sci. Tech.* 13:97–115.

Harder, D. L., C. L. Hurd, and T. Speck. 2006. Comparison of mechanical properties of four large, wave-exposed seaweeds. *Amer. J. Bot.* 93:1426–32.

Heinich, K. 1908. Über die Entspannung des Markes im Gewebeverbande und sein Wachstum im isolierten Zustand. *Jahrbücher für Wissenschaftliche Botanik 1909*, 46:207–69.

Himmelspach, R., R. E. Williamson, and G. O. Wasteneys. 2003. Cellulose microfibril alignment recovers from DCB-induced disruption despite microtubule disorganization. *Plant J.* 36:565–75.

Hofmeister, W. 1859. Über die Beugungen saftreicher Pflanzentheile nach Erschütterung. *Ber. Verh. Ges. Wiss. Leipzig* 11:175–204.

———. 1860. Über die durch die Schwerkraft bestimmten Richtungen von Pflanzentheilen. *Ber. Verh. Ges. Wiss. Leipzig* 12:175–213.

Hoiczyk, E., and A. Hansel. 2000. Cyanobacterial cell walls: News from an unusual prokaryotic envelope. *J. Bacteriol.* 182:1191–99.

Holbrook, N. M., M. W. Denny, and M. A. R. Koehl. 1991. Intertidal "trees": Consequences of aggregation on the mechanical and photosynthetic properties of sea-palms *Postelia palmaeformis* Ruprecht. *J. Exp. Mar. Biol. Ecol.* 146:39–67.

Ishikawa, A., T. Okano, and J. Sugiyama. 1997. Fine structure and tensile properties of ramie fibers in the crystalline form of cellulose I, II, III, and IV. *Polymer* 38:463–68.

Jacobs, M. R. 1938. The fibre tension of woody stems, with special reference to the genus *Eucalyptus. Commonw. For. Bur. Austral. Bull.* 22:1–39.

———. 1945. The growth stresses of woody stems. *Commonw. For. Bur. Austral. Bull.* 28:1–67.

Jeffree, C. E. 1996. Structure and ontogeny of plant cuticles. In *Plant cuticles: An integrated functional approach*, edited by G. Kerstiens, 33–82. Oxford: BIOS Scientific Publishers.

Johnson, A. S., and M. A. R. Koehl. 1994. Maintenance of dynamic similarity and environmental stress factor in different flow habitats: Thallus allometry and material properties of a giant kelp. *J. Exp. Bot.* 195:381–410.

King, M. J., and J. F. V. Vincent. 1996. Static and dynamic fracture properties of the leaf of the New Zealand flax (*Phormium tenax*) (Phormiaceae; Monocotyledones). *Proc. Roy. Soc. London*, ser. B, 263:521–27.

Köhler, L., T. Speck, and H.-C. Spatz. 2000. Micromechanics and anatomical changes during early ontogeny of two lianescent *Aristolochia* species. *Planta* 210: 691–700.

Kolattukudy, P. E. 1980. Cutin, suberin, and waxes. In *Lipids: Structure and function*, edited by P. K. Stumpf, 571–646. New York: Academic Press.

———. 1996. Biosynthetic pathways of cutin and waxes, and their sensitivity to environmental stresses. In *Plant cuticles: An integrated functional approach*, edited by G. Kerstiens, 83–108. Oxford: BIOS Scientific Publishers.

Kübler, H. 1959. Studies on growth stresses in trees. 2. Longitudinal stresses. *Holz Roh-Werkst.* 17:44–54.

Kutschera, U. 1989. Tissue stresses in growing plant organs. *Physiol. Plant.* 77: 157–63.

Kutschera, U., and K. J. Niklas. 2007. The epidermal-growth-control theory of stem elongation: An old and a new perspective. *J. Plant Physiol.* 164:1395–1409.

Lee, D. R. 1981. Elasticity of phloem tissues. *J. Exp. Bot.* 69:1224–30.

Lin, T.-T., and R. E. Pitt. 1986. Rheology of apple and potato tissue as affected by cell turgor pressure. *J. Text. Stud.* 17:291–313.

López-Casado, G., A. J. Matas, E. Dominguez, J. Cuartero, and A. Heredia. 2007. Biomechanics of isolated tomato (*Solanum lycopersicum* L.) fruit cuticles: The role of the cutin matrix and polysaccharides. *J. Exp. Bot.* 58:3875–83.

Matas, A. J., E. D. Cobb, J. A. Bartsch, D. J. Paolillo Jr., and K. J. Niklas. 2004. Biomechanics and anatomy of *Lycopersicum esculentum* mill. outer fruit walls and enzyme-treated samples. *Amer. J. Bot.* 91:352–60.

Müller, N. J. C. 1880. *Handbuch der Allgemeinen Botanik. Erster Teil: Anatomie und Physiologie der Gewächse.* Heidelberg: Carl Winters Universitätsbuchhandlung.

Niklas, K. J. 1988. Dependency of the tensile modulus on transverse dimensions, water potential, and cell number of pith parenchyma. *Amer. J. Bot.* 75:1286–92.

———. 1989. Mechanical behavior of plant tissues as inferred from the theory of pressurized cellular solids. *Amer. J. Bot.* 76:929–37.

———. 1991. Bending stiffness of cylindrical plant organs with a "core–rind" construction: Evidence from *Juncus effusus*. *Amer. J. Bot.* 78:561–68.

———. 1992. *Plant biomechanics: An engineering approach to plant form and function.* Chicago: University of Chicago Press.

———. 2004. The cell walls that bind the tree of life. *BioScience* 54:831–41.

Niklas, K. J., F. Molina-Freaner, C. Tinoco-Ojanguren, C. J. Hogan Jr., and D. J. Paolillo Jr. 2003. On the mechanical properties of the rare endemic cactus *Stenocereus eruca* and the related species *S. gummosus*. *Amer. J. Bot.* 90:663–74.

Niklas, K. J., and D. J. Paolillo Jr. 1997. The role of the epidermis as a stiffening agent in *Tulipa* (Liliaceae) stems. *Amer. J. Bot.* 84:735–44.

———. 1998. Preferential states of longitudinal tension in the outer tissues of *Taraxacum officinale* (Asteraceae) peduncles. *Amer. J. Bot.* 85:1068–81.

Niklas, K. J., and H.-C. Spatz. 2010. Worldwide correlations of mechanical properties and green wood density. *Amer. J. Bot.* 97:1587–94.

Nobles, D. R., D. K. Romanovicz, and R. M. Brown Jr. 2001. Cellulose in cyanobacteria: Origin of vascular plant cellulose synthase? *Plant Physiol.* 127:529–42.

Passioura, J. B., and J. S. Boyer. 2003. Tissue stresses and resistance to water flow conspire to uncouple the water potential of the epidermis from that of the xylem in elongating plant stems. *Funct. Plant Biol.* 30:325–34.

Peters, W. S., and A. D. Tomos. 2000. The mechanical state of "inner tissues" in the growing zone of sunflower hypocotyls and the regulation of its growth rate following excision. *Plant Physiol.* 123:605–12.

Petracek, P. D., and M. J. Bukovac. 1995. Rheological properties of enzymatically isolated tomato fruit cuticle. *Plant Physiol.* 109:675–79.

Probine, M. C., and R. D. Preston. 1962. Cell growth and the structure and mechanical properties of the wall in internodal cells of *Nitella opaca*. II. Mechanical properties of walls. *J. Exp. Bot.* 13:111–27.

Richmond, T. A., and C. R. Somerville. 2000. The cellulose synthase superfamily. *Plant Physiol.* 124:495–98.

Robinson, W. C. 1920. The microscopical features of mechanical strains in timber and the bearing of these on the structure of the cell wall in plants. *Phil. Trans. Roy. Soc. London*, ser. B, 210:49–82.

Schopfer, P. 2006. Biomechanics of plant growth. *Amer. J. Bot.* 93:1415–25.

Spatz, H.-C., H. Beismann, F. Brüchert, A. Emanns, and T. Speck. 1997. Biomechanics of *Arundo donax*. *Phil. Trans. Roy. Soc. London*, ser. B, 352:1–10.

Spatz, H.-C., C. Boomgarden, and T. Speck. 1993. Contribution to the biomechanics of plants. III: Experimental and theoretical studies of local buckling. *Bot. Acta* 106:254–64.

Spatz, H.-C., L. Köhler, and T. Speck. 1998. Biomechanics and functional anatomy of hollow stemmed sphenopsids: I. *Equisetum giganteum* (Equisetaceae). *Amer. J. Bot.* 85:305–14.

Stephens, R. C. 1970. *Strength of materials, theory and examples.* London: Edward Arnold.

Taiz, L., and E. Zeiger. 2002. *Plant physiology.* 3rd ed. Sunderland, MA: Sinauer Associates.

Thompson, D. S. 2001. Extensiometric determination of the rheological properties of the epidermis of growing tomato fruit. *J. Exp. Bot.* 52:1291–1301.

Timell, T. E. 1969. The chemical composition of tension wood. *Svensk papperstidn.* 72:173–81.

———. 1986. *Compression wood in gymnosperms.* Vol. 2, 983–1262. Heidelberg: Springer Verlag.

Timoshenko, S. P., and J. N. Goodier. 1970. *Theory of elasticity.* 3rd ed. New York: McGraw-Hill.

Vincent, J. F. V. 1989. Relationship between density and stiffness of apple flesh. *J. Sci. Food Agr.* 47:443–62.

Virgin, H. 1955. A new method for determination of the turgor of plant tissues. *Physiol. Plant.* 8:954–63.

Wegst, U. G. K. 2006. Wood for sound. *Amer. J. Bot.* 93:1439–48.

Wiedemann, P., and C. Neinhuis. 1998. Biomechanics of isolated plant cuticles. *Bot. Acta* 111:28–34.

Wolfe, J., and P. L. Steponkus. 1981. The dynamics of area changes in the protoplast plasma membrane. In *Proceedings of the 13th International Botanical Congress (Sydney, Australia),* 245.

Wood Handbook. 1999. Madison, WI: Forest Products Laboratory, USDA Forest Service.

Wymer, C., and C. Lloyd. 1996. Dynamic microtubules: Implications for cell wall patterns. *Trends Plant Sci.* 1:222–28.

CHAPTER NINE

Experimental Tools

As long as I've gone this far, I can't just leave it after I've found out so much about it. I have to keep going to find out ultimately what is the matter with it in the end.
—Richard P. Feynman, *"Surely You're Joking, Mr. Feynman!"*

Thus far, we have alluded to a variety of experimental protocols and techniques in various contexts, such as the use of the Scholander pressure chamber to measure water potential and the patch clamp technique to determine electrical currents. In this chapter, we review these and other techniques that have been used, both traditionally and more recently, to understand plant physics. This review cannot be comprehensive, especially because new techniques are being developed and used every year, while more traditional techniques are being put to nontraditional uses every day. Our objective, therefore, is more limited: it is to provide a starting point for those who are interested in learning some basic approaches and thus to provide them with a springboard to explore experimental techniques in greater detail using the primary literature as a qualified guide.

9.1 Anatomical methods on a microscale

A number of microscopic techniques have been widely applied in anatomical studies of plants. Microscopic observations, combined with a wealth of staining techniques, allow observers to distinguish among different tissues. As reviewed by Burgert (2006), the structure of the plant cell wall can be elucidated by extending microscopy into the UV and infrared and by the use of fluorescence microscopy as well as Raman scattering. Polarized light microscopy (Niklas and Paolillo 1997) and small-angle X-ray scattering

(Lichtenegger et al. 1999) have been used to determine the microfibrillar angle of cellulose fibers in the S2 layer of the cell wall.

Scanning electron microscopes

Structural details are revealed by scanning electron microscopes (SEM). For most biological samples, conventional SEM has the disadvantage that specimens must be coated with gold or carbon in a high vacuum. In addition, the observation chamber has to be evacuated, such that only dried or cryogenically frozen samples can be observed. In contrast, the environmental scanning electron microscope (ESEM) allows the observation of uncoated specimens at pressures in the range of 1×10^2 to 5×10^3 Pa and humidities up to 100%, making it a valuable tool for investigating living plant tissues. The details of this method are provided by Donald (2003). Like the SEM, the ESEM works by scanning a highly collimated electron beam across the sample and collecting the electrons emitted by the sample. The electrons emitted fall into two classes: high-energy backscattered electrons and secondary electrons resulting from inelastic collisions of the incident primary beam electrons with the sample. A gaseous environment (usually water vapor) is maintained around the sample. The gas has a key role in the signal detection. Due to their low energy, the secondary electrons have a high probability of interacting with gas molecules and ionizing them to generate additional tertiary electrons, thus amplifying the signal. Due to their lower energy, secondary and tertiary electrons can be collected preferentially by a positively charged detector. As many positive ions as daughter electrons are generated in ionizing collisions. These ions drift toward the sample and serve to compensate for negative charges on the sample surface, which explains why samples in the ESEM do not have to be coated with a conducting layer.

As in the SEM, the signal reflects the topography of the sample. But since the emission of secondary electrons depends on the interaction of the primary beam with atoms in the upper layer of the sample, the signal also reflects its molecular constitution to a depth of roughly 10 nm, depending on the molecular composition of the upper layer and the energy of the beam.

Förster resonance energy transfer (FRET)

Soon after the end of World War II, Theodor Förster (1949) developed a theory describing resonance energy transfer between two chromophores with overlapping excitation-emission spectra. This mechanism is termed Förster resonance energy transfer (FRET). When both chromophores are fluorescent, the term "fluorescence resonance energy transfer" is often used instead, although the energy is not actually transferred by fluorescence. The donor chromophore, initially in its electronic excited state, may transfer energy to an acceptor chromophore through nonradiative dipole-dipole coupling. Donor and acceptor must be 10 nm or less from each other for this phenomenon to occur with sufficient probability to be observable. Thus, FRET can provide a yardstick of molecular dimensions (Stryer and Haugland 1967). It is particularly useful for studying the kinetics of protein folding, proteolysis, and molecular interactions (Gadella et al. 1999). In all these cases, pairs of chromophores such as derivatives of the green fluorescent protein (GFP) or cyanide dyes must be coupled to the protein, RNA, or DNA under study.

The spatial resolution of fluorescent microscopy and FRET can be extended far below the diffraction limit into the range of 10 to 20 nm by applying stochastic optical reconstruction microscopy (Rust et al. 2006). This technique enables subcellular localization in living cells. It is particularly useful for FRET studies on single molecules (smFRET) (Roy et al. 2008). The molecule must be immobilized to a support in a way that retains its functional activity. FRET using one pair, a triplet, or two pairs of chromophores allows the study of intramolecular dynamics such as protein folding or conformational changes as well as the kinetics of intermolecular processes such as binding, proteolysis, and catalysis. Processes such as the action of motor proteins and the unwinding of DNA double strands during replication are also being studied. Moreover, if combined with optical or magnetic tweezers, smFRET can be used to record conformational changes under the influence of specifically applied forces.

Atomic force microscopy

The ultimate resolution is reached in the atomic force microscope (Morris et al. 1999). Here, the surface of a specimen is scanned with a microscale cantilever equipped with a tip with a radius of a few nanometers. Forces between the atoms in the specimen surface and in the tip lead to tiny deflections of the cantilever, which are detected by optical methods.

9.2 Mechanical measuring techniques on a macroscale

In principle, mechanics deals with forces, distances, and time. Universal testing machines (fig. 9.1), which measure these quantities over several orders of magnitude with high precision, are widely in use. Most plant materials deform under forces smaller than those that deform steel. Therefore, the compliance of the machine parts, particularly those that are used to deform biological specimens, is usually negligible.

Uniaxial tension and compression tests

Tension tests (fig. 9.2A) are performed by fixing the two ends of a sample in clamps, often using appropriate glues, and recording the tensile force as a function of the distance the clamps are moved apart. However, clamps or other supports restrict deformations at the ends of a specimen and can increase the apparent elastic modulus of the material. End-wall effects resulting from clamps or other means of supporting a specimen are often neglected, leading to spurious values of the elastic modulus E. Thus, it is

FIGURE 9.1. A universal testing machine for measuring forces and recording distances and time.

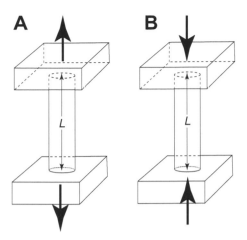

FIGURE 9.2. Schematic drawing of a tension test (A) and a compression test (B). The direction of the applied force indicated by large arrows. The length L of the specimen is measured as the distance between clamps.

always advisable to perform uniaxial tensile tests on specimens that are at least ten times longer than they are wide. This aspect ratio is the minimum ratio for which end-wall effects can be more or less neglected. Another source of uncertainty is possible specimen slippage in the clamps. The preferred method, therefore, is to observe the distance between two markers on the specimen by optical means. If, in addition, the distance between two markers orthogonal to the direction of force is followed by optical means, Poisson's ratios can be measured directly.

Another one-dimensional test is the compression test (fig. 9.2B). Here, the aspect ratio of the sample has to be chosen to prevent buckling. Measurements on wet biological samples are difficult to interpret because water may be pressed out of tissues or, if that is prevented by some means, the stress-strain relationship may be dominated by the hydraulic properties of water.

Biaxial tests

Uniaxial tests should be avoided for biological materials that naturally experience biaxial or triaxial strains, such as the epidermis and the storage parenchyma found deep within large tuberous organs. Two-dimensional straining prevents Poisson's effect—the lateral contraction of the material and the reorientation of fibrous components within it under uniaxial tension. Poisson's effect decreases the apparent elastic modulus and increases the ultimate strains of an anisotropic material. Thus, anisotropic materials will typically yield lower elastic moduli and larger ultimate strains when tested under uniaxial tension than under biaxial tension. An example of biaxial tension is the expansion of cherry skin, mounted like a membrane, bulging under hydraulic pressure (Bargel et al. 2004).

Finally, it is important to remember that the proportional limits measured in tension and in compression for the same material may have very different values. Therefore, it is always advisable to test a material under loading conditions that reflect those it normally sustains. For instance, a transverse compression test (fig. 9.3) can measure the forces on a section of a hollow cylinder as it is deformed from a circular to an elliptical cross section (see section 5.11), simulating the deformation experienced by a hollow stem or leaf in bending.

Bending

Most bending tests are performed in three-point bending (fig. 9.4A). The specimen rests on two supports and is loaded in the middle between the supports. Depending on the aspect ratio (of the span between the supports to the dimension of the specimen in the direction of the force), both bending and shearing contribute to the deflection to varying degrees (see

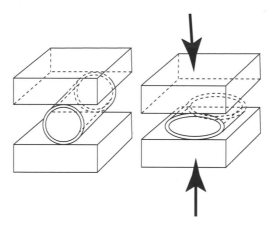

FIGURE 9.3. Schematic drawing of a transverse compression test of a hollow cylinder. The direction of the applied force is indicated by arrows.

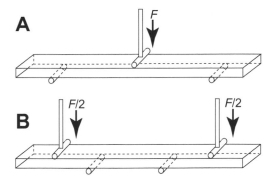

FIGURE 9.4. Schematic of (A) three-point and (B) four-point bending. Note that both tests use the same total applied force (*F*).

section 5.5). If shear contributions are to be avoided, four-point bending is preferable (fig. 9.4B). Here, the specimen rests on two supports and is loaded symmetrically by two forces applied from outside the two supports. The deflection is usually measured at the midpoint between the two supports. A method for testing living trees in bending is presented in box 9.1.

BOX 9.1 **An example of applied biomechanics: Tree risk assessment**

The stability of trees under static loads, particularly wind loads, is a major concern in forestry. In urban environments, risks to human welfare necessitate assessments of tree safety against gravitational loads and wind loads (Mattheck and Breloer 1994; Schwarze 2008). A combination of tree pulling and wind tunnel experiments can reveal the causal relationship between tree characteristics and susceptibility to stem breakage or uprooting (Peltola 2006). Morphological criteria are an important basis for assessing a tree's health (Roloff 1999). In some cases, however, it proves necessary to apply physical methods to monitor the extent of imperfections that often go unseen, such as fungal rot and termite damage. A review of diagnostic methods is given by Rust and Weihs (2007). A variety of methods for risk assessment are available that, for convenience, may be categorized as one-dimensional, two-dimensional, and three-dimensional methods.

An example of a one-dimensional method is the fractometer method, which measures the bending moment necessary to fracture a sample of wood usually

BOX 9.1 **(Continued)**

taken in the radial direction with respect to the trunk. Another one-dimensional method is the resistograph method, which monitors the integrity of wood by measuring the resistance against drilling a hole with a very small diameter, once again, usually in the radial direction with respect to the trunk. Both methods are invasive because they require drilling holes or taking samples at various places, which may potentially damage the tree. The high-speed version of the resistograph method is the least damaging of these two methods. However, it should be noted that the interpretation of a resistograph scan requires a great deal of expertise.

Two-dimensional methods such as electric resistance (Weihs et al. 2007), sound (Gilbert and Smiley 2004), and ultrasound tomography (Nicolotti et al. 2003) have also been developed. Particularly important for the wood industry is the nondestructive X-ray tomographic method (Brännström et al. 2007), which is used to assess wood quality and to detect internal imperfections such as knots or cracks. Measurements with these two-dimensional methods at several positions along the axis lead to a three-dimensional image of the internal structure of the tree or log under investigation.

An example of a three-dimensional method is the pulling test, which provides measurements of an entire tree. The pulling test provides a measure of the force necessary to pull a tree slightly out of its resting position. An inclinometer is attached to the base of the trunk to measure the degree to which the tree is tilted, and extensometers are attached at various heights of the trunk to measure the degree of bending (Sinn 2003). This technique works with small deflections from the resting position and is therefore nondestructive (unless the tree is ready to fall down anyway). The inclinometer readings give an estimate of the strength of the root anchorage and, if extrapolated to maximum experienced wind loads, an assessment of the danger of overturning. The extensometer readings in relation to the pulling force give the flexural stiffness and thus an indication of the integrity of the trunk. Although the breaking strength cannot be inferred from the modulus of elasticity, the extrapolation of the data to maximum experienced wind loads (together with empirical data on the bending strength of green wood from a large number of species; see table 8.2) leads to an assessment of the danger of trunk breakage. A critical overview of the methods for tree risk assessment, particularly with respect to fungal infection and wood decay, is given by Schwarze (2008).

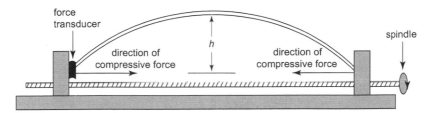

FIGURE 9.5. Schematic of a device to measure Euler buckling and Brazier buckling. Two supports are moved toward each other using a motor-driven spindle. The horizontal component of the force is measured with a force transducer; the degree of bending is measured (usually by optical means) as the distance indicated by h.

Neither three- nor four-point bending is advisable for testing hollow plant organs. Deformations of the cross section at the supports (due to load-induced ovalization) and, in three-point bending, at the load point as well may lead to seriously erroneous measurements of deflection. Even more important, deformation at the load point will lead to a local reduction of the second moment of area and will thus compromise the evaluation of the force-deflection relationship. A more appropriate method is an Euler buckling test, which can be extended to examine Brazier buckling (fig. 9.5). Here, the forces are applied from both ends of the specimen. The ends should be mounted in supports that permit rotation of the specimen orthogonal to the direction of the applied force, which allows for the free alignment of the specimen's bending line. With this technique, continued loading causes the mechanical behavior of the specimen to undergo a transition from Euler to catastrophic Brazier buckling (Spatz et al. 1998). Until the onset of inelastic behavior is reached, the process is reversible.

Oscillations

Recording the frequency of oscillation of a plant organ or a whole plant allows the determination of the modulus of elasticity with some accuracy. The damping ratio yields information on its viscoelastic properties in addition to its aerodynamic resistance. Extensometers, inclinometers, or accelerometers (provided their weight is low compared with that of the specimen being tested) can be attached to an organ. Data should be read at a rate at least ten times higher than the oscillation frequency, which requires fast data logging. An alternative is an electromagnetic motion tracking system. Optical recordings have the advantage of not interfering with the plant's movement and are therefore useful for the study of

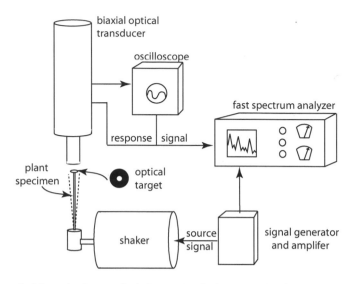

FIGURE 9.6. Schematic of one method of measuring the frequency of oscillation of plant materials. A biaxial optical transducer is used to track the motion of a target placed on the tip of a specimen that is vibrated with white noise. The response signal is then analyzed to determine the primary and higher frequencies of oscillation, which can be used to calculate the elastic modulus of the specimen. (Adapted from Niklas and Moon 1988.)

small plants such as grasses (fig. 9.6). Video recording and computer-aided frame by frame evaluation is suitable if the oscillation period is on the order of 1 second or more (Niklas and Moon 1988).

Sound velocity

An indirect way to quantify the modulus of elasticity E in a solid homogeneous specimen (see section 4.3) over a large frequency range is the measurement of sound velocity c, which is easily accomplished with a sound source, two microphones at some distance from each other along the specimen, and a fast recording device. If the density of the material is known, E can be calculated from the relationship

(9.1) $c = \sqrt{E/\rho}\,.$

This technique works well for samples of homogeneous materials. Its application to specimens composed of heterogeneous materials should be approached with caution, however.

For fluids, the relevant elastic modulus is primarily the volumetric elastic modulus K (see eq. [4.14]). The sound velocity is given by

(9.2a)
$$c = \sqrt{\frac{K + 4/3G}{\rho}}.$$

Since the shear modulus G in a fluid is much smaller than K, equation (9.2a) simplifies to

(9.2b)
$$c \approx \sqrt{K/\rho}.$$

Hardness

Hardness is the ability of a material to resist plastic deformation. Therefore, it is not an intrinsic material property precisely defined in terms of the fundamental units of mass, length, and time. It is operationally defined as the result of a normalized measurement procedure. Typically, a steel ball (Brinell test) or a diamond with the shape of a four-sided pyramid (Vickers test) is pressed into the specimen. The force divided by the area of an indentation left after the indenter is withdrawn is a qualitative measure of hardness.

Torsion

The mechanical response to shear stresses is most easily measured in torsion (see section 5.7). If the polar moment of inertia and the length of the object are known, the shear modulus can be calculated according to equation (5.18) from the quotient of the applied torque and the angle of twist. Greater accuracy can be achieved if a torsion pendulum is used (fig. 9.7). From the frequency of its oscillation, the shear modulus is obtained. The decay of its amplitude yields the loss modulus of the material (Köhler et al. 2000), which measures the torsional energy that is typically dissipated as heat. The disadvantage of this method is the limited frequency range in which measurements can be carried out.

Work of fracture

The work of fracture (see section 4.15) can be measured in a variety of ways, one of which is to quantify the energy required to drive a sharp wedge through a specimen. Using a wedge implies that part of the energy

FIGURE 9.7. A torsion pendulum. A moving part with arms carrying two weights is suspended from a thin steel wire. The specimen is fixed between a stationary clamp and a clamp on the moving part. A mirror reflects a laser beam to a detector (not shown).

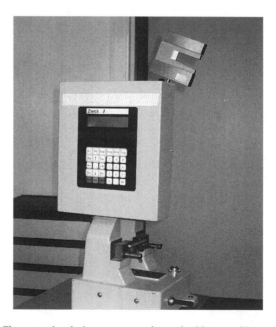

FIGURE 9.8. A Charpy testing device to measure the work of fracture. Upon release, the pendulum fractures the specimen and continues to swing to a height depending on the energy lost in the fracture process.

is used in bending and is regained as the crack propagates ahead of the wedge (see section 5.6). Using a razor blade instead of a wedge minimizes bending (Vincent 1990).

The Charpy method of testing is widely used and standardized (fig. 9.8). Here, the sample rests on two L-shaped supports with a gap between them. A pendulum is released, which hits and fractures the sample. Due to the loss of energy during fracture, the pendulum does not swing to the same height from which it was released. Commercial Charpy testing devices convert the difference in height into the energy loss and, dividing it by the cross section of the sample, into readings of work of fracture. If, as is often done, a small cut is introduced into the sample, the cross section of the uncut portion of the sample is taken.

9.3 Mechanical measuring techniques on a microscale

Most experimental approaches to characterizing mechanical properties at the cellular and subcellular levels have been developed for animal tissues. Their application to plant material has been reviewed by Geitmann (2006).

Microtensile tests

Two important prerequisites for tensile tests on single cells or single fibers (Thygesen et al. 2007) are gentle preparation of the specimen and precise measurements of its dimensions. Laser microdissection (Angeles et al. 2006) allows for the well-defined dissection of microscopically small intact specimens from selected plant tissues. A commercial system for this purpose uses a nitrogen laser with a wavelength of 355 nm, focused to a spot of about 1 μm in diameter, pulsed at a rate of 1 to 200 Hz with a pulse duration of 1.0 nanosecond and an energy output of around 100 μJ. This system allows for precise cuts of 10 μm in width without overheating adjacent cells.

Samples prepared in this way are laser catapulted into microvessels and subsequently transferred to a microtensile testing device (Burgert et al. 2003) equipped with a force transducer with a resolution of approximately 0.5 mN and a video extensometer to record strains. The power of this method becomes most evident in its combination with other analytical methods such as X-ray diffraction (Köhler and Spatz 2002; Keckes et al.

2003; Kölln et al. 2005), Raman spectroscopy (Gierlinger et al. 2006), and ESEM (see section 9.1). For a recent review, see Burgert (2006).

Nanoindentation

A valuable method for studying the mechanical properties of plant material on a microscopic or even submicroscopic scale is nanoindentation (Gindl et al. 2004; Geitmann 2006), which is typically employed to measure the hardness and elastic properties of the cell wall. In contrast to macroscopic hardness tests, the indentations are so small that their area may not be measurable under the microscope. Their direct measurement has to be replaced by the measurement of the indentation depth.

In a nanoindentation test, an indenter with a fine diamond tip, in the shape of a four-sided pyramid (Vickers tip) or a three-sided pyramid (Berkovich tip), is driven electromagnetically into the specimen. The resolution of the indentation is on the order of 1 nanometer, and the forces applied are measured with an accuracy on the order of 1 nanoNewton. If the geometry of the tip is known, the indentation depth can be converted into the indentation area. The hardness value is obtained by dividing the maximum load by the area of the indentation left after the indenter has been withdrawn.

Strictly speaking, the definition of hardness applies to isotropic materials. Therefore, care has to be taken in the interpretation of the data from nanoindentation tests on anisotropic biological materials. In addition, viscoelastic behavior may compromise the interpretation of hardness as resulting from permanent plastic deformations. It is advisable to perform the tests at different time scales.

During the loading phase of indentation, elastic, viscoelastic, and plastic deformations will all take place in the material. However, the very beginning of the unloading phase can be ascribed to elastic behavior. Oliver and Pharr (1992) have proposed a method to obtain the modulus of elasticity E directly from the unloading curve using the formula

$$(9.3) \qquad E = \frac{\pi}{2} \frac{1}{\sqrt{A_c}} (1 - v^2) \left(\frac{dF}{dh} \right)_{F_{max}},$$

where A_c is the contact area, v is Poisson's ratio, and $(dF/dh)_{F_{max}}$ is the slope of the force-deflection curve upon unloading from the maximum force F_{max}. The contact area A_c is a function of the contact height h_c, depending on the geometry of the indenter. The contact height is given by the formula

(9.4) $$h_c = h_{max} - 0.75 \left(F \bigg/ \frac{dF}{dh} \right)_{max}.$$

Again, equation (9.3) applies to isotropic materials under conditions where viscoelasticity plays a minor role.

Microcompression

The mechanical properties of the cell wall can be studied by compressing a cell between a flat plate and the flat end of a glass fiber and recording the forces and the deformation of the cell (compare box 8.3). For isolated tomato cells, forces on the order of several hundred μN are needed to deform the cell by 10% (Wang et al. 2006). The compression time should be small enough to avoid water loss, typically about 50 ms. The shapes of force-deflection curves derived using this technique are in good agreement with a theoretical model describing changes of shape upon compression and the corresponding stresses and strains in the cell wall, assuming linear elastic behavior.

A higher resolution can be achieved with a micro-ball tonometer (Lintilhac et al. 2000), a device that is used to measure tension or pressure. A micro-ball tonometer is a small, rigid sphere mounted on a cantilever that is pressed onto the surface of a cell. The deflection of the cantilever can be recorded optically. With known mechanical characteristics, the deflections can be converted to forces. The deformation of the surface as a function of the applied load yields information on the elastic properties of the cell wall in combination with the turgor pressure of the cell (Wei et al. 2001).

Optical tweezers

Stresses on single cells can be exerted by optical tweezers. This technique uses a laser to manipulate minute objects. Particles with a higher refractive index than their surroundings are moved in a pressure gradient created by light toward the focal point and kept there by a net force on the order of a few hundred picoNewtons. If silica beads are attached at diametrically opposite points on a cell, such that one is trapped in the laser beam and the other is fixed to a support and moved relative to the other bead, the cell can be deformed. Unfortunately, these forces are not sufficient to deform a plant cell with an intact cell wall. However, by inserting objects of high refractive index into the cell and moving them within the cytoplasm, the technique can be used to investigate the viscosity and viscoelasticity of the cytoplasm and its organelles. For example, the effects of the cytoskeletal structure on the movement of statoliths have been studied in *Chara* rhizoids by this method (Leitz et al. 1995).

Bead rheometry

Another technique for studying the viscous and viscoelastic properties of the cytoplasm is magnetic bead rheometry, in which small ferromagnetic particles taken up by or inserted into a cell are moved in a magnetic field while being tracked optically (Bausch et al. 1999). A similar method not yet employed for plant cells is magnetic rotational microrheometry (Wilhelm 2003), in which magnetic nanoparticles are taken up by the cell into the endocytotic compartments. The endosome movement in a magnetic field reflects properties associated with the configuration of the cytoskeleton. Even in the absence of external forces, Brownian motion of particles within a cell can be monitored by scattered light from a focused low-power laser. This method, called laser tracking microrheology, has been successfully applied to estimate cytoplasmic shear moduli (Yamada 2000).

Scanning acoustic microscopy

Ultrasound can be used to study the spatial distribution of the mechanical properties of samples of biological materials. In principle, the resolution of acoustic microscopy can be brought down to about the wavelength of the sound, which at a frequency of 1.0 GHz is roughly 1.5 μm in water at 20°C. The first acoustic microscope was built by Lemons and Quate (1974). Several different types of acoustic microscopes exist today, of which only the scanning acoustic microscope working in the reflection mode is described here (Bereiter-Hahn et al. 1979). The sound frequency range used by this apparatus lies between 50 MHz and 1.0 GHz. Lower frequencies have the advantage of greater penetration in depth, but less spatial resolution.

Sound pulses are generated by a piezoelectric transducer, and a system of sapphire lenses focuses the sound to the appropriate depth in the sample. Air does not transmit sound at high frequencies, so water has to be used as a coupling liquid between the sapphire lenses and the sample. An

incident sound pulse is reflected whenever a change in the acoustic impedance Z is encountered. This impedance is a function of the local density ρ
and sound velocity c. By applying equation (9.2b) for a fluid medium, impedance can also be expressed as a function of the local volumetric elastic
modulus K by the formula

(9.5)
$$Z = \rho c = \sqrt{K\rho} \ .$$

The intensity of the reflected sound is given by

(9.6)
$$\frac{I_r}{I_0} = \frac{(Z_2 - Z_1)^2}{(Z_2 + Z_1)^2},$$

where I_r is the reflected intensity, I_0 is the incident intensity, and Z_2 and Z_1
are the impedances of the adjacent portions of the sample 1 and 2.

The advantage of using the reflection mode is that the same system of
transducer/receiver and focusing sapphire lenses is used for both sound
emission and the detection of the reflected signal. This arrangement avoids
the need for a complicated alignment of a transducer and a separate re-
ceiver, which is required using the transmission mode.

Each emitted sound pulse has a duration ranging between 4 and 25 ns.
The echo between any two pulses consists of three signals. The first echo
to appear is due to the reflection from the front surface of the specimen;
the second echo, which is the one of interest, results from possible discontinuities within the specimen; and the third echo comes from the reflection
at the back surface. By scanning in the x- and y-directions and by changing the focal point in the z-direction (i.e., depth within the specimen), a
three-dimensional image of the acoustic impedance, and therefore of the
mechanical properties of the sample, can be generated from the multitude
of echoes.

9.4 Scholander pressure chamber

The water potential of a plant sample can be determined by measuring the
magnitude of the external pressure uniformly applied over the sample's
surface that drives water from the living cells into the xylem tissue at atmospheric pressure (Scholander et al. 1965). The entire sample is placed into
a dark, humid pressure chamber so that the sample's cut end is exposed to

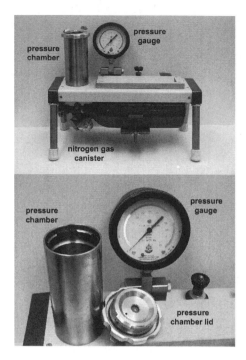

FIGURE 9.9. A portable Scholander pressure chamber system, showing the general configuration of the device. A specimen such as a leaf or stem is placed in the pressure chamber, with its cut end exposed to atmospheric pressure by extending through a small hole, and nitrogen gas is used to increase the pressure in the chamber until xylem water is extruded from the cut end extending through a small opening at the top of the chamber (see also fig. 3.3).

atmospheric pressure and thus to view (fig. 9.9). In order to minimize the gravitational potential, the sample can be oriented horizontally. After transpiration stops, the pressure within the chamber is gradually increased (typically at a rate of ≤ 0.03 MPa/s) until a xylem water meniscus appears at the cut end. At this point, the water potential is zero, and therefore the negative value of the applied pressure $-\psi_{bp} = \psi_s + \psi_m$. As confirmed by experiments in which branches were centrifuged and thus subjected to negative pressures (Holbrook et al. 1995), the water potential of the sample ψ_w is the negative balancing pressure $-\psi_{bp}$ (i.e., the difference between the pressure within the chamber and the atmospheric pressure). As pointed out by Tyree and Zimmermann (2002), the negative balancing pressure is a true measure of the water potential only if the osmotic potential of the xylem water is zero; that is, if xylem water can be regarded as an extremely diluted aqueous solution.

9.5 Pressure probe

Turgor pressure in a cell can be measured directly by inserting a micropipette into the cell (Hüsken et al. 1978; Tomos and Leigh 1999) (fig. 9.10). The micropipette is filled with water and then oil such that the tip inserted into the cell contains the water component. An attached sensor records pressures, which are typically in the range of 0.1–1 MPa. By using a metal plunger manipulated by a screw to displace water into the cell, the turgor pressure in the cell can be changed artificially. The position of the water-oil interface in the micropipette provides a measure of the amount of water pressed into or taken from the cell. Because this method is invasive, care must be taken to avoid artifacts. For instance, the method does not allow the measurement of large negative pressures in water-conducting vessels because the insertion of the micropipette will induce cavitation.

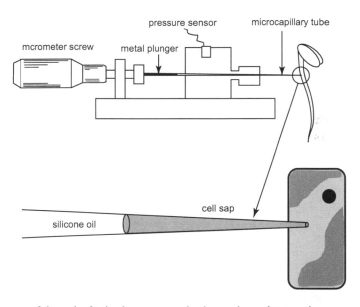

FIGURE 9.10. Schematic of a simple pressure probe that can be used to record turgor pressure in a plant cell. A micrometer screw can be used to inject water into or withdraw water from the cell. The pressure exerted by or on evacuated cell sap or the cell contents can be registered and used to measure water potential.

9.6 Recording of electric potentials and electrical currents

Various techniques for recording ion fluxes and electric potentials are described in Volkov's (2006) recent book on plant electrophysiology. Plant cell walls and turgor pressure may pose some difficulties for measuring potentials between the inside and the outside of a cell. A standard technique involves impaling the cell with an electrode and recording the voltage difference between it and a reference electrode in the solution immediately surrounding the cell. Sharp-pointed tungsten wires, platinized to avoid hydrogen overvoltages, can serve as electrodes. Alternatively, glass microcapillaries, drawn out to a very fine tip and filled with a saturated KCl solution, are often used for this purpose. K^+ and Cl^- in saturated aqueous solution have nearly the same mobility; thus, the electric potentials recorded are not biased by diffusion potentials (see eq. [7.7c]) generated in the microcapillary. A silver wire, coated with AgCl and placed in the body of the microcapillary, serves for a reversible transition of ionic to electric conductance. Amplifiers with high input impedance must be used.

9.7 Patch clamp techniques

The patch clamp technique was first used to record the ion current passing through a single channel in the membrane of a muscle fiber (Neher and Sakmann 1976). Such recordings are made by pressing a heat-polished glass pipette with a tip diameter on the order of 1 μm against a clean membrane surface. When suction is applied to the interior of the pipette, a tight seal forms between the pipette tip and the membrane, as characterized by a gigaohm seal resistance. Both cell-attached configurations and excised patches of membrane allow for the measurement of single-channel currents with a resolution of less than 1 picoampere. Single-channel currents are recorded as rectangular pulses of varying duration. Their frequency of occurrence and their duration depend on the type of channel recorded as well as on experimentally controlled variables such as the chemical composition of the solutions on both sides of the membrane and the voltage across the membrane.

As reviewed by Hedrich and Schröder (1989), patch clamping of plant cells poses some difficulties. Mature plant cells are characterized by a cell wall and a plasma membrane in series with the vacuolar membrane. Therefore, great care is required to properly identify the target membrane and the composition of the solutes in the various cell compartments that are the subjects of enquiry. Even in simple systems such as guard cells or suspensions of cultured cells, removal of the cell wall by lytic enzymes to obtain a clean membrane surface is required for patch clamp recording. With the caveat that protoplasts may not be functionally equivalent to the intact cells, recording from protoplasts provide an alternative. Even then, sealing patch pipettes to the plasma membrane of protoplasts is more difficult than sealing them to vacuolar membranes or animal cells.

9.8 Biomimetics

Over millions of years of evolution, natural selection has fine-tuned the relationship between organic structure and function. By turning to the treasure box provided by nature, engineers can extract ideas for solving technical problems. Although *biomimetics*, or synonymously, *bionics*, is a relatively new field of research, the concept is ancient. The Chinese tried to develop synthetic silk over 3,000 years ago, and Joseph Paxton is believed to have based the structure of the Crystal Palace on a lily pad. Perhaps the most famous example is Velcro, which George Mestral invented after studying the burrs that stuck to his dog's coat. He observed that the burrs consisted of hundreds of tiny hooks that locked in the dog's soft fur. With the help of French weaving experts, he mimicked the burrs and fur to produce Velcro, which was patented in 1952.

As expressed by Julian Vincent, together with Werner Nachtigall, one of the pioneers in the field, biomimetics is the abstraction of good design from nature. Important in this definition is the word "abstraction." As mentioned before, all biological structures emerge evolutionarily from the "bottom up," while most technical products result from "top down" procedures. In addition, since natural structures often have complex dimensions that are incompatible with their application as technical products, biomimetics is not simply a process of copying nature. Biomimetics is characterized by independent and creative development inspired by natural models. Current biomimetic research is helping to develop a great many new products, such as fabrics that respond to a wearer's needs. Its true potential is only now being realized. Biomimetics, however, is not a one-way street from biologist to engineer. Analysis of the mechanics of living organisms from an engineer's point of view can also help biologists to gain a better understanding of their biological design.

A paradigmatic example of biomimetics is the application of the so-called lotus effect (Barthlott and Neinhuis 1997). The leaves of the lotus *Nelumbo nucifera*, a symbol of purity in Buddhism, possess self-cleaning properties. The hydrophobic leaf surface is characterized by protrusions of the epidermis, called papillae, measuring 10–20 µm in height at distances of 10–20 µm apart, which are covered with epicuticular waxes. Due to its cohesion, water cannot penetrate between the papillae, but instead forms nearly spherical droplets on the outer surface of the leaf (fig. 9.11). By the same token, fungal spores and most dust particles are too large to settle between the papillae. The adhesion of these particles to water is greater than to the epicuticular waxes, so they adhere to water droplets, which eventually drain off the leaves and thus clean them.

At first sight, it seemed counterintuitive that surfaces with roughness on the micrometer scale should be easier to clean than smooth surfaces, but once the physicochemical basis of the lotus effect was understood, a number of applications were found, ranging from facade paints and roof tiles to fabrics with diverse surface coatings. Leaves can replenish their epicuticular waxes, but waxes wear away too quickly for most technical applications. Therefore, special fluorochemical or silicone treatments on structured surfaces containing microscale particulates are being used.

The first and most successful product with superhydrophobic self-cleaning properties was the facade paint Lotusan, which was first put on

FIGURE 9.11. Droplets of beaded water on the surface of a *Canna* leaf, illustrating the "lotus effect."

the market in 1999. Lotusan has been applied on more than half a million buildings worldwide so far. Further applications have been marketed, such as the self-cleaning glass installed in the sensors of traffic control units on German autobahns. A spray for generating self-cleaning films on various substrata has also been developed. Lotus-effect superhydrophobic coatings applied to microwave antennas can significantly reduce rain fade and the buildup of ice and snow.

The pine cone has provided inspiration for the development of a smart fabric, capable of reacting to the activity of the wearer. Pine cones open and close in response to changes in humidity, opening as they dry to release their seeds. This movement is due to the large ovuliferous scales, which comprise two easily distinguishable tissues, one on the outer surface and one on the inner surface of the cone. Just as a bimetallic strip bends with heat, the ovuliferous scales bend with increased humidity because the outer tissue expands more readily than the inner tissue, which results in the cone closing (Dawson et al. 1997). A fabric with a similar ability to react to humidity could be used in the desert, as it would provide cooling during the day and warmth at night when temperatures drop, alleviating the need to carry extra clothing. Taking inspiration from the pine cone, a smart material has been made of two bonded fabric layers interspersed with tiny flaps that open with increased humidity. Thus, as the wearer sweats, the flaps open, increasing airflow and cooling the wearer. The material is currently being developed for fashion and sportswear, and commercial prototypes should be available soon.

The rough horsetail (*Equisetum hyemale*) gives an example of an ultralight structure with remarkable mechanical stability (Speck et al. 1998). A cross section of the hollow structure of the aerial stem shows an outer ring of nonsclerenchymatous tissue (the hypodermal sterome) and an inner ring of double-layered endodermis, both of which function as strengthening tissues (Spatz and Emanns 2004) (see fig. 5.9). The two rings are connected by a number of structural elements having the appearance of I-beams. Between the two rings is a layer of parenchyma and chlorenchyma enclosing remarkably large vallecular canals, which greatly reduce the weight of the aerial stem without compromising its mechanical stability. In this respect, *Equisetum* stems are the ultimate cellular solids!

Starting from these observations, researchers developed an ultralight structure with excellent mechanical properties (Milwich et al. 2006). A helical arrangement of fibers, inspired by the helical arrangement of bundles of cellulose fibrils in wood fibers and of fiber bundles in the stems of the giant reed *Arundo donax*, was also incorporated into the final product to provide stability and vibration damping. The resulting "technical plant stem" is an innovative fibrous compound material that is stiff, strong, and light, with high energy damping and benign fracturing behavior—features that improve on currently available materials. This extremely versatile material is currently being developed for use in the automotive and aircraft industries.

Inspired by the structure of bones and his observations of secondary growth in trees, Mattheck (1992) developed a method to optimize load-bearing structures. Using finite element methods (see section 10.5), parts of an initially proposed design that experience little or no stress upon appropriate loading are eliminated. In a reiterative process, a new design is found that combines load-bearing capacity with minimal weight. The procedure is widely used in the design of the chassis of cars and other technical structures in which weight is a concern.

The nonautonomous elastic deformation in the *Strelicia regina* flower provides a model for deployable systems such as facade-shading laminae (Poppinga et al. 2010). The 2 to 3 cm thick peel of pummelos (*Citrus maxima*) serves as an antetype for shock-absorbing metal foams (Fischer et al. 2010).

A particularly interesting system of self-cleaning sewage pipes was developed at the FITR Research Institute at Weimar, Germany. Although the water-conducting vessels of terrestrial plants operate at entirely different Reynolds numbers, their structure provided the inspiration to arm the bottom of sewage pipes with staggered structural bodies, which serve as vortex generators and thus prevent sedimentation of particles and ultimately clogging of the pipes. Further developments include flexible tubes with burlings, which can be introduced into existing sewage pipes to provide an inner lining serving the same purpose.

As a last example of extracting ideas from nature for solving technical problems, we turn to self-healing processes, which are common in nature and essential for the survival of all higher organisms, but still uncommon in technical products. Repair processes in the Dutchman's pipe (*Aristolochia macrophylla*) were observed nearly a century ago by Haberlandt (1924). Internal growth in this plant leads to rupturing of the subepidermal sclerenchymatous ring. When this occurs, turgid parenchyma cells intrude into the fissure and seal it. When the fissure extends deeply into the sclerenchymatous ring, the intruding cells start to divide. In some fissures, these cells remain parenchymatous, while in others, the walls of the most peripheral sealing cells increase in thickness and start to lignify. In this way, the mechanical function of the sclerenchymatous ring can be partially restored.

Self-repair is a desideratum for all hydrostatically inflated structures, as they are vulnerable to tearing and puncturing, which can cause a drop in their internal pressure and a subsequent reduction in, or even loss of, their load-carrying capacity. This is also true, although to a lesser extent, for Tensairity structures (i.e., bridges, pontoons, and similar structures composed of lightweight combinations of inflated struts, cables, and other pneumatic elements; see Luchsinger et al. 2004). An obvious solution to this problem would be a cellular structure; control of the individual elements could be quite complex, however. Alternatively, provisions for automatic sealing of defects are being developed. For example, Speck et al. (2006) created a pressurized cellular material that mimics the self-repair mechanism in plants. This biomimetic material consists of an inner coating beneath an outer membrane. When the membrane is punctured, the repair layer seals the hole in a manner similar to the parenchyma cells in *Aristolochia*. In their preliminary experiments, pressurized polyurethane foam was used for the repair layer. In a laboratory setup, air leakage through a hole of 5 mm diameter was reduced by three orders of magnitude.

These few examples of problem solving using plants as sources of inspiration illustrate the potential of the biomimetic approach. As plants have evolved to live in very different environments, to make optimal use of limited resources, to use intricate methods of fertilization, and to develop defense mechanisms against herbivores and fungal attack, they will be used by scientists and engineers as concept generators in future research, possibly in an increasingly complex way.

Literature Cited

Angeles, G., J. Berrio-Sierra, J.-P. Joseleau, P. Lorimer, A. Lefebvre, and K. Ruel. 2006. Preparative laser capture microdissection and single-pot cell wall material preparation: A novel method for tissue-specific analysis. *Planta* 224:228–32.

Bargel, H., H.-C. Spatz, T. Speck, and C. Neinhuis. 2004. Two dimensional tension tests in plant biomechanics—sweet cherry fruit skin as a model system. *Plant Biol.* 6:432–39.

Barthlott, W., and C. Neinhuis. 1997. Purity of the sacred lotus or escape from contamination in biological surfaces. *Planta* 202:1–7.

Bausch, A. R., W. Möller, and E. Sackmann. 1999. Measurement of local viscoelasticity and forces in living cells by magnetic tweezers. *Biophys. J.* 76:573–79.

Bereiter-Hahn, J., C. H. Fox, and B. Thorell. 1979. Quantitative reflection contrast microscopy of living cells. *J. Cell Biol.* 82:767–79.

Brännström, M., J. Oja, and A. Grönlund. 2007. Predicting board strength by X-ray scanning of logs. *Scand. J. For. Res.* 22:60–70.

Burgert, I. 2006. Exploring the micromechanical design of plant cell walls. *Amer. J. Bot.* 93:1391–1401.

Burgert, I., K. Frühmann, J. Keckes, P. Fratzl, and S. E. Stanzl-Tschegg. 2003. Microtensile testing of wood fibres combined with video extensometry for efficient strain detection. *Holzforschung* 57:661–64.

Dawson, C., J. F. V. Vincent, and A.-M. Rocca. 1997. How pine cones open. *Nature* 390:668.

Donald, A. M. 2003. The use of environmental scanning electron microscopy for imaging wet and insulating materials. *Nature Mater.* 2:511–16.

Fischer, S. F., M. Thielen, R. R. Loprang, R. Seidel, C. Fleck, T. Speck, and A. Bührig-Polaczek. 2010. Pummelos as concept generators for biomimetically inspired low weight structures with excellent damping properties. *Adv. Eng. Mater./Adv. Biomater.* 12: B658–B663.

Förster, T. 1949. Experimental and theoretical investigation of the intermolecular transfer of electronic excitation energy transfer. *Z. Naturforschung A* 4:321–27.

Gadella, T. W. J., G. N. M. van der Krogt, and T. Bisseling. 1999. GFP-based FRET microscopy in living plant cells. *Trends Plant Sci.* 4:287–91.

Geitmann, A. 2006. Experimental approaches used to quantify physical parameters at cellular and subcellular levels. *Amer. J. Bot.* 93:1380–90.

Gierlinger, N., M. Schwanninger, A. Reinecke, and I. Burgert. 2006. Molecular changes during tensile deformation of single wood fibres followed by Raman spectroscopy. *Biomacromolecules* 7:2077–81.

Gilbert, E. A., and E. T. Smiley. 2004. PICUS sonic tomography for the quantification of decay in white oak (*Quercus alba*) and hickory (*Carya* spp.). *J. Arboricult.* 30:277–81.

Gindl, W., H. S. Gupta, T. Schoberl, H. C. Lichtenegger, and P. Fratzl. 2004. Mechanical properties of spruce wood cell walls by nanoindentation. *Appl. Phys. A Mater.* 79:2069–73.

Haberlandt, G. 1924. *Physiologische Pflanzenanatomie.* 6th ed. Leipzig: Engelmann.

Hedrich, R., and J. I. Schröder. 1989. The physiology of ion channels and electrogenic pumps in higher plants. *Annu. Rev. Plant Physiol.* 40:539–69.

Holbrook, N. M., M. J. Burns, and C. B. Field. 1995. Negative xylem pressures in plants: A test of the balancing pressure technique. *Science* 270:1193–94.

Hüsken, D., E. Steudle, and U. Zimmermann. 1978. Pressure probe technique for measuring water relations of cells in higher plants. *Plant Physiol.* 61:158–63.

Keckes, J., I. Burgert, K. Frühmann, M. Müller, K. Kölln, M. Hamilton, M. Burg-
hammer, S. V. Roth, S. E. Stanzl-Tschegg, and P. Fratzl. 2003. Cell-wall recovery
after irreversible deformation of wood. *Nature Mater.* 2:810–14.

Köhler, L., and H.-C. Spatz. 2002. Micromechanics of plant tissues beyond the linear elastic range. *Planta* 215:33–40.

Köhler, L., T. Speck, and H.-C. Spatz. 2000. Micromechanics and anatomical
changes during early ontogeny of two lianescent *Aristolochia* species. *Planta*
210:691–700.

Kölln, K., I. Grotkopp, M. Burghammer, S. V. Roth, S. S. Funari, M. Dommach,
and M. Müller. 2005. Mechanical properties of cellulose fibres and wood: Ori-
entational aspects in situ investigated with synchrotron radiation. *J. Synchrotron Radiat.* 12:739–44.

Leitz, G., E. Schnepf, and K. Greulich. 1995. Micromanipulation of statoliths in
gravity-sensing *Chara* rhizoids by optical tweezers. *Planta* 197:278–88.

Lemons, R. A., and C. F. Quate. 1974. Acoustic microscope—scanning version.
Appl. Phys. Lett. 24:163–65.

Lichtenegger, H., A. Reiterer, S. E. Stanzl-Tschegg, and P. Fratzl. 1999. Variation
of cellulose microfibril angles in softwoods and hardwoods—a possible strategy
of mechanical optimization. *J. Struct. Biol.* 128:257–69.

Lintilhac, P. M., C. Wei, J. J. Tanguay, and J. O. Outwater. 2000. Ball tonometry: A
rapid nondestructive method for measuring cell turgor pressure in thin walled
plant cells. *J. Plant Growth Regul.* 19:90–97.

Luchsinger, R. H., M. Pedretti, and A. Reinhard. 2004. Pressure induced stability:
From pneumatic structures to Tensairity. *J. Bionic Eng.* 1:141–48.

Mattheck, C. 1992. *Design in der Natur. Der Baum als Lehrmeister.* Freiburg: Rombach Verlag.

Mattheck, C., and H. Breloer. 1994. *Handbuch der Schadenskunde von Bäumen.*
2nd ed. Freiburg: Rombach Verlag.

Milwich, M., T. Speck, O. Speck, T. Stegmaier, and H. Planck. 2006. Biomimetics
and technical textiles: Solving engineering problems with the help of nature's
wisdom. *Amer. J. Bot.* 93:1455–65.

Morris, V. J., A. A. Gunning, and A. R. Kirby. 1999. *Atomic force microscopy for
biologists.* Singapore: World Publishing Company.

Neher, E., and B. Sakmann. 1976. Single-channel currents recorded from membrane of denervated frog muscle fibers. *Nature* 260:779–802.

Nicolotti, G., L. V. Socco, R. Martinis, and L. Sambuelli. 2003. Application and
comparison of tree tomographic techniques for detection of decay in trees.
J. Arboricult. 29:66–78.

Niklas, K. J., and F. C. Moon. 1988. Flexural stiffness and modulus of elasticity of flower stalks of *Allium sativum* as measured by multiple resonance frequency spectra. *Amer. J. Bot.* 75:1517–25.

Niklas, K. J., and D. J. Paolillo Jr. 1997. The role of the epidermis as a stiffening agent in *Tulipa* (Liliaceae) stems. *Amer. J. Bot.* 84:735–44.

Oliver, W. C., and G. M. Pharr. 1992. A new improved technique for determining hardness and elastic modulus using load sensing indentation experiments. *J. Mater. Res.* 7:1564–82.

Peltola, H. M. 2006. Mechanical stability of trees under static loads. *Amer. J. Bot.* 93:1501–11.

Poppinga, S., J. Lienhard, T. Masselter, S. Schleicher, J. Knippers, and T. Speck. 2010. Biomimetic deployable systems in architecture. In *IFMBE Proceedings* 31, edited by C. T. Linn and J. C. H. Goh, 40–43. Heidelberg: Springer.

Roloff, A. 1999. Tree vigor and branching pattern. *J. For. Sci.* 45:206–16.

Roy, R., S. Hohng, and T. Ha. 2008. A practical guide to single-molecule FRET. *Nature Methods* 5:507–16.

Rust, M. J., M. Bates, and X. Zhuang. 2006. Sub-diffraction-limit imaging by stochastic optical reconstruction microscopy (STORM). *Nature Methods* 3:793–95.

Rust, S., and U. Weihs. 2007. Geräte und Verfahren zur eingehenden Baumuntersuchung. *Jahrbuch der Baumpflege* 2007:215–29.

Scholander, P. F., E. D. Bradstreet, E. A. Hemmingsen, and H. T. Hammel. 1965. Sap pressure in vascular plants. *Science* 148:339–46.

Schwarze, F. W. M. R. 2008. *Diagnosis and prognosis of the development of wood decay in urban trees.* Rowville, VIC, Australia: ENSPEC.

Sinn, G. 2003. *Baumstatik, Stand- und Bruchsicherheit von Bäumen an Straßen, in Parks und der freien Landschaft.* Braunschweig: Thalacker Medien.

Spatz, H.-C., and A. Emanns. 2004. The mechanical role of the endodermis in *Equisetum* plant stems. *Amer. J. Bot.* 91:1936–38.

Spatz, H.-C., L. Köhler, and T. Speck. 1998. Biomechanics and functional anatomy of hollow stemmed sphenopsids: I. *Equisetum giganteum* (Equisetaceae). *Amer. J. Bot.* 85:305–14.

Speck, T., R. Luchsinger, S. Busch, M. Rüggeberg, and O. Speck. 2006. Self-healing processes in nature and engineering: Self-repairing biomimetic membranes for pneumatic structures. In *Design and Nature* III, edited by C. A. Brebbia, 105–14. Southampton: WIT Press.

Speck, T., O. Speck, A. Emanns, and H.-C. Spatz. 1998. Biomechanics and functional anatomy of hollow stemmed sphenopsids: III. *Equisetum hyemale. Bot. Acta* 111:366–76.

Stryer, L., and R. P. Haugland. 1967. Energy transfer: A spectroscopic ruler. *Proc. Nat. 'Acad. Sci. USA* 58:719–26.

Thygesen, L. G., M. Eder, and I. Burgert. 2007. Dislocations in single hemp fibres—investigations into the relationship of structural distortions and tensile properties at the cell wall level. *J. Mater. Sci.* 42:558–64.

Tomos, A. D., and R. A. Leigh. 1999. The pressure probe: A versatile tool in plant cell physiology. *Annu. Rev. Plant Physiol. Plant Mol. Biol.* 50:447 72.

Tyree, M. T., and M. H. Zimmermann. 2002. *Xylem structure and the ascent of sap.* Berlin: Springer Verlag.

Vincent, J. F. V. 1990. *Structural biomaterials.* Princeton, NJ: Princeton University Press.

Volkov, A. G., ed. 2006. *Plant electrophysiology.* Berlin: Springer Verlag.

Wang, C., J. Pritchard, and C. R. Thomas. 2006. Investigation of the mechanics of single tomato fruit cells. *J. Text. Stud.* 37:597–606.

Wei, C., P. M. Lintilhac, and J. J. Tanguay. 2001. An insight into cell elasticity and load bearing ability: Measurement and theory. *Plant Physiol.* 126:1129–38.

Weihs, U., D. Bieker, and S. Rust. 2007. Zerstörungsfreie Baumdiagnose mittels "Elektrischer Widerstandstomographie." *Jahrbuch der Baumpflege* 2007: 230–41.

Wilhelm, C. 2003. Rotational magnetic endosome microrheology: Viscoelastic architecture inside living cells. *Phys. Rev.* 67:1–12.

Yamada, S., D. Wirtz, and S. C. Kuo. 2000. Mechanics of living cells measured by laser tracking microrheology. *Biophys. J.* 78:1736–47.

Theoretical Tools

Theory relates unmeasurable quantities to measurable quantities.
—Attributed to Carl Wagner

Throughout this book, we have made use of theoretical concepts such as transport equations, thermodynamic relations, the theory of elasticity, and the theory of fluid flow. In this chapter, we present some other theoretical tools that have been used, both traditionally and more recently, to study plant physics as well as a discussion of the potential and the limitations of modeling. This review cannot be comprehensive, especially because, with the availability of increasingly powerful computers and computer programs, new tools are being developed and used every year. Our objective, therefore, is limited: it is to provide examples of theoretical approaches used to understand plant physics.

10.1 Modeling

It is almost tautological to say that most biological structures are complex. Even the smallest part of a unicellular plant is chemically and structurally extremely heterogeneous. For example, although the cell walls of land plants may appear uniform when viewed with a light microscope, when fractured and viewed at high magnification, a cell wall is seen to be a composite material with a polylaminate ultrastructure consisting of crystalline cellulose microfibrils embedded in a chemically complex matrix composed of a broad spectrum of carbohydrates, lipids, and proteins. In addition, some cell walls contain lignin and, in the case of some algae, silica. In light of this complexity, it is essential to reduce biological systems to more manageable subjects of study by means of modeling. Much of biophysical analysis rests on modeling organisms according to physical laws and processes while either neglecting or simplifying attending biological phenomena. Thus, a typical biophysical model is a highly stylized and simplified conceptual representation of a limited aspect of biological reality. It is proposed as a theoretical construct that permits us to test certain of our assumptions about how a recognizably more complex system works. In one sense, a model tests our perception of a phenomenon more than it tests the phenomenon itself.

There are two extreme mind-sets about modeling: one is skeptical, while the other is naïve. One distrusts any attempt to reduce biological complexity to a more manageable form, while the other embraces reductionism, possibly to an extreme. Regardless of where any person may fall along this spectrum of opinion, it is important to recognize that modeling is an unavoidable and fundamental aspect of any experimental design because it is a requisite for examining the interrelationships among a large number of variables, which would otherwise be an unmanageable enterprise. Additionally, modeling is an intrinsic part of how scientists analyze their experimental results. All forms of statistical inference are based on mathematical models of one form or another; for example, experimental variables are frequently regressed against one another and correlated by means of a regression equation that presumes a specified frequency distribution as a null hypothesis. Thus, even the most skeptical cannot avoid using models of one sort or another.

In one respect, the only good model is one that fails the test of reality, because a model that gives the right answers can do so for the wrong reasons, whereas a model that yields predictions that conflict with reality immediately requires us to evaluate our assumptions about how reality works. Good models allow us to reject our preconceptions; poor models delude us into believing we have identified causalities correctly.

How we model any system depends on our perspective, and each model reflects to some extent a certain degree of tunnel vision. Consider a transverse section through a typical dicot leaf. Our attention might be first drawn to the diversity of cell shapes, sizes, and geometries. With the appropriate qualifications, this perspective could lead us to propose a taxonomic model that distinguishes one dicot species from another. Yet another perspective might lead us to consider light interception and thus draw our attention to the spatial distribution of chloroplasts. If so, we might model the leaf as a multilaminated "photovoltaic device" with a vertical gradient of light-harvesting particulates whose concentration per unit volume declines from the top to the bottom of the leaf. Or we might notice large air spaces within the leaf and develop a model addressing the various resistances of carbon dioxide, oxygen, and water vapor transport through the various layers of the leaf's architecture. Alternatively, the presence of a cuticle and guard cells flanking stomata might draw our attention to the evolution of the earliest land plants and their need to conserve water, yet permit the diffusion of CO_2 and O_2 into and out of photosynthetic tissues. The locations of mechanical tissues or the distributions of plasmodesmata and xylem might lead us to construct other models, some dealing with structural stability and others dealing with the hydraulics of the leaf. Finally, we might propose an integrated model, one that views all of the functions performed by a leaf, ranging from the requirements for light interception, gas diffusion, and mechanical support to the evolution of water conservation and the hydraulics of mass fluid transport.

This leaf example illustrates that the same biological structure can be viewed in many ways, each with the potential to add to our understanding as well as to deflect our ability to see that the whole is greater than the sum of its parts. It is tempting to argue that any initial bias in our perspective will eventually lead us to a holistic view, but even the integrated model is the product of a particular perspective (i.e., the physiology of photosynthesis), which is incomplete because it neglects other aspects of plant metabolism as well as the processes that attend the growth and development of a typical foliage leaf.

10.2 Morphology: The problematic nature of structure-function relationships

Before we continue our discussion of some of the theoretical approaches to the investigation of plant form-function relationships, it seems prudent to entertain an important philosophical question: namely, how do we know the function of an organic structure? This is not a trivial question, nor is it always easy to answer. Indeed, we obliquely addressed this topic in section 1.7. Evolution has endowed us with the ability to recognize often subtle differences among organic structures. Indeed, this is one reason why the perception and description of structures stood long in the foreground of traditional biology and why morphology was the sole basis for systematics and taxonomy before the advent of molecular techniques of analysis. However, evolution has also endowed us with the ability for abstraction and what some may see as a predilection to postulate a priori structural-functional relationships, which in turn steers our experimental approaches to understanding biological phenomena.

Consider the differences in how engineers and biologists approach their respective occupations. An engineer is typically made aware of the functional obligations that a structure or device must fulfill as well as many of the details about the environment in which it will function. With this a priori understanding of functional obligations and the workplace, the task of designing begins. Functional design constraints, factors of safety, and costs can be reconciled and calculated and the device can be constructed. At this point, the engineer can stipulate the building materials and the dimensions required for the structure's fabrication. In many cases, a prototype will be built and tested in a variety of ways to ensure that the structure functions appropriately, at which point the full-scale production of the device can proceed.

In contrast, consider the situation of a biologist confronted with an organic structure that has not been analyzed before. The structure can be examined in a variety of ways, at the subcellular, cellular, or (if present) higher levels of organization. If the structure is living and intact, it can be observed to infer how it relates to the organism's general body plan and the environment in which the organism lives. If it is a fossil and fragmented, the general body plan must be inferred, as must the environment in which the structure functioned. In either case, the function of the structure must be deduced before a form-function relationship can be examined experimentally. The important concern here is that the question of whether this relationship is correct can be answered with varying degrees of success. Sometimes the function of an organic structure or device is comparatively easy to infer. Few (if any) would argue against the proposition that the function of a heart is to pump blood, or that the function of a lung is to pump air. Tracheids and vessel members obviously conduct water. But it is not always the case that the function of an organic structure is obvious.

Consider, for example, trichomes (i.e., hairlike epidermal extensions that can be found on leaves, stems, and roots). Depending on the species, these structures can function to deter herbivory, temporarily confine or direct a pollinator to a limited space, insulate against extremes in temperature, serve as triggers for leaf closure, absorb or repel water, or reflect light. Indeed, for some species trichomes serve as a climbing device (Bauer et al. 2011). So what is their function, and how can we be sure that a particular kind of trichome isn't performing more than one? Unless we are able to answer such questions, how can we be sure that the experimental techniques we use to examine a structure are appropriate? If a structure serves to resist tensile forces, will measuring its elastic modulus in compression lead to a better understanding of its biology? Unless a leaf functions as a photosynthetic structure (and many kinds of leaves do not), what is the purpose of measuring its reflectance or pigment concentration?

Unfortunately, there is no canonical protocol that can assure us that a form-function relationship has been assessed correctly. Each case has to be evaluated on its own merits, and each must be reevaluated as we learn more about a particular organic structure. All that can be said with any reasonable certainty is that few structures in biology perform a single function—most are multitaskers—and that no form-function relationship can be properly evaluated without considering carefully the biotic and abiotic environment in which it occurs.

10.3 Theoretical morphology, optimization, and adaptation

Theoretical morphology is a discipline concerned with the analysis of evolutionary constraints. Its objective is to mathematically simulate the full range of morphologies that organisms can attain and to discover why certain morphologies exist while others, although geometrically or structurally possible, do not. This approach begins with the generation of a morphospace (i.e., an N-dimensional domain of all mathematically conceivable morphologies for the particular group of organisms or structures under investigation). This step is followed by analyses that identify (1) functional versus nonfunctional morphologies within the morphospace, (2) functional morphologies that are developmentally attainable, and (3) developmentally attainable morphologies that have been achieved at some time during the evolutionary history of the organisms or structures being modeled (fig. 10.1).

This approach has a long tradition in zoology, particularly in paleontology (see McGhee 1999). It has been underutilized in the study of plants, which is surprising, given the comparative ease with which plants can be simulated and functionally evaluated with the aid of computers. For example, it is easy to simulate a simple stem bearing leaves differing in their shape and in their divergence angle θ (i.e., the "phyllotactic" angle between two successively formed leaves when the stem is viewed along its

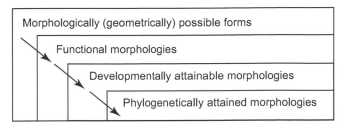

FIGURE 10.1. Concept map for a typical analysis in theoretical morphology. A morphospace consisting of all morphologically possible forms is generated, and the functional morphologies within the space are identified. The boundary separating functional from nonfunctional morphologies defines a geometric or structural design constraint. Within the more limited set of functional possibilities, developmentally attainable forms are identified, and within this still smaller set of possibilities, those that have been expressed phylogenetically are identified. An additional step in the analysis (not shown) is quantifying the relationship between the ability of a morphology to perform its functional obligations and the morphology's reproductive fitness.

length) to assess the effects of divergence angle and leaf shape on the ability to intercept sunlight (Niklas 1988). The resulting "phyllotactic" morphospace reveals an optimal $\theta = 137.5°$, but it also shows that altering leaf shapes, internodal distances, or the leaf divergence angle can "flatten" the fitness landscape, making the phyllotactic pattern largely unimportant in terms of light interception (fig. 10.2).

Optimization models can also be used to determine whether evolutionary patterns resulting from natural selection conform to theoretical predictions. These models have three components:

1. The generation of a morphospace
2. The quantification of how well each phenotype performs specified functions
3. A simulated "adaptive walk" in which the morphospace is searched for phenotypes that perform those functions better

This approach is illustrated in figure 10.3. In this example, a morphospace was generated for all geometrically conceivable ancient vascular plants, and that morphospace was searched for successively fitter phenotypes. The search criterion was the optimization of both mechanical stability and the ability to intercept sunlight. Starting with the ancient vascular land plant fossil *Cooksonia*, the adaptive walk progressively located the series of morphologies shown in the upper panel until it reached a phenotype that was surrounded by others that were less successful at performing those two functions simultaneously. The series of morphologies identified by this adaptive walk conforms in many ways to the predictions of the telome theory of Walter Zimmermann for the evolution of the megaphylls (Zimmermann 1930).

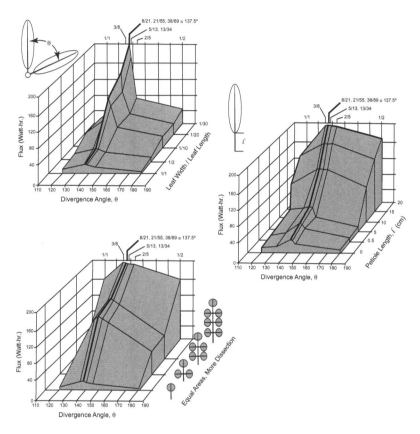

FIGURE 10.2. Three portions of a phyllotactic morphospace, showing the relationships between leaf divergence angle (θ), leaf shape—defined here in terms of (A) leaf width divided by leaf length, (B) petiole length, and (C) leaf dissection—and the ability of a single vertical stem bearing different-shaped leaves with different divergence angles to intercept light (quantified in terms of total flux in watt-hours). In each case, the divergence angle is plotted in degrees (foreground axis) and as phyllotactic fractions (part of the Fibonacci series) (rear axis). In each case, the single stem bears the same total leaf area and the light interception "fitness landscape" manifests an "adaptive ridge" (shown as a thick line). However, changes in leaf shape can "smooth" the adaptive landscape in ways that make differences in θ unimportant. (Data taken from Niklas 1988.)

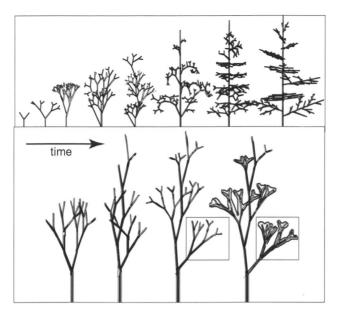

FIGURE 10.3. An example of an "adaptive walk" through a morphospace for early vascular leafless plants. The criterion for the search algorithm was the identification of phenotypes that optimized the simultaneous performance of light interception and mechanical stability. The successively located phenotypes in this adaptive walk, shown in the upper panel, conform in many ways with the predictions of the telome theory of W. Zimmermann (1930), illustrated in the lower panel. (Adapted from Niklas and Kerchner 1984.)

10.4 Size, proportion, and allometry

The relationships among size, geometry, and shape are never simple, even when dealing with inorganic objects, but they assume remarkably complex proportions when we treat organic structures because, in theory, size, geometry, and shape can change independently as an organism or part of an organism grows and develops. One of the important tools used to cope with this complexity is *allometry*, which is traditionally defined as the study of the change in proportion of the various parts and functions of an organism as a consequence of growth. This definition is incomplete, however, because allometric analyses have also been used to study the changes in the proportions of phyletically related organisms as a consequence of evolutionary modification. Allometry has also been used to determine whether general rules governing organic proportions hold true across all or most forms of life.

To begin this topic, consider the relationship between the surface area A and the volume V of any solid object. How do these two variables change as the object increases in size? We know that, for any geometry, surface area is typically measured by multiplying two linear reference dimensions, and we know that volume is measured by multiplying three linear reference dimensions. If the principle of geometric similarity is applicable, we obtain the proportional relationships $A \propto \ell^2$ and $V \propto \ell^3$, where ℓ denotes the reference dimension, from which it follows mathematically that $A \propto V^{2/3}$. In words, the surface area of an object scales as the $2/3$ power of the object's volume.

This "$2/3$ power rule," first articulated by Galileo Galilei in the seventeenth century, was based on his study of solid mechanics. Nevertheless, it is not a "rule" in the strict sense of the word because it can be violated by organisms that are capable of changing either geometry or shape as they grow in size. It is also violated if we compare organisms that differ in size, shape, and geometry. For example, various species of unicellular algae are spheroidal, cylindrical, or disklike, and within each of these geometric categories, shape can change as a function of size. As a consequence, the allometry of A with respect to V is approximated by the proportional relationship $A \propto V^{3/4}$, rather than $A \propto V^{2/3}$. The difference between $2/3$ and $3/4$ may appear trivial numerically. In each case, it is clear that increases in A fail to keep pace with increases in V for the entire ensemble of algae. However, the $3/4$ exponent indicates that A is increasing more rapidly with respect to increases in V as compared with the $2/3$ exponent. The slight difference between the exponent predicted by simple geometry and the exponent observed in nature (i.e., $2/3$ and $3/4$, respectively) may confer a significant advantage on algae; for example, the half-time for passive diffusion scales as V/A (see eq. [2.11]). Plant allometry is covered in detail elsewhere (Niklas 1994).

Allometric relationships require a normalization constant to compute the absolute values of variables such as Y_o for known values of variables such as Y_a. With this in mind, the general formula for an allometric relationship is

(10.1) $$Y_o = \beta Y_a^a,$$

where Y_o and Y_a are variables plotted on the ordinate and abscissa, respectively; β denotes the normalization ("allometric") constant; and α is the scaling exponent (Huxley 1932; Gould 1966). When $\alpha = 1$, equation (10.1) describes a relationship that plots as a straight line on both linear and logarithmic axes (i.e., an isometric relationship). When $\alpha \neq 1$, equation (10.1) describes a linear function only on logarithmic axes. For allometric relationships, equation (10.1) is typically log-transformed to determine the numerical values of β and α:

(10.2) $$Y_o = \log \beta + \alpha \log Y_a.$$

This formula shows that α is the slope of the log-log linear relationship and that $\log \beta$ is the "elevation" (Y-intercept) of the log-log linear curve.

Although the linearization of data by means of logarithmic transformation has become a conventional practice in allometric studies, the parameters estimated in this way do not invariably provide the best fit to a regression model compared with minimizing the squared residuals for the actual function by means of nonlinear regression protocols. Analyses of residuals are required to determine whether log-log linear or log-log nonlinear functions optimize the goodness of fit.

The objective of most allometric analyses is to determine the numerical values of $\log \beta$ and α. When a predictive relationship is sought, simple ordinary least squares regression (OLS) analysis is used. When the objective is to establish a functional relationship between Y_o and Y_a, as is generally the case, OLS regression analysis is ill equipped for the purpose, in part because it is based on the assumption that Y_a is measured with little or no error. A number of regression protocols have been used to determine functional allometric relationships (Sokal and Rohlf 1980), and considerable controversy revolves around which of those methods gives the best results (Smith 1980; Harvey 1982; Prothero 1986; Rayner 1985; Warton et. al. 2006). This issue is not trivial because different regression methods can produce different numerical values for α and $\log \beta$, even for the same data set.

Space precludes a detailed discussion of the merits and deficits of these different regression methods. However, reduced major axis (RMA, also called standardized major axis, or SMA) regression analysis has emerged as a "standard" allometric technique over the past few years. Statistical software is available to perform RMA regression analyses, but access to this software is not critical because OLS regression summary statistics provide all the data required to compute α and $\log \beta$ and their corresponding 95% confidence intervals (box 10.1).

The following analysis of an extensive data set published in a synoptic

BOX 10.1 **Comparison of regression parameters**

The relationships between the regression parameters of reduced major axis (RMA) analysis and ordinary least squares (OLS) analysis are given by the formulas

(10.1.1)
$$\alpha_{RMA} = \frac{\alpha_{OLS}}{r}$$

and

(10.1.2)
$$\beta_{RMA} = \overline{\log Y_o} - \alpha_{RMA} \overline{\log Y_a},$$

where α_{RMA} is the RMA scaling exponent, α_{OLS} is the OLS regression slope, r is the OLS correlation coefficient, $\log \beta_{RMA}$ is the allometric constant of the log-transformed relationship, and $\overline{\log Y}$ denotes the mean value of $\log Y$ (Sokal and Rohlf 1980). The corresponding 95% confidence intervals of these two regression parameters are computed using the formulas

(10.1.3)
$$\alpha_{RMA} \pm t_{N-2} \left(\frac{MSE}{SS_a} \right)^{1/2}$$

and

(10.1.4)
$$\beta_{RMA} \pm t_{N-2} \left[MSE \left(\frac{1}{N} + \frac{\overline{\log Y_a}^2}{SS_a} \right) \right]^{1/2},$$

where MSE is the OLS mean square error, SS_a is the OLS sums of squares for $\log Y_a$, N is the sample size, and $t_{N-2} = 1.96$ when $N \geq 200$.

---

worldwide survey of different forests (Cannell 1982) illustrates the general format of an allometric analysis. The survey compiled measurements for the basal stem diameter of trees, D (in m), and for total dry leaf mass per tree, M_L (in kg). OLS regression of $\log_{10}$-transformed values of M_L against $\log_{10}$-transformed values of D yields $r^2 = 0.789$, $\alpha_{OLS} = 1.77$, and $\beta_{OLS} = 2.09$, with a regression model error of 0.087 ($n = 587$). Inserting these values and those for the means of $\log M_L$ and $\log D$ into equations (10.1.1)–(10.1.4) (see box 10.1) yields $\alpha_{RMA} = 2.0$ (with 95% confidence intervals of 1.95 and 2.03) and $\log \beta_{RMA} = 2.3$ (with confidence intervals of 2.2 and 2.4). Taking the antilog of 2.3, we see that M_L (in kg) = 200 D^2 (in m²), which indicates that, on average, the total dry leaf mass of a tree scales as the square of the basal stem diameter (i.e., $M_L \propto D^2$).

This scaling relationship opens up a number of plausible avenues of interpretation that are not mutually exclusive, given that the stems of trees function both hydraulically and mechanically. One possible interpretation of this scaling relationship is that the amount of water lost by leaves during transpiration is proportional to total leaf area, and the amount of water supplied to leaves is proportional to the cross-sectional area of the basal-most stem (the trunk). Determining whether the $M_L \propto D^2$ scaling relationship is mainly due to hydraulic or mechanical limitations requires much more detailed analyses, although the available data suggest that mechanical explanations are not very convincing (Niklas et al. 2009). For example, taking the data for total dry leaf mass M_L and total dry stem mass M_S

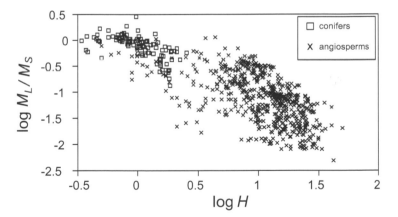

FIGURE 10.4. Bivariate plot of $\log_{10}$-transformed data for the quotient of dry leaf mass and dry stem mass (M_L/M_S) versus tree height (H, original units in m) for conifers and angiosperms. The data show that with few exceptions, the quotient M_L/M_S does not exceed unity ($\log M_L/\log M_S = 0$) for the smallest trees and decreases across species with increases in tree height. The data also show that, on average, conifer stems support more leaf mass than their angiosperm counterparts of equivalent height (because most conifers are evergreen, whereas most angiosperm species are deciduous). (Data from Cannell 1982.)

provided by Niklas and Spatz (2004), we see that M_L/M_S for a 35 m tall an-giosperm tree is about 10^{-2}, whereas for a 7.5 m tall tree, $M_L/M_S \approx 5 \times 10^{-2}$ (fig. 10.4). Even if we estimate the total wet leaf mass, the M_L/M_S values of average-sized angiosperm trees are typically less than unity. Naturally, M_L becomes more mechanically meaningful if we introduce the effects of wind drag and the possibility of snow loading, which might explain why many angiosperm species shed their leaves during the snow season.

10.5 Finite element methods (FEM)

Finite element analysis (FEM) is a numerical method for the approxi-mation of solutions of partial differential equations or integral equations in complex domains. The mathematical foundations of FEM go back to the beginning of the 1940s. The method originated as a matrix formula-tion during the 1950s and 1960s in the context of structural analyses of airframes. It has since been developed into a standard tool in mechanical engineering, fluid dynamics, thermodynamics, and electrodynamics and is even used today in weather forecasting programs.

The delineation of the mathematics of FEM is beyond the scope of this book. Here, we confine ourselves to the analysis of mechanical elastostatic problems. The approach can be subdivided into several steps:

1. The object of interest, which can be a prismatic bar in the simplest case or an entire tree, is subdivided into a three-dimensional mesh of elements, which may be of cuboid or tetrahedral form or of any other geometric shape. A combina-tion of differently shaped elements can be used if adequate to the problem to be solved.

2. The corners of the elements, referred to as nodes, are denoted by numbers 1 to N.

3. Boundary conditions can be formulated as restrictions on the displacement of a number of nodes, as in the case of rigid restraints on part of the object. Other types of boundary conditions are point forces assumed to act solely on specific nodes or distributed forces assumed to act on a number of nodes.

4. For linear elasticity and isotropic materials, each element is characterized by its modulus of elasticity E and its Poisson's ratio v. For a three-dimensional prob-lem, the stress-strain relationship, including shear stresses and shear strains, is given by $\sigma = C\varepsilon$, where σ and ε are vectors and C is a 6×6 matrix. For anisotropic materials such as wood, three moduli of elasticity and six Poisson's ratios deter-

mine mechanical behavior. For nonlinear elasticity, the stress-strain relationship is a function of the strain applied (see chap. 4).

The differential equations that govern the relationship between external forces applied and displacements of the nodes from their positions in the unloaded condition can be derived from the conservation of energy E_{total}, which states that, within the volume Vol of the domain,

(10.3)
$$E_{total} = \sum_i F_i u_i - \int_{Vol} \frac{\varepsilon\sigma}{2} dVol = 0,$$

where F_i is the force acting at node i and u_i the displacement of node i. The integral summarizes the contribution of the strain energies in every element. The principle of variation states that at equilibrium,

(10.4)
$$\frac{\partial E_{total}}{\partial u_i} = 0.$$

The forces acting on each of the nodes are external forces and forces resulting from displacements of adjacent nodes, with corresponding stresses in the respective elements.

Rather than solving the differential equations for the displacements u_i exactly, the solutions are approximated by piecewise first-, second-, or even higher-order functions with unknown coefficients, which have to fulfill the condition of continuity at each of the nodes to account for the cohesiveness of the material. For some applications the functions must also be continuously differentiable, which requires at least second-order spline functions.

Taken together, the foregoing leads to a formulation of the system of equation (10.3) in the form of a quadratic matrix with a size of the number of nodes N multiplied by the order of the approximating functions. In an appropriate mesh, the number of nodes adjacent to—and therefore in interaction with—a particular node i is small. Therefore, most of the entries in the matrix will be zeros. Commercially available FEM programs can solve a system of equations (10.3) even if the number of nodes goes into the millions.

Since the solutions are only numerical approximations, an important concern is whether the inaccuracies add up to give meaningless results, or whether the solutions converge. This can be tested by choosing a finer mesh with a larger number of nodes, by using higher-order approximating functions, or by applying a combination of both methods.

The solution of the calculations for an elastostatic problem is a vector u with the displacements u_i for each node i, usually displayed graphically. Correspondingly, the stresses in the object can be visualized. For example, FEM methods were used to calculate stresses in a bifurcation of a tree. By comparing it with other possible forms of a bifurcation, Mattheck (1992) could demonstrate that the tree realized a structure with a uniform distribution of stresses—the optimal design to minimize the danger of failure.

Other mechanical problems, such as the eigenwert problem of oscillating elastic structures such as a tree with all its branches, can be solved using the same formalism (Rodriguez et al. 2008). In general terms, finite element methods provide a terrific reduction of complexity. FEM can therefore be used wherever a very large number of interdependent entities have to be considered to obtain a reasonable conception of the physical reality even of plant development (Niklas 1977).

10.6 Optimization techniques

In mathematics, optimization refers to finding the solution closest of all feasible solutions to a preset objective. In its simplest case, the problem is reduced to finding a minimum or maximum value of a function by changing variables within a set defined by specified conditions. More complex problems require the use of calculus of variations, stochastic programming, or game theory. Since a general treatment of optimization techniques is beyond the scope of this book, the topic is only exemplified.

Two simple cases of optimization were treated in box 5.7, in which the optimal geometry of tree trunks subjected to wind loads was calculated under the boundary conditions of constant stresses. As another example, the optimal allocation of biological resources by trees under different environmental conditions can be treated by employing certain simplifications (box 10.2). A more advanced method of optimization for traits subject to certain constraints is known as the Lagrange multiplier method. Using this method, McCulloh et al. (2009) showed that in the majority of compound leaves investigated, the branching vascular system follows a law, originally proposed by Murray (1926) to describe a branching network, that maximizes hydraulic conductance. The analysis is detailed in box 10.3.

BOX 10.2 **Optimal allocation of biological resources**

For simplicity, consider a model tree of total height H, with a trunk of constant radius R and length H and a crown with diameter D, proportional to H. Note that the following allometric relations are also valid for more complicated forms of trunk and crown:

(10.2.1)   Total mass: $M \propto R^2 \cdot H$

(10.2.2)   Light capture: $L \propto D^3 \propto H^3$

(10.2.3)   Water conduction: $W \propto R^2$

If we take the total tree mass M as proportional to the expenditure of biological resources E, we can split the expenditure into two parts:

(10.2.4)   Expenditure for light capture: $E_L \propto M / L \propto R^2 / H^2$

(10.2.5)   Expenditure for water conduction: $E_W \propto M / W \propto H$

Biological resources are allocated in different proportions between the two tasks. The optimum of biological resource expenditure is equivalent to the minimum of E as a function of H,

(10.2.6)   minimum$(E) = $ minimum$(c_1 \cdot R^2 / H^2 + c_2 \cdot H)$,

where c_1 and c_2 are different normalization constants whose numerical values depend on environmental as well as biological conditions.

The simplest procedure to find the optimum tree height H as a function of the trunk radius R is to choose appropriate values of the constants c, differentiate equation (10.2.6) with respect to H, and set the derivative to zero:

(10.2.7a) $\dfrac{\partial E}{\partial H} = \dfrac{\partial}{\partial H}\left(c_1 \dfrac{R^2}{H^2} + c_2 H\right) = -2c_1 \dfrac{R^2}{H^3} + c_2 = 0$

or

BOX 10.2 **(Continued)**

$$(10.2.7b) \qquad\qquad H_{opt} = \left(\frac{2c_1}{c_2}\right)^{\frac{1}{3}} R^{\frac{2}{3}}.$$

Note that equation (10.2.7b) has the same form as the Euler-Greenhill formula (see eq. [5.29]). The relationship between optimal tree height and trunk radius complies with a $\frac{2}{3}$ scaling exponent that is independent of the particular choice of c_1 and c_2. For the specific example shown in figure 10.2.1, the constants are chosen to obtain a relation for wind-sheltered trees ($U = 0$) exactly like the Euler-Greenhill relation for trees with a density-specific modulus of elasticity $E/(g\rho) = 10^6$ m, where E is the modulus of elasticity, g the acceleration due to gravity, and ρ the density of the fresh wood.

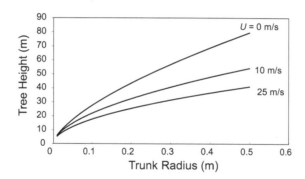

FIGURE 10.2.1. Simulation of the relationship between tree height and trunk radius based on an optimization approach. The constants in equation (10.2.11) are $c_1 = 1$ and $c_2 = 10^{-6}$ (m^{-6}). The upper curve refers to wind-sheltered trees (average wind speed $U = 0$). The two lower curves (representing windy conditions at $U = 10$ m/s and $U = 25$ m/s) are obtained when $c_3 = 2 \times 10^{-14}$ (s^2/m^5).

An extension of the optimization formalism can lead to a description of growth processes under particular environmental conditions. As an example, the influence of average wind loads on the relationship between tree height and trunk radius can be modeled. Stability against wind loads can be represented by the allometric relation

$$(10.2.8) \qquad\qquad S \propto R^3 /(H^3 \cdot U^2),$$

BOX 10.2 **(Continued)**

where U is the absolute value of the average wind speed. The expenditure of biological resources for stability is given by

(10.2.9) $$E_S \propto M / S \propto H^4 \cdot U^2 / R.$$

Correspondingly, the total expenditure of biological resources is given by

(10.2.10) $$E = E_L + E_W + E_S = c_1 \cdot R^2/H^2 + c_2 \cdot H + c_3 \cdot H^4 \cdot U^2/R.$$

Using a calculation analogous to equation (10.2.7a), the minimum of E as a function of H can be found by differentiation:

$$\frac{\partial E}{\partial H} = \frac{\partial}{\partial H}\left(c_1 \frac{R^2}{H^2} + c_2 H + c_3 H^4 \frac{U^2}{R} \right) = -2c_1 \frac{R^2}{H^3} + c_2 + 4c_3 H^3 \frac{U^2}{R} = 0.$$

(10.2.11a)

The solution of this equation is

$$H_{opt} = \left(\frac{1}{8c_3 U^2} \left(R\sqrt{c_2^2 + 32 c_1 c_3 R U^2} \right) - c_2 R \right)^{1/3} \quad \text{with } U > 0.$$

(10.2.11b)

With appropriate choices of the constants c_1, c_2, and c_3, this relation can be represented graphically (see fig. 10.2.1).

BOX 10.3 **Lagrange multipliers and Murray's law**

The Lagrange multiplier method provides a strategy for finding the maximum or minimum of a function subject to certain constraints. For the optimization problem of minimizing $f(x, y)$ subject to $g(x, y) =$ constant, a new variable λ, called a Lagrange multiplier, is introduced, and the Lagrange function is defined as

(10.3.1) $$L(x, y, \lambda) = f(x, y) + \lambda \cdot g(x, y).$$

BOX 10.3 **(Continued)**

The minimum of the function is found by forming the partial derivatives and solving the equations

(10.3.2) $$\frac{\partial L}{\partial x} = \frac{df}{dx} + \lambda \cdot \frac{\partial g}{dx} = 0, \quad \frac{\partial L}{\partial y} = \frac{df}{dy} + \lambda \cdot \frac{\partial g}{dy} = 0, \text{ and } \frac{\partial L}{\partial \lambda} = 0.$$

McCulloh et al. (2009) analyzed the minimization of hydraulic resistance by the structures of vascular networks in compound leaves under the constraint that the total volume of the branching network is constant. Generalizing their approach, the hydraulic resistance for a network with N branching levels can be expressed according to the Hagen-Poiseuille law (see box 6.2) as

(10.3.3) $$\Omega = \frac{8k}{\pi} \mu \sum_{m=0}^{N} \frac{l_m}{r_m^4 n_m},$$

where μ is the dynamic viscosity, k is a numerical correction factor for deviations from an ideal Hagen-Poiseuille flow, l_m is the length of a conduit, r_m its radius, and n_m is the number of branches at the branching level m. The volume is given by

(10.3.4) $$V = \pi \sum_{m=0}^{N} r_m^2 l_m n_m = V_{total},$$

where V_{total} is a constant (see above).

For the partial derivatives, we obtain

(10.3.5) $$\frac{\partial \Omega}{\partial r_m} = -\frac{32k}{\pi} \mu \frac{l_m}{r_m^5 n_m}, \text{ and } \frac{\partial V}{\partial r_m} = 2\pi r_m l_m n_m.$$

From

(10.3.6) $$\frac{\partial \Omega}{\partial r_m} + \lambda \cdot \frac{\partial V}{\partial r_m} = 0,$$

BOX 10.3 **(Continued)**

---

it follows that

(10.3.7)
$$\lambda = \frac{16\,k}{\pi^2}\,\frac{\mu}{r_m^6\,n_m^2}.$$

Partial differentiation with respect to r_0 yields

(10.3.8)
$$\frac{\partial\Omega}{\partial r_0} = -\frac{32\,k}{\pi}\,\mu\,\frac{l_0}{r_0^5\,n_0}, \qquad \frac{\partial V}{\partial r_0} = 2\,\pi\,r_0\,l_0\,n_0$$

and

(10.3.9)
$$\frac{\partial\Omega}{\partial r_0} + \lambda\cdot\frac{\partial V}{\partial r_0} = -\frac{32\,k}{\pi}\,\mu\,\frac{l_0}{r_0^5\,n_0} + \frac{32\,k}{\pi}\,\mu\,\frac{r_0\,l_0\,n_0}{r_m^6\,n_m^2} = 0,$$

which simplifies to

(10.3.10) $r_m^6\,n_m^2 = r_0^6\,n_0^2$ or $r_m^3\,n_m = r_0^3\,n_0$ with $0 \le m \le N$,

which can also be written as

(10.3.11)
$$r_m^3 \propto n_m^{-1}.$$

This third-order relation of the diameter of the conduit to the number of branches at any branching level is called Murray's law.

Reintroducing equation (10.3.10) into equations (10.3.4) and (10.3.3) yields

(10.3.12)
$$V = \pi\,r_0^3\,n_0\,\sum_{m=0}^{N}\frac{l_m}{r_m} = V_{total}$$

and

(10.3.13)
$$\Omega = \frac{8\,k}{\pi}\,\frac{\mu}{r_0^3\,n_0}\,\sum_{m=0}^{N}\frac{l_m}{r_m} = \frac{8\,k}{\pi^2}\,\frac{\mu}{r_0^6\,n_0^2}\,V_{total}.$$

BOX 10.3 **(Continued)**

Inspection of this final equation leads to the interesting result that the hydraulic resistance in the vascular network considered here is independent of the length of the conduits at every branching level, which is consistent with the anatomical findings of McCulloh et al. (2009).

In the examples treated here, the goal of optimization could be formulated mathematically. This is not always the case, and experiments have to be performed to check whether a change in a relevant parameter leads to an improvement of performance under given constraints. Because this can be time-consuming and expensive in complex systems, a strategy of advancement is usually developed. Typically the route with the steepest gradient is being followed. More advanced is the downhill simplex method for optimization (Nelder and Mead 1965). It should be noted that both procedures run the risk that only a local optimum, and not the absolute optimum, will be reached.

A different strategy developed by Rechenberg (1973) follows the route of biological evolution in an iterative procedure introducing "mutations" by making changes of varying size in relevant parameters chosen at random. Tests of performance under specified conditions provide the basis for selection for products with improved performance and against products with poorer performance. In its simplest case, the strategy involves only two members, "parent" and "mutant." If the mutant shows improved performance, it will become a parent in the next iteration. The advantages of this evolution strategy become apparent if it is extended to populations of many parents and if combinations of advantageous mutations ("recombination") are allowed. This optimization procedure is less likely to get "stuck" in a local optimum.

Literature Cited

Bauer, G., M.-C. Klein, S. N. Gorb, T. Speck, D. Voight, and F. Gallenmüller. 2011. Always on the bright side: the climbing mechanism of *Galium aparine*. *Proc. R. Soc. B* 278:2233–39.

Cannell, M. G. R. 1982. *World forest biomass and primary production data.* London: Academic Press.

Gould, S. J. 1966. Allometry and size in ontogeny and phylogeny. *Biol. Rev.* 41: 587–640.

Harvey, P. H. 1982. On rethinking allometry. *J. Theor. Biol.* 95:37–41.

Huxley, J. S. 1932. *Problems of relative growth.* New York: MacVeagh.

Mattheck, C. 1992. *Design in der Natur. Der Baum als Lehrmeister.* Freiburg: Rombach Verlag.

McCulloh, K. A., J. S. Sperry, F. C. Meinzer, B. Lachenbruch, and C. Atala. 2009. Murray's law, the "Yarrum" optimum, and the hydraulic architecture of compound leaves. *New Phytol.* 184:234–44.

McGhee, G. R. 1999. *Theoretical morphology: The concept and its applications.* New York: Columbia University Press.

Murray, C. D. 1926. The physiological principle of minimum work. I. The vascular system and the cost of blood volume. *Proc. Nat. Acad. Sci. USA* 12:207–14.

Nelder, J. A., and R. Mead. 1965. A simplex method for function minimization. *Comput. J.* 7:308–13.

Niklas, K. J. 1977. Applications of finite element analyses to problems in plant morphology. *Ann. Bot.* 41:133–53.

———. 1988. The role of phyllotactic pattern as a "developmental constraint" on the interception of light by leaf surfaces. *Evolution* 41:1–16.

———. 1994. *Plant allometry: The scaling of form and process.* Chicago: University of Chicago Press.

Niklas, K. J., E. D. Cobb, and H.-C. Spatz. 2009. Predicting the allometry of leaf surface area and dry mass. *Amer. J. Bot.* 96:531–36.

Niklas, K. J., and V. Kerchner. 1984. Mechanical and photosynthetic constraints on the evolution of plant shape. *Paleobiology* 10:79–101.

Niklas, K. J., and H.-C. Spatz. 2004. Growth and hydraulic (not mechanical) constraints govern the scaling of tree height and mass. *Proc. Nat. Acad. Sci. USA* 101:15661–63.

Prothero, J. 1986. Methodological aspects of scaling in biology. *J. Theor. Biol.* 118:259–86.

Rayner, J. M. V. 1985. Linear relations in biomechanics: The statistics of scaling functions. *J. Zool.* 206:415–39.

Rechenberg, I. 1973. *Evolutionsstrategie, Optimierung technischer Systeme nach Prinzipien der biologischen Evolution.* Stuttgart: Friedrich Fromann Verlag.

Rodriguez, M., E. de Langre, and B. Moulia. 2008. A scaling law for the effects of architecture and allometry on tree vibration modes suggests a biological tuning to modal compartmentalization. *Amer. J. Bot.* 95:1523–37.

Smith, R. J. 1980. Rethinking allometry. *J. Theor. Biol.* 87:97–111.

Sokal, R. R., and F. J. Rohlf. 1980. *Biometry: The principles and practice of statistics in biological research.* San Francisco: Freeman.

Warton, D. I., I. J. Wright, D. S. Falster, and M. Westoby. 2006. Bivariate line-fitting methods for allometry. *Biol. Rev. Camb. Phil. Soc.* 81:259–91.

Zimmermann, W. 1930. *Die Phylogenie der Pflanzen: ein Überblick über Tatsachen und Probleme.* Jena: Gustav Fischer.

Glossary

abaxial. Oriented away from the axis. Compare with **adaxial.**

abiotic environment. The component of the total environment of an organism provided by physical factors, individually or in concert. Compare with **biotic environment.**

abscisic acid (ABA; also known as dormin). A plant hormone involved in many important developmental processes, including seed dormancy and stomatal opening and closing.

abscission. The shedding of plant parts (leaves, flowers, stems, or any other structure). Typically occurs after the formation of an abscission zone.

achene. A dry indehiscent fruit produced by a single, uniovulate carpel.

action potential. A transient alteration of the membrane potential across an excitable membrane generated by the activity of voltage-gated ion channels embedded in the membrane.

actuator disk. A term from the momentum theory of rotors, whereby a rotor is treated as an infinite number of airfoils producing a discontinuous but uniformly distributed pressure rise.

adaptation. Any characteristic or property of an organic process, an organ, or an organism that contributes to survival; any process that maintains or improves organic function.

adaxial. Oriented toward the axis. Compare with **abaxial.**

adventitious. Referring to plant structures developing from unusual sites, such as roots originating on stems or leaves instead of on other roots.

aerenchyma. Parenchymatous tissue with large intercellular spaces resulting from the breakdown (lysogeny) or shearing and ripping (schizogeny) of cells.

alamethicin. An antibiotic peptide produced by the fungus *Trichoderma viride* containing the amino acid 2-aminoisobutyric acid that can act as a monovalent cation ionophore.

algae (singular **alga).** A term referring collectively to all eukaryotic plants that are not embryophytes, thereby distinguishing a grade of plant organization rather than a monophyletic group of plants.

allometry (heterogony of Julian Huxley). A constant relative growth pattern often expressed in terms of the formula $y = \beta x^a$, where y and x are two interdependent variables (e.g., length and weight or weight and elastic modulus), β is the value of y when x is of some arbitrary magnitude, and α is the ratio of the growth rate of y to that of x.

allowable stress (safe stress). See **working stress.**

alternation of generations (diplobiontic life cycle). A life cycle comprising more than one multicellular type of organism. In the embryophytes, the alternation of generations involves a diploid multicellular individual (the sporophyte) that produces meiospores and a haploid multicellular individual (the gametophyte) that produces gametes.

angiosperms. Flowering plants; a monophyletic group. Plants in which the ovule is borne within a closed structure (carpel).

anisotropic. Having different material properties (e.g., elastic modulus) when measured along the principal axes of symmetry. Anisotropic materials include axisymmetric materials (those that have two principal axes of symmetry) and orthotropic materials (those that have different mechanical properties along all three of their principal axes of symmetry). Compare with **isotropic.**

annual growth layer. A layer of secondary tissue produced in the plant body during one growth season. Annual growth layers are produced by the vascular cambium ("wood") and the phellogen ("cork") and are typically referred to as "annual rings" when seen in transverse section.

anomocytic. Referring to stomata that lack differentiated subsidiary cells surrounding their guard cells.

anther. The pollen-bearing part of the stamen.

anticlinal. Referring to a plane of cell division in a layer of cells that is perpendicular to an adjacent layer of cells. Compare with **periclinal.**

apical meristem. A lenticular cluster of cells found at the tips of roots and stems that initiates vegetative and reproductive organs by cellular division.

apoplast. The portion of the plant body not composed of the living protoplast (symplast); typically refers to the complex of cell walls that is continuous throughout the plant body. Compare with **symplast.**

autogyroscopic. Referring to an object that rotates about one of its axes as a result of its relative motion with respect to the fluid it is immersed in.

autotroph. An organism that synthesizes organic materials from inorganic precursors. The source of energy may be either light, via photosynthesis (photoautotroph), or chemical reactions independent of light (chemoautotroph). Compare with **heterotroph.**

axil. The angle between the adaxial surface of an organ and the organ it is attached to, as with a leaf axil or branch axil.

axillary bud (axillary meristem). Juvenile apical meristem found in a leaf axil. May be dormant for a time or may commence development immediately after initiation. Gives rise to either a vegetative shoot (a branch) or a flower (a determinate, reproductive shoot).

axis. In botany, a cylindrical stemlike organ whose homology with the three principal plant organs (stem, leaf, root) is not specified (as in early land plants whose organographic constructions are unclear). The longitudinal dimension of an elongated plant structure. In engineering or mathematics, one of the three Cartesian coordinates that specify the three-dimensional geometry of an object.

axisymmetric. Having equivalent material properties (e.g., elastic moduli) when measured along two of the three principal axes of symmetry.

bark. A nontechnical term applied to all primary and secondary tissues external to the vascular cambium of a plant organ exhibiting secondary growth. Outer bark refers to nonliving tissues external to the vascular cambium; inner bark refers to living tissues (typically just the phloem) internal to the vascular cambium.

beam. A structural support member placed in bending. Compare with **column; shaft.**

bending moment. The product of a bending force and the length of the radius arm at right angles to the direction of application of the bending force.

bending stresses. Stresses caused by bending, of which there are three kinds: tensile ($\sigma+$), compressive ($\sigma-$), and shear bending stresses (τ). Tensile and compressive bending stresses increase in intensity toward the perimeter of cross sections; shear bending stresses increase in intensity toward the centroid axis.

biaxial stresses (tension or compression). A pair of coaxial forces operating orthogonally to each other. See also **coaxial.**

bilateral symmetry (zygomorphy). The condition of having two geometrically complementary sides so that the structure can be symmetrically divided by a single longitudinal plane into two mirror images.

bilins. A category of pigments consisting of linearly arranged, open-chain tetrapyrroles, found in the antennae (phycobilisomes) of cyanobacteria and red algae as well as in the bile of mammals.

biomass. The total weight of living matter of an organ, an organism, or a population of organisms.

biomimetics (bionics). The abstraction of good design from nature; the development of materials or designs inspired by natural materials or structures.

bionics. See **biomimetics.**

biotic environment. The component of the total environment of an organism provided by other organisms or the interrelations among them. Compare with **abiotic environment.**

bluff-bodied. Having a shape that lacks streamlining (e.g., a sphere or cylinder). Compare with **streamlined.**

body force. Any force operating within the volume of an object, such as gravity. Compare with **surface force.**

bordered pit. A type of perforation that permits water to flow between cells with secondary walls, in which the secondary wall arches over the pit cavity and narrows toward the opening to the cell lumen. The middle lamella of a bordered pit is typically modified into a torus and margo, which can operate as a safety valve if a pressure drop occurs in one of the cells.

boundary condition. A time-independent or physical constraint. Boundary conditions are classified as either geometric or kinetic. An example of the former is the way a beam is anchored yet free to vibrate; an example of the latter is the force or moment applied to a beam.

boundary layer. The layer of fluid (gas or liquid) immediately surrounding an object, in which viscous forces predominate in dictating fluid dynamic behavior. The dimensions (thickness) of the boundary layer around the same object vary as a function of the ambient velocity (speed and direction of flow) of the fluid. Boundary layers also differ among objects differing in their geometry or absolute size but experiencing the same ambient flow conditions.

bract. A modified leaf that subtends or surrounds a plant organ. Generally provides protection.

Brazier buckling. A short-wave mode of mechanical failure in very long, thin-walled tubes resulting from localized crimping when a large bending moment is applied. It is assumed that buckling occurs in a plane of symmetry of a cross section; that is, that no torsional buckling is involved.

breaking load. The load that results in the mechanical failure of a structure; the strength of a structure expressed in units of weight.

breaking stress (breaking strength). The stress level at which a nonductile material breaks; the strength of a material expressed in units of force per area (stress).

bryophytes. Nonvascular embryophytes encompassing the mosses (Musci), the liverworts (Hepaticae), and the hornworts (Anthocerotae). The gametophyte is the morphologically dominant and free-living generation in the life cycle.

bud. A short vegetative or reproductive shoot bearing a densely packed series of leaves and intervening stem segments (internodes).

bulb. A modified perenniating shoot with short internodes and fleshy, scalelike leaves.

bulk modulus. Symbolized by K; the ratio of the uniformly applied hydrostatic pressure to the relative change in volume. The reciprocal of K is called **compressibility.**

calmodulin. A small (~148 amino acids long), acidic, calcium-binding protein found in all eukaryotic cells that binds to and regulates a number of protein targets important to plant development.

cambium. A region of cells configured as a more or less cylindrical layer of embryonic, growing cells; typically found in stems and roots. The derivative cells are commonly produced in two directions and are arranged in radial files. The term "cambium" should be applied only to the two lateral meristems, the vascular cambium and the phellogen (the "cork cambium"). See also **phellogen; vascular cambium.**

cantilever (cantilevered beam). A nonvertical beam anchored at one end and free to deflect at the other. See also **beam.**

carotenoids. Orange tetraterpenoid pigments that function as antenna complex pigments and photoprotective agents.

carpel. The leaflike organ in flowering plants that produces one or more ovules.

Cauchy strain (engineering strain or conventional strain). The ratio of the difference between a deformed (l) and original dimension (l_0) to the original dimension: $\varepsilon = (l - l_0)/l_0 = \Delta l/l_0 = l/l_0 - 1$.

cavitation. The formation of water vapor bubbles in columns of water subjected to tensile stresses that exceed the tensile strength of water, resulting in the formation of an embolism in the conduit conducting water.

cellular solid. Any fabricated or biological material (whose relative density is equal to or less than 0.3) consisting of a geometrically arranged solid phase and a gas or, less commonly, liquid phase. The solid phase may be arranged in the form of walls (complete partitions, in the case of closed-walled cellular solids) or in the form of strutlike or beamlike interconnected elements (incomplete partitions, in the case of open-walled cellular solids). Cellular solids have a complex pattern of mechanical behavior that differs from that of their solid phase and varies as a function of the stress or strain level.

cellulose. A polysaccharide consisting of long, straight chains of β-D-glucose residues joined by 1,4 links, in which cellobiose is the repeating polymeric unit. The main component of cell walls in many plants.

cell wall. The more or less rigid shell secreted by and enveloping the protoplast. The chemical composition of the wall varies from one plant group to another. Typically, however, in vascular and nonvascular land plants, the cell wall is composed of cellulose, other carbohydrate polymers, and proteins.

centrifugal. Developing or produced successively farther from the center of an organ.

centripetal. Developing or produced successively closer to the center of an organ.

centroid axis. The longitudinal axis of a structure defined by the center of mass in all successive cross sections. The geometric center of a plane area. The sum of all the elements of area over a plane area multiplied by the distance from any axis through the centroid axis must be zero.

chemical potential. A measure of the capacity of any substance to do work. When expressed in terms of units of energy per unit volume, the chemical potential of water is referred to as water potential (ψ_w).

chlorenchyma. Parenchymatous tissue containing numerous chloroplasts, as in the mesophyll of photosynthetic leaves.

chlorophyll. Green, light-absorbing chlorin pigments found in almost all photosynthetic eukaryotes, synthesized from succinyl-CoA and glycine.

chloroplast. An organelle in which photosynthesis is carried out.

coaxial. Having a common axis, specifically in relation to two opposed, externally applied forces operating along the same axis through a material or structure. Coaxial forces directed inward place a material in uniaxial compression; a pair of coaxial forces directed outward place a material in uniaxial tension. A pair of coaxial forces, in which each set operates orthogonally to the other set, can produce biaxial tension or biaxial compression.

coenocyte. An organism formed by cytoplasmic growth and nuclear division without the formation of cell wall partitions.

cohesion-tension theory. A theory explaining the ascent of xylem water within tall vertical vascular plants in terms of the high cohesive forces developed among neighboring water molecules, the adhesion of water molecules to the inner surfaces of capillary-like tracheary elements, and transpirational "pull" on columns of water within those tracheary elements.

coleoptile. The cylindrical, tubular sheathlike organ enveloping the epicotyl of the embryo of grasses (Poaceae) and other monocots; sometimes considered the first leaf of the epicotyl.

collenchyma. A mechanical tissue composed of more or less elongated living cells with unevenly thickened, nonlignified primary walls. Common in regions of primary growth in stems and leaves.

column. A structural support member primarily experiencing compressive axial loading. Compare with **beam; shaft.**

companion cell. A cell type found in the phloem of flowering plants, produced by an asymmetric division of a procambium cell that also produces a sieve tube member.

compliance. Symbolized by D; for elastic materials, the reciprocal of Young's modulus. For viscoelastic materials, the ratio of strain to stress. For a viscoelastic material or any material exhibiting creep, compliance is the reciprocal of the tangent modulus.

composite material. A material with an infrastructure sufficiently heterogeneous that its material properties cannot be adequately predicted from the properties of any of its constituents. Composite materials may have a periodic infrastructure, in which constituents have a geometric spatial regularity, or they may have a nonperiodic infrastructure, in which constituents have no clearly defined spatial regularity.

compound middle lamella. The two primary walls and intervening middle lamella of two neighboring cells when the walls and middle lamella are not distinguishable.

compressibility. The reciprocal of the bulk modulus K.

compression wood. Reaction wood typically found in conifers and some angiosperms, formed on the sides of branches and stems subjected to bending stresses; characterized by a dense structure and extensive lignification; capable of expansion. Compare with **tension wood.**

compressive (normal) **strain.** Symbolized by ε^-; normalized deformation of a material or structure subjected to compression.

compressive (normal) **stress.** Symbolized by $\sigma-$; the normal compressive force component divided by the surface area through which the force component operates.

conductance. The reciprocal of resistance.

conventional strain. See **Cauchy strain.**

convergence (convergent evolution). The evolutionary acquisition of similar characteristics in nonhomologous organs or taxonomically disparate organisms.

cork cambium. See **phellogen.**

corpus. That part of the shoot apical meristem that gives rise to the procambium and the ground meristem.

cortex. In dicots and many nonflowering plants, the ground tissue found external to the vascular tissue.

cotyledon. An embryonic leaflike organ produced by angiosperms and gymnosperms; markedly different in form from leaves subsequently produced by the seedling. In dicots, the cotyledon often stores nutrients.

couple. A pair of forces acting in parallel and separated by a distance. The moment of a couple or torque is the product of one of the two forces and one-half the perpendicular distance between the two forces.

creep. Mechanical behavior typically exhibited by viscoelastic materials, characterized by changes in the magnitude of strain under a constant level of stress.

critical crack length. See **Griffith length.**

critical load. Symbolized by W_{cr} or P_{cr}; the compressive load that will result in the elastic buckling of a column.

cross section. See **transverse section.**

cuticle. A layer of hydrocarbons, fatty acids, and other materials on the outer walls of epidermal cells; serves a variety of functions, one being to reduce the rate of water loss from plant tissues.

cutin. One of two waxy polymers, consisting of omega $C16$ and $C18$ hydroxy acids and their derivatives, covering the aerial surfaces of some nonvascular plants and the majority of vascular plants.

cytoplasm. The portion of the living cell bounded by the plasma membrane and excluding the nucleus and visible vacuoles.

cytosol. The fluid phase of the cytoplasm; the cytoplasmic matrix.

damping. The absorption of vibrational energy by a structure. Generally, three forms of damping are recognized: material (or mass) damping, which results when vibration energy is dissipated by the material of a structure: fluid damping, which results when energy is absorbed by the fluid surrounding the structure; and another damping that results from the impact and scraping of articulated components making up the structure.

deformation. The displacement of a structure from its equilibrium position; any distortion resulting from an externally applied force.

degree of elasticity. The ratio of the elastic (recovered) deformation to the total deformation when a material is loaded to a given stress level and then unloaded.

dehiscence. The spontaneous rupture or opening of a plant structure, such as an anther or a fruit, releasing reproductive bodies.

derivative cell. A cell resulting from the division of a meristematic cell that subsequently differentiates and matures into a nonmeristematic cell within the plant body.

determinate growth. A type of development in which the number of lateral organs (branches and leaves) produced by an apical meristem is limited. Compare with **indeterminate growth.**

dichotomy (dichotomization). The bifurcation of the apical meristem of a plant axis into two equal-sized apical meristems leading to two axes, neither one developmentally dominant.

dicots (dicotyledons). Nontechnical term for angiosperms whose embryos have two cotyledons. Dicots are thought to be a monophyletic group within the angiosperms. Compare with **monocots.**

dielectric constant. A numerical expression of the ability of a substance to become polarized in an applied electrical field.

differentiation. The developmental phase during which physiological and morphological changes lead to the specialization of cell, tissue, organ, or plant body.

diplobiontic life cycle. See **alternation of generations.**

dipole. A molecule that has two opposite electric charges somewhere along its atomic structure.

distal. Farthest from a structure's point of origin or site of attachment to another structure. Compare with **proximal.**

drag. The component of the total hydro- or aerodynamic force acting on an object that resists motion in the direction of the velocity vector defining the relative motion of the object with respect to a fluid. Drag has three components: normal pressure drag results from the resolved components of pressure forces normal to the surface of an object with motion relative to a fluid; friction drag results from the resolved components of the pressure forces tangential to the surface of an object with motion relative to a fluid; vortex drag (also called induced drag) results from the formation of trailing vortices—that is, leeward vortices that result from the flow of a fluid from a region of high pressure to a region of low pressure.

drag coefficient. A dimensionless quantity that reflects the fluid resistance associated with the relative motion of an object with respect to a fluid.

ductile. Referring to materials that yield (undergo plastic deformations) once their proportional limits are exceeded. Ductility is the capacity of elastic materials to deform without fracturing, as in metals that can be drawn out into wires.

dynamic viscosity. Symbolized by μ; the product of the kinematic viscosity (v) and the density (ρ) of a fluid.

early wood (spring wood). The secondary xylem (wood) formed at the initiation of a growth layer; characterized by lower tissue density and larger cell diameters than in subsequently formed secondary xylem in the same growth layer (late wood).

eigenfrequency. After an initial excitation, the frequency with which an object oscillates in the absence of external forces.

elastic behavior. The ability to restore deformations instantly when the level of stress drops to zero. Compare with **plastic behavior.**

elastic hysteresis. The amount of energy that a material dissipates internally during a loading-unloading cycle; shown in a stress-strain diagram as an elastic hysteresis loop where the trajectories of the loading and unloading portions of the plot do not coincide.

elastic limit (proportional limit). The stress level at which elastic behavior is lost. Beyond the elastic limit, many elastic materials either yield plastically or break. Thus, either the yield stress or the breaking stress is reached after the elastic limit has been exceeded.

elastic moduli. The moduli that collectively define the behavior of a material in its elastic range of behavior; that is, E, G, and K, together with the Poisson's ratio v.

elastic modulus (Young's modulus, modulus of elasticity). Symbolized by E; the quotient of normal stress to normal strain measured within the elastic range.

electrical capacitance. The ability of an object to hold an electric charge; a measure of the amount of electrical energy stored for a given electric potential.

electroneutrality. A physicochemical principle stating that in every aqueous region that is large as compared with atomic dimensions, the number of positive electric charges carried by cations is essentially equal to the number of negative electric charges carried by anions.

embolism. The formation of water vapor bubbles in water-conducting xylem cells resulting in a drop in hydraulic conductivity (typically due to water stress or a freeze-thaw cycle).

embryophytes (formerly Embryophyta). Nontechnical term for plants that retain multicellular embryos within sporophytic tissues; the group of plants encompassing the bryophytes and the tracheophytes (vascular plants).

endodermis. The innermost layer of the cortex, found in the roots and stems of most vascular plants, which forms a sheath around the vascularized region of the organ and restricts the flow of water through the symplast.

end-wall effects. The mechanical consequences of supports at the ends of a specimen; particularly evident in anisotropic materials stretched in the direction of their greater stiffness. Supports increase the apparent stiffness of a specimen by locally restricting deformations. Since the strains are reduced for each level of stress, the elastic modulus appears higher than its actual value. End-wall effects in uniaxial tensile tests can be reduced by ensuring that a specimen has a length:radius ratio greater than 10 and by measuring stresses and strains along the midspan of the specimen.

energy. The capacity to exert a force over some distance. Potential energy results from the position of one body with respect to another or relative to the positions of parts of the same body; kinetic energy results from the motion of a body or from the relative motions of parts of a body.

engineering strain. See **Cauchy strain.**

engineering stress. See **nominal stress.**

epidermis. The outer layer of cells covering the primary plant body, produced by the protoderm.

epiphyte. A plant that grows on the aerial portions of another plant.

euglenoids. A lineage of algae derived from primary endosymbiotic events involving a green algal endosymbiont and a heterotrophic host cell. Most species are unicellular and live in fresh water. Some species are phagocytotic.

eukaryote. An organism whose chromosomes are contained within a nucleus and which possesses membrane-bound organelles. Compare with **prokaryote.**

Euler buckling. Elastic instability; a deflection of a slender column from its original form under an applied axial compressive load.

exogenous. Arising in superficial tissue, as with an axillary bud or a leaf primordium.

extensin. A family of approximately twenty kinds of hydroxyproline-rich glycoproteins found in plant cell walls that are believed to participate in the expansion of the cell wall.

extension ratio. The ratio of a deformed (l) to an undeformed (l_0) dimension: l/l_0. Equivalent to the stretch ratio (λ) of rubbery materials.

factor of safety. In solid mechanics, a dimensionless measure of the ability of a structure to withstand a mechanical force. Typically calculated as the breaking stress divided by the working stress a structure can withstand. The critical buckling height of a plant divided by the height also serves as a factor of safety.

fiber. An elongated, tapered cell with a lignified or nonlignified secondary wall; may or may not have a living protoplast at maturity.

Fibonacci series. Any series of numbers resulting from the successive addition of the last two numbers: 0, 1, 1, 2, 3, 5, 8, 13, 21, etc. Fibonacci series occur in phyllotactic patterns. Named in honor of Leonardo of Pisa, son (Filius, abbreviated to "Fi") of Bonacci.

finite element method (FEM). A numerical method for the approximation of solutions of partial differential equations or integral equations in complex domains by subdividing the domains into a large number of finite elements.

flexural stiffness (flexural rigidity). Symbolized by EI; the product of the elastic modulus and the second moment of area. Measures the ability of a structure to resist bending.

fluid. A general term encompassing all gases and liquids.

fluidity. Symbolized by α; the reciprocal of viscosity.

force. A physical quantity that changes the state of rest or motion of a free material object, measured by the acceleration of a mass (Newton's second law). In a fixed body, it causes stress and leads to deformation. Force has both magnitude and direction, making it a vector quantity.

fracture. The ultimate irreversible failure of a material object under a critical load, often separating the object into two or more pieces. Fracture is usually initiated by a crack in the material.

free energy (Gibbs free energy). The energy within a system that can do work.

frequency. In uniform circular motion or in any periodic motion, the number of revolutions or cycles completed per unit time.

gametophyte. The multicellular phase in the life cycle that produces gametes. Compare with **sporophyte.**

gas. A state of matter in which molecules or atoms are unrestricted (ideal gas) or nearly unrestricted (real gas) by cohesive forces so that any given quantity has no definite shape or volume.

Goldman equation. An equation that relates the electric potential across a membrane separating two compartments to the concentration of ions in either compartment and their ability to pass through the membrane.

gravitational potential. The effect of the force of gravity on the chemical potential of a system. In reference to xylem water or soil water, the product of the density of water, the acceleration of gravity, and elevation (a positive value in the vertical above the ground surface; a negative value in the vertical below the ground surface).

gravitropism. A turning or growth movement by a plant in response to gravity.

Griffith length (critical crack length). Symbolized by L_G; the limiting length of a crack or flaw within a structure that, when exceeded, will self-propagate a fracture and, when shorter, will remain stable and not propagate a fracture when the structure is subjected to stress.

ground meristem. One of the three primary meristems derived from an apical meristem; gives rise to the ground tissue system.

ground tissue. Any tissue other than the epidermal, peridermal, or vascular tissue.

growth layer. See **annual growth layer.**

growth stresses. Stresses that result from the differentiation and maturation of cells (typically cells derived from the cambium). In trees, growth leads to tensional stresses in the periphery of the stem or axis and compressive stresses in the inner structure.

guard cell. An epidermal cell surrounding a stoma (opening) that regulates the stoma's diameter by means of hydrostatic changes in its size and geometry; the mechanical device whereby CO_2 and O_2 exchange between the plant and the external atmosphere is regulated.

gymnosperms. A diverse group of seed plants that do not produce ovules within ovaries.

Hagen-Poiseuille equation. An equation that relates the rate of flow of a fluid through a hollow tube or capillary to its inner radius, the pressure gradient, and the viscosity of the fluid.

hardness. Operationally defined as resistance to fracture or plastic deformation due to friction from a sharp object (scratch hardness), resistance to plastic deformation due to a constant load from a sharp object (indentation hardness), or height of the bounce of an object dropped on the material (rebound hardness).

harmonic motion. A periodic oscillatory motion in a straight line such that the restoring force is proportional to the magnitude of the displacement.

heartwood. The innermost growth layers of secondary xylem that have ceased to transport water and which store metabolites; functions principally in mechanical support and storage. Compare with **sapwood.**

hemicelluloses. A chemically heterogeneous group of alkali-soluble polysaccharides (including galactans, glucans, glucomannans, mannans, and xylans) found in the matrix of plant cell walls that are neither pectinaceous nor lignified fractions of the cell wall.

Henchy strain (natural strain, true strain). Symbolized by ε_t; the natural logarithm of the extension ratio $\varepsilon_t = \ln(l/l_o)$.

heterospory. The formation of spores that produce unisexual gametophytes: megaspores that develop into megagametophytes and microspores that develop into microgametophytes.

heterotroph. An organism that acquires carbon by ingesting organic materials. Compare with **autotroph.**

Hookean material. A linearly elastic material; a material that behaves according to Hooke's law.

Hooke's law. The ratio of stress to strain is linearly proportional within the elastic range of behavior: $\sigma = \varepsilon E$, where E (the elastic modulus or Young's modulus) is the proportionality factor.

horsetails. Nontechnical term for the Sphenopsida, a group of seedless vascular plants characterized by whorls of leaves at nodes, hollow internodes, internodal anatomy with carinal and vallecular canals, and sporangia borne in a conelike structure.

hydroid. A cell type, found in some species of bryophytes, that is functionally analogous to the tracheary element; transports or stores water.

hydrophilic. Referring to typically charge-polarized molecules (or portions of molecules) capable of binding water molecules and thereby able to dissolve in water. Compare with **hydrophobic.**

hydrophobic. Referring to typically nonpolar molecules (or portions of molecules) that are insoluble in water. Compare with **hydrophilic.**

hydrostatic pressure. The pressure acting on an object or a finite element within a fluid. Technically, the product of the distance from the upper surface of a fluid, the density of the fluid, and the acceleration due to gravity g.

hygroscopic. Referring to a substance, material, or structure with the ability to absorb and retain moisture from its surrounding environment (e.g., salt, sugar, some spore walls, and some plant tissues).

hypodermis. A subepidermal layer of tissue in internodes that is distinct from other cortical tissues found in the stem.

hysteresis (loop). A system whose behavior is path-dependent (the stress vs. strain pathways resulting from a loading-unloading cycle).

IAA. Abbreviation for indole-3-acetic acid, a plant growth hormone (auxin).

indeterminate growth. A type of development in which the number of lateral organs (e.g., branches and leaves) produced by an apical meristem is unrestricted. Compare with **determinate growth.**

inflorescence. A collection of flowers sharing the same subtending stem (peduncle).

initial cell. A meristematic cell that gives rise to two cells, one remaining in the meristem, the other added to the plant body.

instantaneous elastic modulus. See **tangent modulus.**

integument. A layer of tissue enveloping the ovule that forms the seed coat during the development of the ovule into a seed.

internode. A stem axis between two nodes.

inviscid fluid. A hypothetical fluid that has no viscosity and therefore can support no shear stress.

ion channel. A pore-forming transmembrane protein that allows the passive flow of ions across a membrane down a gradient of electrochemical potential. Compare with **ion pump.**

ion pump. A transmembrane protein that actively transports ions across a membrane against a gradient of electrochemical potential. Compare with **ion channel.**

isotropic. Having material properties that, when measured in any direction, are equivalent. Compare with **anisotropic.**

kinematic viscosity. Symbolized by v; the ratio of dynamic viscosity (μ) to density (ρ).

laminar flow. Fluid flow conditions characterized by lack of macromolecular mixing among adjacent layers of fluid. Compare with **nonlaminar flow.**

lateral meristem. Cells of the vascular cambium and the phellogen (cork cambium); any meristem parallel to the sides of a plant axis that gives rise to a secondary tissue.

late wood (summer wood). Secondary xylem formed during the latter part of the deposition of an annual growth layer. Late wood is typically denser and composed of smaller cells than early (spring) wood.

leaf. The principal lateral (appendicular) organ of the plant stem. The leaves of vascular plants are often classified as either microphylls or megaphylls.

lift. The component of the total aerodynamic force operating perpendicular to the drag force.

lignification. The impregnation of a cell wall with lignin.

lignin. An organic polymer with a complex three-dimensional configuration of variable composition whose monomeric units include monosaccharides, phenolic acids, aromatic amino acids, and alcohols. Typically found in secondary plant cell walls, particularly those of the xylem tissue.

linear elasticity. The property of an elastic material such that stress and strain are proportionally related by a single constant (the elastic modulus E) within the elastic range of behavior: $\sigma = \varepsilon E$. Compare with **nonlinear elasticity.**

liquid. A state of matter characterized by a high volumetric modulus of elasticity but a low shear modulus. In contrast to solids, liquids share with gases the ability to flow. In contrast to gases, liquids have a density comparable to that of solids. On the molecular level, liquids have a much less ordered structure than solids; therefore, glass is considered a liquid.

lodging. The mechanical failure (typically in bending) of the main stem (typically of cereals such as wheat, corn, and barley) resulting from the loads produced by ripening seeds or excessive wind-induced drag.

lycopods. Nontechnical term for the Lycopsida (Lycophyta), a group of pteridophytes possessing, among other distinguishing features, lateral, reniform sporangia and stems with centripetal xylem differentiation.

margo. The portion of the middle lamella spanning a bordered pit that supports the torus and permits water to pass from one adjoining cell to another. See also **torus.**

material damping. Damping of oscillations that results when vibration energy is dissipated in the material of a structure due to internal friction.

matric potential. Symbolized by ψ_m; the effects of cell surface areas, colloids, and capillarity on the water potential of a system. Always has a negative value.

mechanical tissue (supporting tissue). Any tissue composed of thick-walled cells; any primary (collenchyma) or secondary (sclerenchyma) tissue that confers strength on the plant body.

mechanoperception. The ability to sense and response to mechanical stimuli.

megagametophyte. The gametophyte in a heterosporous diplobiontic life cycle that produces the gamete functionally equivalent to the egg. Compare with **microgametophyte.**

megaphyll. One of two types of leaves produced by vascular plants; as used in this book, any leaf produced by a vascular plant other than a lycopod. Compare with **microphyll.**

megaspore. The meiospore in a heterosporous diplobiontic life cycle that will give rise to the megagametophyte. Compare with **microspore.**

membrane potential. The electric potential across a membrane separating the inside and the outside of a cell or two cellular compartments. In the absence of an imposed electrical voltage difference, the membrane potential is due to an inequality of ion concentrations between the two sides of the membrane.

meristem. A tissue that retains the embryonic capacity to produce new protoplasm and cells by mitotic cell division.

mesophyll. The photosynthetic parenchymatous tissue of a leaf that composes the ground tissue found between the upper and lower epidermal layers.

microfibril. A threadlike, cellulosic component of the cell wall visible only at the level of resolution possible with the electron microscope.

microfibrillar angle. The angle subtended between the longitudinal axis of a cell and the longitudinal axis of a cellulose microfibril within a cell wall layer.

microgametophyte. The gametophyte in a heterosporous diplobiontic life cycle that produces the gamete functionally analogous to sperm. Compare with **megagametophyte.**

microphyll. One of two kinds of leaves produced by vascular plants; as used in this book, any leaf produced by a lycopod. Compare with **megaphyll.**

micropyle. The small tube or pore that remains from the incomplete closure of the integument or integuments of an ovule, through which pollen gains access to the megagametophyte.

microspore. The meiospore in a heterosporous diplobiontic life cycle that will give rise to the microgametophyte. Compare with **megaspore.**

middle lamella. The intercellular layer of chiefly pectinaceous material found between the primary cell walls of adjoining cells.

modulus. A mechanical parameter that is the ratio of a stress component to the analogous strain component; for example, elastic modulus, shear modulus.

modulus of rupture. Symbolized by M_R; the tensile stress at fracture of a material measured in bending.

molal solution. A solution containing one mole (gram-molecular weight) of a solute per kilogram of solvent.

molar solution. A solution containing one mole (gram-molecular weight) of a solute per liter of the solution.

mole (mol). A mass numerically equivalent to the molecular weight of the substance expressed in grams.

momentum. Quantity of motion measured as the product of mass and velocity.

monocots (monocotyledons). A nontechnical term for angiosperms whose embryos have a single cotyledon. Monocots are thought to be a monophyletic group within the angiosperms. Compare with **dicots.**

monophyletic group (natural group or natural taxon). A group of organisms of any taxonomic rank believed to be descended from a common ancestral group of organisms of equal or lesser taxonomic rank.

monostromatic. Referring to a form of tissue or body plan construction consisting of a single layer of cells.

morphology. A discipline of biology dealing with the macroscopic form and structure of organisms and their parts.

nabla operator. Symbolized by ∇; depending on the mathematical context, a vector differential operator used to denote the standard derivative of a function with a one-dimensional domain, or the gradient of divergence or the curl of a vector field.

natural selection. According to the Darwinian theory of evolution, the principal mechanism responsible for evolution. Competition is believed to result in the death of some individuals and the survival of others whose genetic composition confers a competitive advantage. The survivors pass their genetic advantages on to the next generation, resulting in evolutionary change over sequential generations.

natural strain. See **Henchy strain.**

Navier-Stokes equation. A nonlinear partial differential equation describing the velocity of a fluid at any given point in space and time.

Nernst-Planck equation. Also called the Nernst equation; describes the net flow of a substance through any cross section (which may or may not be that of a membrane), which, according to irreversible thermodynamics, is proportional to a driving force (i.e., a gradient of chemical potential).

neutral plane. Also called the neutral axis. The plane of zero stress in the cross section of a structure. The neutral and centroid axes precisely coincide for beams composed of homogeneous materials, provided the axial load is zero and the beam sustains only a bending load.

Newtonian fluid. Any fluid for which the viscosity (the ratio of stress to the rate of shear strain) is a constant for a specific temperature and pressure. Newtonian fluids have additional features; for example, they are isotropic materials with a Poisson's ratio of 0.5. Compare with **non-Newtonian fluid.**

nodal diaphragm. A transverse septum of tissue found at the node of a stem possessing hollow internodes.

node. In botany, the part of a stem where one or more leaves are attached. In dynamic beam theory, a point on a structure that does not deflect during vibration in a given mode.

nominal stress (engineering stress). Symbolized by σ_n; the stress calculated as the normal force component divided by the original cross-sectional area through which the force component is applied.

nonlaminar flow. Fluid flow conditions in which layers of the fluid are stirred, disrupted, or mixed laterally. Compare with **laminar flow.**

nonlinear elasticity. The condition in which the stresses exhibited by an elastic material are not proportionally related to strains by a single constant: $\sigma \neq \varepsilon\, E$. Rather, the proportionality factor relating a given stress to its corresponding strain is the tangent modulus: $E_t = \partial\sigma/\partial\varepsilon$. Compare with **linear elasticity.**

non-Newtonian fluid. Any fluid for which the viscosity (the ratio of stress to the rate of shear strain) is not constant at a specified temperature and pressure. Some non-Newtonian fluids can be extended to form rods or other geometric configurations and will retain their shape for a time when subjected to a stress; they have a high coefficient of viscous traction (a property equivalent to the elastic modulus of solids). Compare with **Newtonian fluid.**

normal strain components. Symbolized by ε_x, ε_y, or ε_z; strain components resulting from normal stresses that operate perpendicular to each of the three principal axes of symmetry (x, y, z) of a material element.

normal stresses. Symbolized by σ_x, σ_y, or σ_z; stresses operating perpendicular to the three principal axes of an element within a material. For materials placed in coaxial tension or compression, only the normal stress components need be considered in terms of evaluating static equilibrium.

ontogeny. The development of an individual organism throughout its lifetime.

organ. A distinct and visibly differentiated multicellular part of a plant, such as a stem, leaf, or root.

orthotropic. Having material properties that differ when measured in all three of the principal axes of symmetry.

osmotic adjustment. A metabolic effect that results in the net accumulation of solutes (carbohydrates, organic acids, and inorganic ions) within the protoplasm of water-stressed plants.

osmotic potential (osmotic pressure). Symbolized by Π; the pressure that must be applied to a membrane to prevent the net movement of water molecules across the membrane.

ovule. The reproductive structure of gymnosperms and angiosperms, composed of a megasporangium surrounded by one or more layers of tissues (integument) that develop from the base of the megasporangium; it possesses a small tube or pore (the micropyle) resulting from the incomplete closure of the integument.

palisade mesophyll. A layer or layers of mesophyll tissue composed of columnar, prismatic cells possessing a large number of chloroplasts, typically found in the leaves of dicots.

pappus. A low-density, parachute-like structure developed from the sepals of a single flower. Aids in long-distance dispersal of the subtending fruit.

PAR. Photosynthetically active radiation; light with wavelengths between 400 and 700 nm.

parenchyma. A tissue composed of more or less isodiametric, thin-walled cells, characterized by cellular division in each of the three principal axes, with plasmodesmata found in most adjacent cell walls.

patch clamp technique. A technique to record the electrical current (of the order of picoamperes) passing through a single ion channel in a membrane.

pectins. A chemically heterogeneous group of acidic polysaccharides (typically long-chain, branched or unbranched polymers of arabinose, galactose, galacturonic acid, and methanol) found mainly external to the primary cell wall. Soluble pectins or soluble pectic acids are straight chains of galacturonic acid residues precipitated as calcium and magnesium pectates, forming the middle lamella.

peduncle. The stem subtending an inflorescence.

periclinal. Referring to a plane of cell division in a layer of cells that is parallel to an adjacent layer of cells. Compare with **anticlinal.**

periderm. Secondary tissue derived from the phellogen (cork cambium) that replaces the epidermis. Consists of phellem (cork) and phelloderm.

petiole. The cylindrical, stalklike axis subtending the lamina of a leaf.

phellem. The external layer of cells produced by the phellogen (cork cambium).

phelloderm. A tissue produced internally by the centripetal cell divisions of the phellogen (cork cambium); a component of the periderm.

phellogen (cork cambium). The lateral meristem that forms the periderm. Produces phellem (cork) centrifugally and phelloderm centripetally by means of tangential cell divisions.

phloem. The tissue of vascular plants that is specialized to conduct cell sap; composed in angiosperms predominantly of sieve tube members and various kinds of phloem parenchyma, fibers, and sclereids.

phloem loading. The translocation of photosynthates produced in mesophyll into phloem. The process includes short-distance symplastic or apoplastic transport into companion cells and long-distance symplastic transport through sieve tube members.

phloem unloading. The translocation of photosynthates from phloem into cells that either metabolize or store them.

photoautotroph. An organism that can convert light energy into chemical energy by means of photosynthesis (e.g., cyanobacteria and plants).

photosynthate. A substance synthesized as a direct result of photosynthesis that contains carbon; typically a sugar.

photosynthesis. The synthesis of organic compounds by the reduction of carbon dioxide in the presence of light energy absorbed by chlorophyll. In plants, water is the hydrogen donor and the source of oxygen released by the photosynthetic process, which is represented by the empirical equation $CO_2 + H_2O \rightarrow (CH_2O)_n + O_2$.

photosynthetically active radiation. See PAR.

phototropism. A directional growth response to light, in which the direction of growth is determined by the direction of the incident light.

phyllotactic. Referring to the arrangement of leaves on the shoot.

pith. Ground tissue found toward the center of dicot stems and some dicot and monocot roots.

plasma membrane (plasmalemma; cell membrane). The single membrane enveloping the protoplasm and appressed to the cell wall.

plasmodesma (plural **plasmodesmata**). A symplastic strand passing through the cell walls of two adjoining cells that connects their protoplasts.

plasmolysis. A process in which the plasma membrane pulls away from the cell wall as a result of cytoplasmic dehydration; the reverse of cytolysis.

plastic behavior. Irreversible molecular or microstructural deformation within a solid resulting from the application of external forces. Compare with **elastic behavior.**

point load. A force acting on a structure at only one point.

Poisson's effect. In uniaxial tension, materials are permitted to contract laterally, whereas in biaxial tension, lateral contraction is restrained. Thus, materials evince larger strains for a given uniaxial stress level than for the strains measured in biaxial tension.

Poisson's ratio. Symbolized by v; the ratio of negative lateral strain to the strain measured in the direction of the applied force. For any planar section having two principal axes of symmetry, two Poisson's ratios may be calculated. For isotropic materials, the two Poisson's ratios are equivalent in magnitude.

polar moment of inertia. The sum of the second moments of area measured in the two orthogonal planes of a cross section. For shafts with circular cross sections, the polar moment of inertia equals the torsional constant J.

pollen (pollen grain). The microspore-microgametophyte of seed plants.

pollination droplet. A viscous liquid, produced from the disintegration of some nuclear cells, exuded from the micropyle of some gymnosperms, to which pollen grains adhere.

porphyrin. A group of organic molecules produced by the synthesis of D-aminolevulinic acid by the reaction of the amino acid glycine and succinyl-CoA in the citric acid cycle.

pressure chamber. A device used to determine water potential by measuring the magnitude of the external pressure uniformly applied over a sample's surface that drives water from the living cells into the xylem tissue at atmospheric pressure.

pressure-flow model. A widely accepted explanation for how photosynthates are translocated in the phloem, first proposed by Münch in 1930. The model states that photosynthate transport from source to sink tissues is driven by a pressure gradient that is osmotically generated. See **Phloem loading; Phloem unloading.**

pressure potential. Symbolized by ψ_p; the effect of any pressure on the water potential of a system. Usually has a positive value, but may be negative in the case of rapidly moving xylem water owing to high transpirational water losses.

pressure probe. A device used to measure the turgor pressure in a cell directly.

prickle. Typically a woody, ascicular epidermal outgrowth that can have multiple functions or no (known) function.

primary cell wall. That portion of the cell wall formed mainly while the cell is increasing in size, in which the orientation of the cellulosic microfibrils varies from random to more or less parallel to the longitudinal axis of the cell. Compare with **secondary cell wall.**

primary growth. That phase of the ontogeny of vegetative and reproductive organs from the time of their meristematic initiation until the completion of their expansion and elongation. Compare with **secondary growth.**

primary meristem. Any of the three meristematic tissues derived from an apical meristem: protoderm, ground meristem, and procambium.

primary tissues. Tissues derived from the apical meristems and the embryo.

procambium (provascular tissue). One of the three primary meristems derived from an apical meristem; gives rise to the primary vascular tissue.

progymnosperms (Progymnospermopsida). An extinct group of free-sporing middle Devonian to lower Carboniferous plants, some of which produced secondary xylem (wood); the group consisted of three orders (Aneurophytales, Protopityales, and Archaeopteridales)

prokaryote. An organism that does not have its nuclear material enclosed in a nuclear envelope and which lacks other membrane-bound cytoplasmic organelles. Prokaryotic organisms include cyanobacteria, nonphotosynthetic bacteria, and mycoplasmas. Compare with **eukaryote.**

propagule. A term encompassing seeds, fruits, and vegetative structures that can propagate an individual plant. Any reproductive structure that is capable of producing a plant either by possessing an embryo or by the vegetative generation of embryonic roots and shoots.

proportional limit. See **elastic limit.**

protoderm. One of the three primary meristems derived from an apical meristem; gives rise to the epidermis.

protoplast. The living contents of the cell; also, the inclusive term for the living fraction of an entire plant; the symplast.

proximal. Closest to a structure's point of origin or site of attachment to another structure. Compare with **distal.**

pseudoparenchyma. A tissue found in some algae that is constructed from interwoven filaments of cells, giving the appearance of true parenchyma.

pteridophytes. Nontechnical term for free-sporing (non-seed-bearing) vascular plants. Encompasses the ferns, horsetails, lycopods, and several extinct seedless plant groups. The pteridophytes were formerly thought to be a monophyletic group and were referred to as the Pteridophyta.

radial. Referring to any plane parallel to a radius through an object or a plane of section through a biological structure.

Raoult's law. A law that states that the partial pressure of a gas at equilibrium with its volatile liquid phase is proportional to the molar fraction of that solvent in the liquid.

reaction wood. Secondary xylem formed by stems in response to mechanical changes from their original orientations. See also **compression wood; tension wood.**

redox (reduction-oxidation) **reaction.** A chemical reaction in which the oxidation num ber (state) of reactants is changed; reduction involves a decrease in the oxidation number, whereas oxidation involves an increase in the oxidation number.

relaxation modulus. The ratio of the time-varying stress to the fixed strain.

relaxation time. Symbolized by T_R; the time required for a given stress to diminish to $1/e$ its original magnitude. Also, the ratio of a material's dynamic viscosity to its elastic modulus.

residual strain. The plastic component of deformation. A perfectly elastic material has no residual strain, provided its proportional limit is not exceeded during loading. Most elastic materials are not perfectly elastic, and many exhibit a residual (plastic) strain after they are unloaded.

resilience. A measure of the elastic energy stored within a body; the capacity to restore original dimensions after an applied load has been removed.

Reuss model (equal stress model). A model assuming that material elements differing in their elastic properties are aligned normal to the direction of application of an external force so that all elements experience stresses of equivalent magnitude.

Reynolds number. Symbolized by Re; the dimensionless parameter that is the product of a reference dimension (parallel to the direction of ambient fluid flow) and the ambient fluid speed, divided by the kinematic viscosity of the fluid. The Reynolds number expresses the ratio of inertial to viscous forces in the fluid flow in reference to an object obstructing fluid flow. Note that fluid flow does not have a Reynolds number except in reference to an object.

rheology. The discipline of engineering and physics studying the flow of liquids, viscoelastic solids, and other materials capable of flowing.

root. One of the three primary plant organ types, characterized by a root cap and by centripetal xylem differentiation. All but the primary root result from endogenous development.

root cap. A thimble-like mass of cells covering the root apical meristem.

sacci. The two bladderlike extensions of the pollen grains produced by some gymnosperms resulting from the separation of the inner and outer spore walls.

safe stress. See **working stress.**

safety factor. See **factor of safety.**

samara. A fruit possessing an extended bladelike wing or membrane. Aerodynamically, any propagule with a wing that generates lift.

saponifiable. Referring to a molecule with one or more ester functional groups that can be hydrolyzed under basic conditions (e.g., phospholipids and triglycerides).

sapwood. The outermost growth layers of secondary xylem, in which some cells are living and others (tracheary elements) conduct water. Compare with **heartwood.**

sclereid. Any sclerenchymatous cell type other than a fiber.

sclerenchyma. A mechanical tissue composed of thick-walled, lignified cells that can provide mechanical support; a collective term for the sclerenchymatous tissues of an organ or the entire plant body. Includes fibers and sclereids.

secondary cell wall. That portion of the cell wall deposited in some cells by the protoplast between the plasma membrane and the primary wall after the primary wall ceases to increase in size. Compare with **primary cell wall.**

secondary growth. A phase of ontogeny that follows primary growth in some plant species. Growth that results from the activity of lateral meristems and is characterized by the deposition of secondary tissues, resulting in an increase in the thickness of organs, typically stems and roots. Compare with **primary growth.**

secondary phloem. The phloem tissue produced by a vascular cambium during secondary growth in a vascular plant.

secondary xylem. The xylem tissue produced by a vascular cambium during secondary growth in a vascular plant. Colloquially referred to as wood.

second moment of area. Symbolized by I; the integral of the product of each elemental cross-sectional area and the square of the distance of each elemental cross-sectional area from the centroid axis. A dimensional parameter that quantifies the distribution of mass in each cross section with respect to the center of mass of the cross section.

seed. A mature ovule containing an embryo or embryos surrounded by a seed coat composed of the mature integument or integuments.

shaft. A support member subjected to a torque, such as a rod supporting a rotating propeller. Compare with **beam; column.**

shear. The result of the tangential application of an external force to the surface of a material or structure.

shear modulus. Symbolized by G; the ratio of the shear stress (τ) to the shear strain (γ) for a material.

shear strain. Symbolized by γ; the tangent of the rotation angle (in radians) resulting when a material is subjected to a tangentially applied force.

shear stress. Symbolized by τ; the shear force component divided by the tangential area over which the shear force component acts.

shell. A very thin elastic structure whose material is confined to a curved surface. A shell evincing negligible flexural rigidity is called a membrane.

sieve tube member. One of a number of cells that collectively comprise a sieve tube in the phloem of angiosperms. When mature and functional, this cell type lacks a nucleus. Its metabolism is controlled by the nucleus of its companion cell. See also **phloem.**

siphonous. Referring to a coenocytic body plan (in a unicellular multinucleated organism) in which some cellular septa may have formed.

S layer. Any secondary cell wall layer. The numerical order of the S layers reflects the growth of the cell wall by apposition; ascending numbers indicate progressively older (inner) cell wall layers.

softwood. The wood of conifers.

solid. A state of matter characterized by the restricted motion of molecules or atoms such that the shape and volume of any portion are restricted relative to those of any other portion.

solute. The constituent of a solution that is dissolved in the solvent. The solute is the smaller of the two amounts.

solute potential. Symbolized by ψ_s; the effects of solutes dissolved in water on the water potential of the system. Always has a negative value.

solvent. A solid, liquid, or gas that dissolves another solid, liquid, or gaseous solute. The solvent is the larger of the two amounts.

spongy mesophyll. A type of parenchyma containing large intercellular spaces filled with air, found in the leaves of some plants; aids gas exchange between the plant body and the external atmosphere.

sporangiophore. A branchlike structure that bears one or more sporangia; the stalk of a sporangium.

sporophyte. The multicellular phase in the diplobiontic life cycle that produces meiospores; the dominant organism in the life cycle of vascular plants. Compare with **gametophyte.**

spring wood. See **early wood.**

statolith. A free-moving intracellular particle or concretion.

stem. One of the three principal vegetative organs of the vascular plant body; produced by the shoot apical meristem and bearing appendicular organs (leaves) produced by exogenous growth.

stoma (plural stomata). An opening in the epidermis that is surrounded by one or more guard cells, through which gases can be exchanged between the plant body and the atmosphere.

strain. Symbolized by ε; a normalized measurement of deformation (or deflection) regardless of the nature of the deformation, although it is typically assumed that a strain is the result of an externally applied load.

strain energy. The component of the total energy within an object or a structure that is stored in the form of molecular deformations. Strain energy is a form of potential energy; in elastic materials, within the proportional limits of loading, the strain energy is used to restore the material's original dimensions when the stress is removed. Within the range of elastic behavior, the strain energy per volume of the material is the area measured in the stress-strain diagram.

stramenopiles (heterokonts). A monophyletic group consisting of unicellular and multicellular eukaryotes, the majority of which are photosynthetic algae that have chloroplasts with four membranes.

streamline. A line in a flow field where the local velocity vector is tangential to every point on the line. Streamlines can be visualized by following the flow of particles carried by the fluid.

streamlined. Having a body profile that minimizes pressure drag. Compare with **bluff-bodied.**

strength. In reference to a structure, the load (breaking load) that will cause the structure to fail. In reference to a material, the stress (breaking stress) that will break the material.

stress. Symbolized by σ; a force normalized with respect to the area through which it acts. The SI unit of stress is Pa or N/m^2.

stress-strain diagram. A plot of stress versus strain for a given material or structure.

summer wood. See **late wood.**

supporting tissue. See **mechanical tissue.**

surface force. Any force operating on the external boundaries of an object, such as hydrostatic pressure. Compare with **body force.**

symplast. The living portion of the plant body; consists of the metabolically functional protoplast. Compare with **apoplast.**

tangential. Referring to the direction of the tangent (normal to the radius) of a structure. Used in reference to the plane in which a structure is sectioned.

tangent modulus (instantaneous elastic modulus). Symbolized by E_T; the ratio of the change in stress to the change in strain measured at any point along a stress-strain diagram. When measured within the linear portion of the stress-strain diagram of a linearly elastic material, the tangent modulus is called the modulus of elasticity (Young's modulus, elastic modulus).

telome. Any distal branch of a dichotomized axis; the fundamental morphological unit of an ancient land plant with vascular tissues.

tensile (normal) **stress.** Symbolized by $\sigma+$; the normal tensile force component divided by the surface area through which the force component operates.

tension wood. The reaction wood of dicots, formed on the surfaces of branches experiencing stresses; characterized by lack of lignification and often by a high content of gelatinous fibers; capable of contraction. Compare with **compression wood.**

thigmomorphogenesis. Any morphogenetic response to mechanical perturbation, typically involving some form of growth inhibition.

thixotropic materials. Materials that behave as thick solids but that are capable of flowing like fluids when mechanically perturbed (as in shaking).

thorn. A modified branch, typically functioning for protection.

tissue. Any group of cells organized into a structural or functional unit. The constituent cells may or may not be similar in size and shape.

tissue system. Tissue (or tissues) in the plant body that is structurally and functionally organized into a unit. The three tissue systems are the dermal, vascular, and fundamental (ground) tissue systems.

tonoplast. The membrane surrounding a cell vacuole.

torsion. The mechanical consequence of a torque moment.

torsional constant. Symbolized by J; a parameter describing the geometric contribution of a shaft toward resisting torsion. For shafts with a circular cross section, the torsional constant equals the polar moment of inertia. For shafts with nonterete cross sections, the torsional constant is less than the polar moment of inertia.

torsional rigidity. Symbolized by C or GJ; the product of the shear modulus (G) and the polar moment of inertia. A measure of the ability of a shaft to resist a moment of torque.

torus. The thickened region of the middle lamella spanning a bordered pit that is suspended by the margo. See also **margo.**

toughness. The energy per unit volume required for fracturing a material. The area under a stress-strain diagram up to failure provides a measure of a material's toughness in terms of the energy absorbed per unit volume.

tracheary element. Any water-conducting cell type found in vascular plants; specifically, a collective term for tracheids and vessel members.

tracheid. A spindle-shaped tracheary element lacking end-wall perforations that transports water. May occur in either the primary or the secondary xylem or in both. Compare with **vessel member.**

tracheophytes. Vascular plants. Includes ferns, horsetails, lycopods, gymnosperms, and angiosperms; excludes mosses, liverworts, hornworts, and the algae.

transverse section (transection, cross section). Any plane section taken normal to the longitudinal axis of a structure.

true strain. See **Henchy strain.**

true stress. Symbolized by σ_i; the stress calculated from the normal force component divided by the instantaneous cross-sectional area through which it operates. True stress should be calculated when the loading conditions vary over time or when the strains are large ($\geq 5\%$).

tunica. The external layer (or layers) of cells in the apical meristem of a stem that divides to give rise to the protoderm, which in turn gives rise to cells that mature and differentiate into the epidermis.

turbulent flow. Fluid flow characterized by random fluctuations of velocity and pressure at any given point in the fluid outside the boundary layer and at Reynolds numbers in excess of 10^5 even within the boundary layer.

turgor pressure. When matric potentials are neglected, the turgor pressure is defined as the difference between the water potential and the solute potential of a cell or tissue.

uniaxial stresses. A single pair of coaxial forces directed either outward (uniaxial tension) or inward (uniaxial compression) with respect to a material or structure.

Ussing-Teorell equation. A quantification of the ratio of inward to outward flux of ions passively transported through the plasma membrane.

vapor pressure. The pressure of a vapor in thermodynamic equilibrium with its condensed phase, liquid or solid, in a closed compartment.

vascular bundle. A threadlike vascular strand consisting of primary xylem and primary phloem.

vascular cambium. The lateral meristem that gives rise to the secondary vascular tissues.

vascular plants. See **tracheophytes.**

vascular tissue. The xylem and the phloem; sometimes used to refer to either the xylem or the phloem. Typically the term is used with no distinction between the primary and the secondary xylem or phloem.

vein. A single vascular bundle in a dorsiventral organ, such as a leaf.

vessel. A tubelike conduit composed of many vessel members stacked end to end.

vessel member (vessel element). A single cellular component of a vessel. Vessel members have perforated end walls and offer little resistance to the flow of water. Compare with **tracheid.**

viscoelastic. Exhibiting the properties of viscosity and elasticity. A viscoelastic material deforms with the application of a force (or forces) and, over time, elastically restores some or all of its deformation when the force is removed.

viscosity. Symbolized by μ; the ability of a fluid to resist shearing deformation. The viscosity of a Newtonian fluid is the ratio of the shear stress to the resulting velocity gradient.

viscous modulus. Symbolized by E''; the ratio of the dynamic viscosity (μ) to the frequency (ω) at which strain is varied: $E'' = \mu/\omega$.

Voigt model (equal strain model). A model assuming that material elements differing in their elastic properties are aligned parallel to the direction of application of an external force so that all elements experience strains of equivalent direction and magnitude.

von Karman (vortex) **street.** A regular arrangement of vortices in approximately two parallel directions, such that vortices paralleling one another are shed alternately.

vortex. A rotating body of fluid.

vortex street. See **von Karman street.**

vorticity. Symbolized by Γ; rotational motion within a fluid body that at any point in the fluid is defined as twice the angular velocity of a small portion of the fluid surrounding that point.

wake. A region of disturbed fluid flow behind an object having a motion relative to the fluid.

water potential. Symbolized by ψ_w; the chemical potential of water or a solution of water and solutes expressed in terms of the units of energy per unit volume of fluid. The water potential of pure water under standard conditions is zero. The water potential of a solution of water and solutes is the ratio of the difference between the chemical potential of the solution and the chemical potential of pure water to the molal volume of water in the solution.

wavelength. Symbolized by λ; the spatial period of a wave; the distance over which the shape of a wave is repeated, or the distance between any two corresponding points of the same wave.

wood. See **secondary xylem.**

working stress (safe stress). Symbolized by σ_w; the stress level for which a structure is designed to operate. Usually well below the yield stress σ_y or the ultimate strength σ_U of the materials used in the structure's fabrication; that is, $\sigma_w = \sigma_y/n$ or $\sigma_w = \sigma_U/n_1$, where n and n_1 are factors of safety, which define the magnitude of the working stress. In the case of structural steel, the yield point is used and n is taken as $n = 2$

or larger; in the case of wood, the ultimate strength is usually taken as the basis for determining the working stress.

xylem. The tissue produced by the procambium and the vascular cambium that is specialized to conduct water and solutes in water; also serves as the principal mechanical support tissue in organs exhibiting secondary growth.

yielding. The initiation of plastic behavior in an elastic, ductile material.

yield strain. The strain corresponding to the yield stress, providing a measure of the maximum level of deformation that can be elastically recovered when the stress level drops to zero.

yield stress. The level of stress that initiates plastic behavior in an elastic, ductile material. The yield stress represents the limiting stress level for which deformations are recoverable when the stress level drops to zero.

Young's modulus. The elastic modulus E; the proportionality constant relating normal stress to normal strain throughout the linear elastic range of behavior of a material.

Z ("zigzag") scheme. A flow diagram depicting the arrangement of the antenna complexes and reaction centers that make up photosystems I and II (PSI and PSII) and the electron transport chain linking PSI and PSII by their midpoint redox potential. The arrangement is shaped like the letter Z lying on its side.

Author Index

Page numbers followed by the letters *f* and *t* indicate figures and tables, respectively.

Subject Index

Page numbers followed by the letters *f* and *t* indicate figures and tables, respectively.